D0432213

PROGRESS IN WATER TECHNOLOGY, VOLUME I

APPLICATIONS OF NEW CONCEPTS OF PHYSICAL-CHEMICAL

WASTEWATER TREATMENT

Vanderbilt University
Nashville, Tennessee
September 18-22, 1972

Edited by

W.W. ECKENFELDER

and

L.K. CECIL

Sponsored by

The International Association
on Water Pollution Research

and

The American Institute of Chemical Engineers

PERGAMON PRESS, INC.

NEW YORK   TORONTO   OXFORD   SYDNEY   BRAUNSCHWEIG

TD 511
A66
1972

PERGAMON PRESS, INC.
Maxwell House, Fairview Park, Elmsford, N.Y. 10523

PERGAMON OF CANADA, LTD.
207 Queen's Quay West, Toronto 117, Ontario

PERGAMON PRESS, LTD.
Headington Hill Hall, Oxford

PERGAMON PRESS (AUST.) PTY. LTD.
Rushcutters Bay, Sydney, N.S.W.

VIEWEG & SOHN GmbH
Burgplatz 1, Braunschweig

Copyright © 1972, Pergamon Press, Inc.
Library of Congress Catalog Card No. 72-8108

All Rights Reserved. No part of this publication may be reproduced,
stored in a retrieval system or transmitted in any form, or by any
means, electronic, mechanical, photocopying, recording or otherwise,
without prior permission of Pergamon Press, Inc.

Printed in the United States of America

0 08 017243 1

PREFACE

During the decade of the '70's emphasis will be placed on new technology and modifications of old technology for municipal and industrial water pollution control. Much of this effort will be in the physical-chemical treatment technology. While many of the processes now being considered are still in the development stage, some of the technology is under design and construction. This conference is directed toward presenting these new concepts with particular reference to their engineering applications.

This conference is the third in a series of specialized conferences on pertinent and developing technology sponsored by the I.A.W.P.R. and the A.I.Ch.E. The first specialized conference was held in Vienna, Austria in 1971 and devoted itself to the design of large wastewater treatment plants. The second conference held in London in 1972 dealt with the problems of phosphorous in the environment. Future conferences are planned on subjects of pertinent interest to water quality control and management. This volume is the first in a series to be published by I.A.W.P.R. as PROGRESS IN WATER TECHNOLOGY. Pertinent discussions related to this conference will be published in WATER RESEARCH.

The organizers of the conference are particularly grateful to the Department of Environmental and Water Resources Engineering of Vanderbilt University for their support and assistance in the organization and conduct of the conference.

Errata

<u>Page 329, Line 7</u> should read as follows:

"at the plate face is 1.5, so to take care of losses up to
the plate"

<u>Page 329, Line 13</u> should read as follows:

"be 40 minutes and 3/4" spacing would be 22.5 minutes."

<u>Page 332, Line 9</u>, Farada should have been spelled "Faraday"

<u>Page 332, Line 18</u> should read as follows:

"(SR = 415,000/ppm * * .9735)"

<u>Page 332, Line 28</u>, the fourth word is "is" rather than "in".

<u>Page 372, Lines 11-12</u> should read as follows:

"13,4 per cent of the total water consumption (4). The
average daily production from the reclamation plant
during this period was 2.3 ml/d which is equivalent to a
capacity utili-"

<u>Page 376, Line 42</u> should read as follows:

"at a rate of 10 per cent containing 25 g/l of solids"

<u>Page 377, Line 4</u> should read as follows:

"Breakpoint chlorination is maintained in tank with 45
minutes detention."

CONTENTS

Municipal Waste Treatment by Physical-Chemical Methods   1
    J.F. Kreissl and J.J. Westrick

Physicochemical Systems For Direct Wastewater Treatment   13
    W.J. Weber, Jr.

Advanced Waste Treatment at Alexandria, Virginia   27
    C.W. Reh, R.L. Hall, and T.E. Wilson

Physical/Chemical Treatment Design for Garland, Texas   41
    D.P. McDuff and W.J.W. Chiang

Ultrahigh Rate Filtration of Municipal Wastewater   59
    J.F. Malina, Jr.

Interrelationships Between Biological Treatment And
    Physical-Chemical Treatment   75
    W.W. Eckenfelder, Jr.

Tertiary Treatment - The Corner Stone of Water Quality
    Protection and Water Resources Optimization   85
    G.J. Stander

The Effective Utilization of Physical-Chemically
    Treated Effluents   95
    O.O. Hart and G.J. Stander

Upgrading Existing Wastewater Treatment Plants   103
    J.M. Smith, A.N. Masse, and W.A. Feige

Some Problems Associated With the Treatment of Sewage
    By Non-Biological Processes   119
    R.W. Bayley, E.V. Thomas, and P.F. Cooper

The Applicability of Carbon Adsorption in The
    Treatment of Petrochemical Wastewaters   133
    D.L. Ford

Treatment of Wastes From Metal Finishing And
    Engineering Industries   147
    R.K. Chalmers

Physical and Biological Interrelationships in
    Carbon Adsorption   159
    W.W. Eckenfelder, T. Williams, and G. Schlossnagle

Regeneration of Activated Carbon   167
    G. Shell, L. Lombana, D. Burns, and H. Stensel

Principles and Practice of Granular Carbon Reactivation   179
    C.E. Smith

Chemical Regeneration of Activated Carbon   187
    J.M. Rovel

Automation and Control of Physical-Chemical Treatment
    For Municipal Wastewater   199
    J.J. Convery, J.F. Roesler, and R.H. Wise

Recent Studies of Calcium Phosphate Precipitation
    In Wastewaters   211
    D. Jenkins, A.B. Menar, and J.F. Ferguson

Contents (cont.)

Logical Removal of Phosphorous   231
       I. Yall, W.H. Boughton, F.A. Roinestad,
   and N.A. Sinclair

Phosphorous Removal by Chemical and Biological
   Mechanisms   243
       W.F. Garber

Use of Surface Stirrers for Ammonia Desorption
   From Ponds   263
       A.M. Wachs, Y. Folkman, and D. Shemesh

Dewatering Physical-Chemical Sludges   273
       D.D. Adrian and J.E. Smith, Jr.

Disposal and Recovery of Sludges From Physical-Chemical
   Processes   291
       C.E. Adams, Jr.

The Role of Freezing Processes in Wastewater Treatment   309
       J.H. Fraser and W.E. Johnson

Waste Water Treatment Through Electrochemistry   325
       C.E. Smith

Ultrafiltration and Microfiltration Membrane Processes
   For Treatment and Reclamation of Pond Effluents in
   Israel   335
       G. Shelef, R. Matz, and M. Schwartz

Membrane Equipment Selection for the Must Hospital
   Water Recycle System   347
       J.A. Heist

Comparative Cost of Tertiary Treatment Processes   357
       R.F. Weston and R.F. Peoples

Process Selection and Cost of Advanced Wastewater
   Treatment in Relation to the Quality of Secondary
   Effluents and Quality Requirements for Various
   Uses   371
       L.R.J. van Vuuren and M.R. Henzen

Interrelationships Between Fresh Water and Wastewater
   Plants   385
       L.K. Cecil

How Usable is Present Technology for Removing
   Nutrients From Wastewater   389
       L.K. Cecil

Applications of New Concepts of Physical-Chemical
Wastewater Treatment
Sept.18-22, 1972

Pergamon Press, Inc.   Printed in U.S.A.

MUNICIPAL WASTE TREATMENT BY PHYSICAL-CHEMICAL METHODS

James F. Kreissl and James J. Westrick
Environmental Protection Agency
National Environmental Research Center
Advanced Waste Treatment Research Laboratory
Cincinnati, Ohio 45268

INTRODUCTION

The present status of physical-chemical (P-C) treatment technology for municipal wastewaters is sufficiently advanced to permit a reasonably reliable assessment of its capabilities and characteristics. A number of pilot studies
on a variety of wastewaters have produced significant data concerning the design and operation of physical-chemical systems. No full-scale plants are
in operation at this time, but start-up of a 0.6 MGD system at Rosemount,
Minnesota, is scheduled for late 1972.

WHAT IS PHYSICAL-CHEMICAL TREATMENT?

Some conjecture exists as to a precise definition for physical-chemical treatment. Although other systems might be considered,the only one discussed here
is chemical clarification followed by dissolved organic removal. The first
stage of treatment is designed to provide efficient suspended and colloidal
solids removal along with phosphorus removal; the second operation removes a
large percentage of the dissolved organic matter. A typical flow diagram of
this basic system is shown in Figure 1.

A number of variations exist for these two operations. In the clarification
step there are two primary considerations: the choice of chemicals to be used
and the choice of physical equipment. Essentially, the chemical options are
limited to alum, lime or iron salts with or without polymer addition. In the
choice of physical equipment, conventional mixing, flocculation and solids
separation must be evaluated versus solids-contact systems. The solids separation process in the conventional sequence can consist of conventional sedimentation, flotation or high-rate sedimentation. In addition, a filtration
step may be included for improved solids removal.

In the dissolved organic removal step the choices are generally limited to
either granular or powdered activated carbon adsorption. However, other
processes such as ozonation have also been proposed for this purpose. The

choice of a granular activated carbon (GAC) design requires decisions concerning series or parallel contacting, pressure or gravity vessels and upflow or downflow operation. Powdered activated carbon (PAC) designs primarily involve the choice between single-stage and two-stage countercurrent contacting.

In addition to these basic operations provisions can be included for nitrogen removal by ammonia stripping, ion exchange or breakpoint chlorination. Other unit processes may also be added to the basic system in order to provide for the specific requirements of any application. For example, disinfection or final filtration steps may be desirable in many locations. Of course, the treatment and disposal of the sludge from the clarification step, the regeneration of spent activated carbon and management of backwash streams represent some of the additional design considerations.

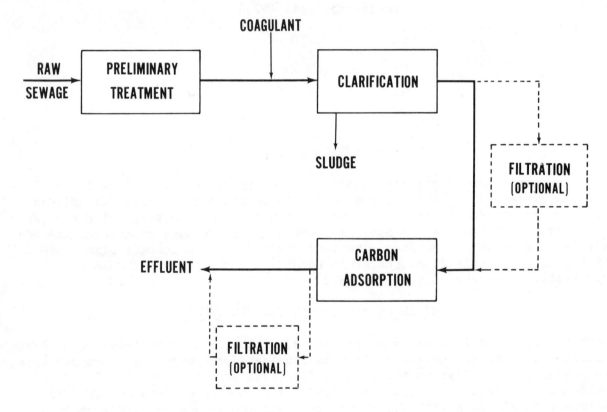

FIG. 1
Basic Flow Diagram of a Physical-Chemical Treatment System

WHY CONSIDER IT?

Like any new technology, physical-chemical treatment must have certain advantages over conventional methods of treatment to warrant full-scale application. A number of advantages inherent in physical-chemical treatment have been cited by other authors (1-4):

1. Reduced land area requirements,
2. More positive control over treatment plant performance,
3. Greater resistance to upset from toxic substances in wastewaters,
4. Capability for removal of numerous toxic materials,
5. Negligible start-up period before normal removals are attained,
6. Improved color removal.

The above advantages may be translated into the following applications:

1. Available land limited in area or prohibitively expensive,
2. Stringent reliability requirements,
3. High proportion of industrial wastes of a toxic nature to biological systems,
4. Heavy metal or pesticide concentrations in the wastewater and stringent effluent requirements for those pollutants,
5. Intermittent or periodic flows as in parks or other recreational areas,
6. Highly-colored industrial wastes which produce an undesirable effluent coloration.

When constraints such as these are facing a designer, the feasibility of physical-chemical treatment is greatly enhanced and should be investigated. It should be noted that physical-chemical systems and the individual processes included therein are proposed as additional weapons in the waste treatment arsenal and not as any form of panacea for all pollution abatement applications. Taken in this light, the particular advantages of a P-C system and its component processes can be used along with conventional processes to create the most flexible and reliable waste treatment plants possible for optimum pollution abatement for any installation.

HOW GOOD IS IT?

In terms of suspended solids, organic matter and phosphorus removals, the larger pilot facilities (>50 gpm) have produced excellent results, as shown in Tables 1 through 5. The Eimco Salt Lake City data, shown in Table 1, is difficult to interpret because a number of parameters were being studied concurrently (5). During the course of the study, the 100 gpm pilot plant utilized solids-contact clarification with each of the three major coagulants, followed by either single-stage or two-stage countercurrent PAC adsorption with a final step of filtration. Essentially, it was determined that alum, ferric chloride and lime could all produce equivalent product quality in the clarification step, and that the powdered carbon systems could be evaluated independent of the chemical used. Therefore, overall removals were not readily apparent for any single combination of processes and variables.

The data generated at Cleveland Westerly by Battelle-Northwest (6) provides the only substantial information on the application of physical-chemical treatment to a strong municipal waste. The 100 gpm pilot facility included conventional clarification prior to filtration and downflow GAC columns. The relatively high organic content of the wastewater was substantially reduced, as shown in Table 2, although severe problems with hydrogen sulfide generation in the activated carbon columns were encountered.

The Environmental Protection Agency-Los Angeles County Sanitary District facility at Pomona, California, is presently operating a physical-chemical treatment system with an excellent product quality (7). The processing sequence for this 50 gpm pilot plant is conventional clarification with either iron or alum followed directly by downflow GAC columns. These data are presented in Table 3 along with the removals of COD, SS and P by the 50 gpm pilot plant and the 10 MGD activated sludge plant during a period of upset caused by a high concentration of copper in raw wastewater.

Another EPA pilot facility, at Lebanon, Ohio, employed physical-chemical

treatment at 75 gpm (8). The process employed single-stage lime treatment
with solids-contact clarification and filtration, followed by pH adjustment
and upflow GAC contacting. The resulting data are presented in Table 4.
Again, substantial removals of the major contaminants were obtained.

The most comprehensive EPA pilot facility is at the Blue Plains Sewage Treatment Plant in Washington, D. C. The physical-chemical system there has been
operating on a diurnal cycle of 3.2 to 1 (9). This system consists of two-
stage lime treatment with solids-contact reactors, filtration and downflow
GAC contacting in series, followed by ion exchange for nitrogen removal.
The six-month averages of performance from March through August of 1970 are
presented in Table 5. The high quality of treatment is evident.

Numerous other pilot-scale studies have been reported (10,11,12,13) with results similar to those described above. In almost every case, the removals
of organics, suspended solids and phosphorus exceed 90 percent.

TABLE 1
Salt Lake City Performance Data[*]

Chemical	Clarification					
	COD		SS		P	
	Raw	Clar.	Raw	Clar.	Raw	Clar.
Alum	158	45	110	30	5.8	1.2
Iron	255	54	94	14	7.8	0.2
Lime	200	62	100	16	2.6	1.1

	PAC Adsorption Effluent	
	Soluble COD	SS
System		
1-stage	16	2
2-stage	10	2

* Conditions: 1) Typical Runs
                 2) Flow = 50-60 gpm
                 3) All values expressed in mg/l

TABLE 2
Average Performance at Cleveland Westerly[*]

Chemical	COD		BOD		SS		P	
	in	out	in	out	in	out	in	out
Iron	500	41	288	29	213	20	6.2	0.4
Lime -								
pH 10.5	498	69	244	41	199	22	4.9	0.8
pH 12	536	48	182	35	144	6	5.3	0.2

[*]Conditions: 1) Flow = 30-100 gpm
                 2) Polymer used for iron and pH 10.5 lime
                 3) All values expressed in mg/l

TABLE 3
Pomona Performance Data*

Normal

Chemical	COD		SS		P	
	Raw	Final	Raw	Final	Raw	Final
Iron	357	27	178	19	11.9	1.6
Alum	285	29	183	9	10.4	1.0
Act. Sl.	–	43	–	18	–	–

Upset Period

System	COD		SS		P	
	Raw	Final	Raw	Final	Raw	Final
Phys.–Chem.	283	30	188	9	10.8	1.1
Act.Sl.	283	80	188	42	10.8	10.8

* All values expressed in mg/l

TABLE 4
Average Performance at Lebanon*

PARAMETER	RAW	LIME CLAR. + FILT.	EFFLUENT
TOC	87	35	9
COD	257	60	29
BOD	98	19	6
SS	110	7	2
P	8	2	2

*Conditions: 1) Averages over 11 M G throughput
                 2) All values expressed in mg/l

TABLE 5
Average Performance at Blue Plains*

PARAMETER	RAW	LIME CLAR.	FILTER EFFL.	FINAL EFFL.
TOC	99	20	18.5	6.3
BOD	122	22	18.0	5.3
COD	305	51	46.2	13.2
SS	158	14	5.5	4.6
P	8.4	0.3	0.2	0.1

*Conditions: 1) 6-month averages
                 2) Flow = 45–140 gpm
                 3) All values expressed in mg/l

WHAT ELSE IS KNOWN?

Pilot studies have shown the ability of physical-chemical systems to produce
an excellent quality of effluent in terms of the primary pollution indicators,
suspended solids, organics and phosphorus. Information on the removal of
various other pollutants of interest and the overall quality of effluents is
somewhat more difficult to determine from the available reports. Also,

minimal design data are presented in these sources. Some of this information
has been compiled herein.

Starting with the chemical clarification step, two studies (5,6) have tested
different overflow rates in order to determine maximum allowable design rates.
The recommendations for peak overflow rates for the three coagulant systems
are shown in Table 6.

TABLE 6
Recommended Peak Overflow Rates

	Peak Overflow Rate, gpd/sf	
Chemical	Eimco (5)	Westerly (6)
Lime	1870	1000
Iron	720	650
Alum	570	–

It should be pointed out that the data from these studies did not show any
significant change in effluent suspended solids at these levels. It is
assumed that other factors, such as ease of operation, contributed to the
choice of these overflow rates. Also, the Eimco study utilized a solids-
contact reactor, while the Westerly study used conventional mixing, flocculation, and sedimentation. The use of polymers may also have a profound effect
on the design overflow rate. In relation to these recommended peak rates,
Bishop, etal, (9) had rain peaks as high as 1450 gpd/sf and obtained excellent clarification with lime. Culp (14) recommends a peak rate of 1400 or an
average rate of 900 gpd/sf for lime with the larger requirement of the two
controlling the design. Swedish full-scale designs of flotation units for
separation of solids after alum coagulation have successfully employed overflow rates of more than 2,500 gpd/sf (15).

Chemical dosage to obtain good clarification varies with the wastewater and
the chemical employed. Lime systems have required 250 to 1500 mg/l as
$Ca(OH)_2$, depending on the desired treatment pH and wastewater alkalinity.
When high pH systems are required, a recarbonation step must be used to
stabilize the wastewater after clarification. Bishop (9) adjusted pH from
11.5 to 10 with an average CO_2 dose of 120 mg/l, while Westerly pH adjustment from 10.5 to 7 required 360 mg/l of CO_2. Bishop (9) and Stander (16)
both required the addition of coagulant to bring down the $CaCO_3$ precipitates.
One of the advantages frequently cited for the use of lime in tertiary or
advanced waste treatment of secondary effluent is the ability to recalcine
and reuse the lime. One investigator has indicated that without a satisfactory method for separating the inert solids from the lime, recalcination and
reuse is not economically feasible (14). Iron requirements for good clarification have varied from 20 to 60 mg/l as Fe, while alum dosages have ranged
from 7 to 20 mg/l as Al.

A question which frequently arises concerns the organic removals taking place
during the chemical clarification of the wastewater. Table 7 summarizes the
reported results from the physical-chemical studies surveyed here. Roughly,
removal of anywhere from 50 to 85 percent of the organic matter can be expected, depending on the clarification efficiency and the nature of the wastewater. Since chemical clarification removes suspended and colloidal solids,
the remaining organic matter will be almost entirely in the dissolved state.

TABLE   7
Percent Removals of Organics by Chemical Clarification

Chemical	BOD	COD	TOC
Iron - clarified	50 - 79	51 - 59	48 - 77
filtered	59 - 79	64 - 70	57 - 85
Alum - clarified	57 - 72	60 - 77	62 - 67
filtered	-	-	-
Lime - clarified	55 - 83	50 - 82	48 - 80
filtered	67 - 85	62 - 85	66 - 81
TOTAL RANGES	50 - 85	50 - 85	48 - 85

If a wastewater has a very high soluble organic content, the major burden of treatment will be shifted toward the activated carbon system. This higher loading will probably cause the carbon to be exhausted more rapidly. This may also mean a higher organic content in the effluent or, at least, require a longer contact time to meet the desired effluent quality. This situation leads to the frequently-speculated problem of applicability for "strong wastes". In fact, it is not the total organic strength of the waste that presents the difficulty but the dissolved strength or organic concentration which remains after the clarification sequence. Thus, total and filtered organic analyses are required in estimating the performance of a physical-chemical system. Because the filtered sample contains some colloidal organics, the filtered organic concentration could be higher than that of the clarification system effluent. For example, the Eimco study indicated soluble COD (SCOD) and soluble TOC (STOC) values were reduced with iron coagulation by 47 and 25 percent, respectively. Oddly enough, lime treatment increased these parameters by about 10 percent. Friedman, et al, (10) reduced STOC by almost 30 percent through the iron clarification system. The result of all this is that the dissolved organic load to the carbon must be determined for each waste by performing a coagulation test in the laboratory. In practical terms a fresh domestic waste, with minimal sewer residence time, may be better suited to physical-chemical treatment than one which has had time for extensive dissolution of the organic constituents.

The final item on the clarification step of the process is the one most discussed, i.e. the sludge problem. Much has been said of the "voluminous quantities" of sludge produced by the chemical clarification step. In terms of sludge weight in pounds of dry solids per million gallons (lb/MG) a conventional activated sludge plant will produce approximately 2,000-2,500 lb/MG of (primary plus waste activated) sludge (4,17). The ranges of reported sludge production by chemical clarification in P-C pilot plants (5,6,15) are 3,100 to 12,100, 1,280 to 2,500 and 930 to 1,780 lb/MG for lime, iron and alum, respectively. The average production levels for these three chemicals are 6,500, 1,740 and 1,120 lb/MG respectively. Sludge volumes of 3,000 to 28,000 gal/MG, with an average of 10,000 gal/MG have been reported for lime clarification (5,6). Similarly, iron clarification has been reported to produce from 9,000 to 20,000 gal/MG of sludge, with an average value of 13,000 gal/MG (5,6). Data on alum sludge volumes is insufficient for any comparative analysis. These volumes can be compared to an average value (primary plus waste activated) of 20,000 to 25,000 gal/MG obtained in activated sludge treatment (17).

More pertinent, however, is the treatability of the sludge generated in

physical-chemical treatment. Some reports (5,6) have indicated that sludge
processing should include thickening, dewatering by centrifugation, vacuum or
pressure filtration and incineration. Isgard (3) suggested a stabilization
step, such as digestion or liming of alum and iron sludges prior to dewatering.

In Table 8, the data generated by the Eimco study are shown for the relative
thickening characteristics of sludges generated by the three major chemicals.
While the alum sludge data was quite limited, it is shown to illustrate that
it will probably be the most dilute sludge of the three. Laboratory vacuum
filtration tests of high pH lime sludge during the Eimco study predicted a
yield of 4.5 lb/hr/sf at 4.5 percent feed solids. The yield of moderate pH
lime sludge at 17.9 percent feed solids was 12.8 lb/hr/sf. Predicted cake
moisture content was 72 percent and 67 percent for the high and moderate pH
sludges, respectively. Iron sludge conditioned with lime and alum sludge
conditioned with polymer exhibited yields of 1.0 and 0.4 lb/hr/sf at feed
solids of 1.6 and 0.7 percent, respectively, with 80 to 85 percent cake moisture. The Westerly study of vacuum filtration and centrifugation of lime
sludges, indicated that polymer dosages of 1 to 2 lb/ton of dry solids were
needed to obtain high solids capture.

TABLE 8
Gravity Thickening of Chemical Sludges (5)

COAGULANT	SS, % in	SS, % out	LOADING lb/day/sf
Lime:			
pH >11.5	4.8	9.4	18
pH <11.0	11.5	17.9	46
Iron	1.5	4.5	12
Alum *	0.36	1.5	6

* Polymer will be needed to improve results

In the discussion of the dissolved organic removal portion of a physical-
chemical treatment sequence, chemical oxidation techniques are bypassed for
the sake of brevity and their limited developmental status, and only activated carbon adsorption is considered. Powdered activated carbon (PAC) has
been studied by some investigators, and their reports (5,12,18,19) reveal
several interesting and valuable pieces of information. Most of this work
has been with 2-stage countercurrent PAC systems, although one study also
tested a single-stage system. Powdered carbon feed rates have varied from
75 to 600 mg/l. Excellent removals of soluble organics have been experienced
in all of these studies. Although numerous differences exist in the systems
used, all of them required an in-depth filter following the PAC system to
remove suspended solids and carbon particles escaping the carbon contactors.
Some of the areas which are in need of further research and development include PAC regeneration and reuse, alternate sources of PAC and optimization
of PAC systems. Regeneration of spent PAC is in the early stages of development but several types of systems appear technically feasible. The pyrolysis
of waste paper, garbage or other waste organic material produces a char which
can be activated. The possibility of combined solid waste disposal-wastewater
treatment systems is intriguing, but additional research is necessary to properly determine their feasibility. Additional research is required in order
to define the optimum conditions for application of PAC systems and to provide
design and performance data necessary for the transition to full scale.

For the most part, granular activated carbon has been utilized in physical-chemical pilot studies. A review of the available reports indicates that only one study has reported an ultimate loading value for carbon. Bishop, et al, (9) reported the exhaustion of the first stage of a two-stage downflow system at a loading of 0.41 lb COD/lb carbon. A value of 0.45 lb COD/lb carbon was reported from the Westerly study, but it is not clear how that value was obtained (6).

Design details of these studies are quite varied, but some recurring items, such as a 30-minute empty-bed contact time, do appear. Weber, et al, (2) have compared upflow and downflow configurations and have found little difference in effluent quality with respect to organic removal. Friedman, et al, (10) used parallel upflow systems to show the value of adding oxygen to the first and third stages of a 4-stage system to maintain aerobic conditions in the columns. The aerobic system produced an effluent of 6.4 mg/l TOC from an influent of 28 mg/l TOC, as opposed to the anaerobic column which had an effluent TOC of 9.3 mg/l. Unfortunately, profuse biological growths resulting from the oxygen addition quickly plug downflow columns (2,6).

The maintenance of aerobic conditions is one way of preventing the production of hydrogen sulfide (H_2S) in the carbon columns, a problem which has assumed major importance in physical-chemical considerations. In attempting to rid their carbon systems of H_2S some researchers have tried surface and air-water backwashing, chlorinated backwash water, high pH backwashing and low-level chlorination of the influent without notable success (6,7,9). The Westerly report (6) indicated that both a high temperature (180°F) backwash for 30 minutes and a large addition of nitrate to the influent achieved some success in eliminating H_2S. Both of these systems were deemed impractical. Bishop (9) has used breakpoint chlorination of the influent with good results.

Most studies have used downflow columns and have employed the 8x30 mesh carbon due to its more desirable hydraulic characteristics. Upflow columns which require significant expansion of the carbon generally utilize the smaller 12x40 mesh for the same reason. Average hydraulic loading rates between 2 and 7 gpm/sf have been used for downflow columns, while upflow columns have been used at 5 to 6 gpm/sf. Successful backwashing systems usually employ surface washing and/or air scour in addition to normal backwashing facilities.

The relative merits of parallel vs series operation, upflow vs downflow configurations and the need for filtration ahead of the carbon columns are not considered here. In many instances the choice will be based on the particular circumstances of any individual application. Further studies in these areas will eventually provide better information upon which to judge the merits of each of these approaches.

In looking at some other aspects of the treatment capabilities of physical-chemical systems, a few "bits and pieces" of information are available. As noted earlier, many heavy metals (such as lead, copper, iron, chromium (III), cadmium, silver, manganese, nickel and zinc) are removed to a great extent in physical-chemical systems (20,21). Kreissl and Cohen (11) reported greater than 99 percent removals of coliforms, fecal coliforms and fecal streptococci by a physical-chemical package plant. Also, removals of color and turbidity as well as organic pollutants such as pesticides are generally in excess of those obtained in conventional treatment.

Nitrogen removal in the basic chemical clarification-carbon adsorption system of physical-chemical treatment is limited to that portion of the organic nitrogen removed along with other organic compounds. Therefore, physical-chemical plants of this type produce effluents that have high concentrations of ammonia nitrogen. This may be particularly undesirable in terms of the load to the receiving stream because of the 4.5 mg/l of oxygen demand per mg/l of NH3-N in the effluent. Physical-chemical nitrogen removal processes such as ammonia stripping, breakpoint chlorination and ion exchange can be added to the basic physical-chemical system, but the associated costs are significant. Since a discussion of these systems could constitute a conference in itself, no further discussion will be undertaken here.

SUMMARY

Although no full-scale physical-chemical treatment plants for municipal wastewaters are in operation at this time, much useful information has been generated by numerous pilot-scale studies. The physical-chemical approach presently has certain basic advantages over conventional biological systems which would warrant its use where specific physical, chemical or economic restrictions exist. The development of physical-chemical systems and individual processes represents a series of new tools to the design engineer for more effective abatement of pollution from municipal wastewaters.

REFERENCES

1. I. J. Kugelman and J. M. Cohen, "Physical-Chemical Processes," paper given to U.S. EPA Technology Transfer Design Seminar, Seattle, Washington (1971).

2. W. J. Weber, C. B. Hopkins, and R. Bloom., Jour. WPCF, 42,83 (1970).

3. E. Isgard, "Chemical Methods in Present Swedish Sewage Purification Techniques." Paper presented to 7th Effluent & Water Treatment Exhibition and Conference (1971).

4. R. L. Culp and G. L. Culp, Advanced Wastewater Treatment, Van Nostrand Reinhold Co., New York (1971).

5. D. E. Burns and G. L. Shell. Final report for U.S. EPA Project No. 17020 EFB (1972).

6. A. J. Shuckrow and W. F. Bonner Final report to Zurn Environmental Engineers (1971).

7. W. Lee, Monthly Progress Reports, Pomona Research Facility (1972).

8. R. A. Villiers, Water & Wastes Engrg., 9,1, 32 (1972).

9. D. F. Bishop, T. P. O'Farrell and J. B. Stamberg, Jour.WPCF, 44 361 (1972).

10. L. D. Friedman, W. J. Weber, R. Bloom and C. B. Hopkins, Water Pollution Control Research Series 17020 GDN (1971).

11. J. F. Kreissl and J. M. Cohen, "Treatment Capability of a Physical-Chemical Package Plant," Water Research (in press).

12. R. L. Beebe and C. F. Garland, Final report for U.S. EPA Project
     No. 17050 EGI (1971).

13. P. F. Atkins, D. A. Scherger and R. A. Barnes, "Ammonia Removal
     in a Physical-Chemical Wastewater Treatment Plant," paper given
     at 27th Purdue Industrial Waste Conference (May 1972).

14. G. L. Culp, "Physical-Chemical Treatment Plant Design," paper
     presented at Technology Transfer Design Workshop, New York, N.Y.(1972)

15. L. Ulmgren, National Swedish Environmental Protection Board
     Publication 1969: 10E (1969).

16. G. J. Stander and L. R. J. Van Vuuren, Jour. WPCF, 41, 355
     (1969.

17. J. E. Smith, "Present Technology of Sludge Dewatering," paper
     presented to 64th Annual Meeting of American Society of Sanitary
     Engineering, Atlanta, Georgia (Oct. 1970).

18. P. V. Knopp and W. B. Gitchel, Proc. 25th Purdue Industrial Waste
     Conference, 25, 687 (1970).

19. A. J. Shuckrow, G. W. Dawson and W. F. Bonner, "Pilot Plant
     Evaluation of P-C Process for Treatment of Raw and Combined
     Sewage Using PAC," paper presented to 44th Annual Conference
     of W.P.C.F. (1971).

20. J. M. Cohen, "Control of Environmental Hazards in Water," paper
     presented to ACS 3rd Central Regional Meeting, Cincinnati,
     Ohio (1971).

Applications of New Concepts of Physical-Chemical
Wastewater Treatment
Sept.18-22, 1972

PHYSICOCHEMICAL SYSTEMS FOR DIRECT WASTEWATER TREATMENT

Walter J. Weber, Jr.
Professor of Environmental and Water Resources Engineering
The University of Michigan
Ann Arbor, Michigan

Introduction

A broad range of physicochemical separation and conversion processes have been studied over the past 15 years for potential applications to municipal and industrial wastewater treatment. Among these have been adsorption, coagulation, chemical oxidation, solvent extraction, ion exchange, distillation, freezing, reverse osmosis, ultrafiltration, electrodialysis, electrochemical degradation, flotation, and foam separation (1,2,3). The process combination of coagulation and precipitation for removal of insoluble impurities followed by adsorption on activated carbon for removal of soluble organic impurities has emerged as the treatment sequence of greatest promise in terms of both technologic and economic feasibility. The purpose of this paper is to summarize the salient features of this physicochemical treatment system in the perspective of its application to water pollution control.

Development of physicochemical processes for higher levels of treatment centered initially on "tertiary" systems designed to follow "primary" sedimentation and "secondary" biological treatment (2,3). There are, however, several fundamental shortcomings to this approach. First, the implementation of tertiary systems depends upon prior implementation of primary and secondary systems. Second, the addition of tertiary processes to primary and secondary processes incurs capital and operating expenses which are likely to discourage such developments. Third, the effective operation of a tertiary process in this format is dependent to a large extent on the consistent and efficient operation of a biological secondary process, which is normally subject to problems arising from transients in waste composition and flow, often requiring at least partial diversion, and from the occasional presence of toxic materials.

13

The concept of applying coagulation-adsorption processes directly to raw wastes rather than to secondary effluents therefore derived partially from considerations regarding the effectiveness and reliability of treatment and partially from the relative economics of "direct" versus "tertiary" treatment systems (4). Direct treatment subsequently has been demonstrated in pilot investigations to be an attractive technical and economic alternative to biological treatment (5,6,7,8,9,10).

With pretreatment of raw sewage by chemical clarification - which results in significant removal of both total and soluble organic matter, phosphates and suspended solids - activated carbon treatment commonly produces a clear effluent of low organic content, suitable to meet most requirements for water reuse and pollution control.

To give some illustration of this, Table 1 summarizes overall treatment results obtained at six different pilot plant installations of physicochemical treatment by coagulation and adsorption. Table 2 gives a partial list of full-scale plants planned or currently under design.

One major factor to which the coagulation-adsorption treatment scheme does not address itself is ammonia nitrogen. If it is necessary to remove ammonia, an additional process such as air stripping at high pH, selective ion exchange, or breakpoint chlorination must be added. Breakpoint chlorination has been shown to be highly effective in this application (10).

Coagulation Systems

Coagulation of wastewater may be accomplished with any of the common water coagulants including lime, iron and aluminum salts, and synthetic polymers. The choice is made on the basis of suitability for a particular waste, availability and cost of the coagulant, and sludge treatment and disposal considerations. For example, iron is sometimes available at no cost as a waste product in the form of pickling liquor, and its presence in sludge presents no particular problems for anaerobic digestion of the sludge. Lime generally provides good clarification, a rapidly settling sludge, and permits the use of a simple method for recovery that also insures destruction of most sewage solids in the resulting sludge. Sludges from lime coagulation can be thickened and incinerated to destroy the sewage organic solids and provide regenerated coagulant by recalcination. Lime, perhaps in combination with an organic polymer, coagulant aid, or small amounts of iron or aluminum salts to aid flocculation and sedimentation and decrease lime requirements, is therefore often the preferred coagulant for raw sewage. Additionally, lime does not contribute undesirable inorganic anions to the treated water, a factor which is of some import relative to potential water reuse applications.

Depending upon the characteristics of the wastewater and the ultimate water quality objectives, lime coagulation may be carried out in a one-stage or a two-stage operation. A one-stage process is generally suitable for efficient solids removal and phosphate

TABLE 1

Operating Results of
Pilot Physicochemical Treatment Plants

Plant	Organic Removal, %	Effluent Concentration
Ewing-Lawrence (New Jersey)	95-98	TOC[1] = 3-5
Blue Plains (Washington, D.C.)	95-98	TOC = 6
Lebanon (Ohio) (powdered carbon)	95	TOC = 11
New Rochelle (New York)	95	COD[2] = 8
Rocky River (Ohio)	93	BOD[3] = 8
Salt Lake City (Utah) (powdered carbon)	91	BOD = 13

(1) TOC - total organic carbon
(2) COD - chemical oxygen demand
(3) BOD - biochemical oxygen demand

TABLE 2

Physicochemical Treatment Plants
Planned or Designed

Site	Capacity, million gallons per day[1]
Rocky River, Ohio	10
Painesville, Ohio	5
Cleveland, Ohio	10
Fitchburg, Massachusetts	15
Waterford, New York	2
Cortland, New York	10
Clay, New York	10
Niagara Falls, New York	60
Garland, Texas	30
Owosso, Michigan	6

(1) 1 million gallons per day = 3785 cubic meters per day

precipitation with waters of high hardness and alkalinities in
excess of approximately 200-250 mg/l as calcium carbonate. However, for waters of alkalinity below about 200 mg/l a two-stage
process frequently gives better results, although even in these
cases a one-stage operation with addition of cationic polymer or
iron or aluminum salts may be satisfactory. In the two-stage
lime process the first stage is accomplished at a high pH, usually
in excess of 11.0-11.5. Carbon dioxide produced during recalcination is then introduced to reduce the pH to approximately 10, and
to precipitate excess calcium ions added in the first stage. This
precipitate is coagulated and settled in the second stage by
addition of small amounts of polymer or metal-ion coagulant.

Adsorption Systems

Adsorption is fundamental to physicochemical treatment for
removal of soluble organic impurities. Adsorption of organic
materials from solution onto solids such as activated carbon, the
adsorbent of choice for water and waste treatment, may be a consequence of either lyophobic behavior on the part of the organic
matter, or of a high affinity for the adsorbent. For the
majority of systems of concern in water and waste treatment, both
factors are involved (11,12,13).

Applications of adsorption for purification of public water
supplies have been well known and widely practiced for years.
Taste and odor producing impurities removed in such applications
are usually present in low concentration, and the need to provide
removal often intermittent or occasional. Conversely, relatively
large amounts of organic impurities are present in wastewaters,
and treatment must be continuous. More efficient utilization of
the capacity of activated carbon is consequently required in
waste treatment applications than is commonly achieved in water
treatment applications. Further, the fact that large amounts
of carbon are used requires a scheme of regeneration and reuse
of this material. These requirements suggest the use of granular carbon continuous contacting systems; granular carbon
because of the relative ease of handling and regenerating this
material relative to powdered carbon.

The most common type of continuous system is one in which
water or wastewater is passed through fixed beds of carbon. In
such systems the wastewater is applied to the beds at rates
generally ranging from 2 gpm/ft^2 to 8 gpm/ft^2 (81-326 l/min/m^2).
In this flow range essentially equivalent adsorption efficiency
is obtained for equivalent contact times. At flow rates below
2 gpm/ft^2 (81 l/min/m^2) adsorption efficiency is reduced, while
at flow rates above 8 gpm/ft^2 (326 l/min/m^2) excessive pressure
drop takes place in packed beds. Contact times employed are in
the range of 30 minutes to 60 minutes on an empty bed basis. In
general, increases in contact time up to 30 minutes yield pro-
portionate increases in organic removal. Beyond 30 minutes the
rate of increase falls off with increases in contact time, and at
about 60 minutes the effects of additional contact time become
negligible. Carbon beds operated at the lower end of the flow
range are generally designed for gravity flow. Systems designed
for higher flow rates must employ pressure vessels if packed beds

are used. A pressure vessel is more expensive to construct than
a gravity flow vessel, but commonly requires less land area, and
provides greater ability to handle fluctuations in flow.

Provision must be made to regularly backwash packed-bed carbon systems because they collect suspended solids and tend to
develop attached biologic growths in this application. Backwashing alone generally relieves clogging due to suspended solids,
but does not completely remove attached biologic growth. It is
advisable to include a surface wash and air scour to be assured
of removal of gelatinous biologic growth.

This attached growth can lead to development of anaerobic
conditions in packed beds. Aeration of the feed is partially
effective in preventing anaerobic conditions, but this also
accelerates biologic growth to the extent that excessive backwash
is required; air-binding can also result. Effective control of
biological growth can be accomplished in most instances by
regular chlorination of the influent to the adsorbers, and/or by
chlorination during regular backwash operations.

Packed beds of granular carbon are well suited for treatment
of solutions containing little or no suspended solids, and under
such circumstances normally operate effectively for extended
periods without clogging or excessive pressure loss. However, the
suspended solids invariably present in municipal and industrial
wastewaters and the potential for biologic growth on the surfaces
of the carbon can present some problems for the use of packed
beds, as noted above. Because solids and biologic activity
usually cause progressive clogging and high head loss in packed
beds, increased interest has developed in the potential of
expanded-bed adsorbers, which have certain inherent operating
advantages over packed-bed adsorbers for treating solutions containing suspended solids. By passing wastewater upward through a
bed of carbon at velocities sufficient to expand the bed, problems
of fouling, plugging and increasing pressure drop are minimized.
Effective operation over longer periods of time results, as has
been demonstrated in comparative laboratory studies and in pilot
field investigations in both "tertiary" and direct physicochemical
applications (7,8,14,15). Another advantage of the expanded bed
is the relatively small dependence of pressure drop on particle
size. It is possible to use carbon of smaller particle size in
an expanded bed than is practical in a packed bed, thus taking
advantage of somewhat higher adsorption rates which obtain for
smaller particles (11).

Perhaps the most significant potential benefit provided by
expanded-bed adsorption systems is the apparent extention of the
operational capacity of activated carbon observed by Weber et al,
(7,8). These researchers found that apparent sorption capacities
in excess of 100 weight-percent as organic matter and 150 weight-
percent as chemical oxygen demand (COD) could be obtained in
expanded-beds of activated carbon in which biologic growth was
allowed to fully develop (8). Because expanded beds require little
maintenance, extended periods of undisturbed operation facilitate
the development and continuous growth of bacteria on the carbon
surfaces. This biologic activity, primary anaerobic,

degrades some of the organic matter which adsorbs on the carbon, functioning to provide in-situ partial regeneration by renewing a portion of the carbon surface for continued adsorption. The direct feeding of settled but uncoagulated (not chemically coagulated) raw sewage to expanded-bed activated carbon systems in which biologic activity was encouraged, followed by chemical coagulation of the carbon-treated wastewater, has been investigated and described (16). The results of these studies indicated no real advantage to this sequence of treatments, and considerable disadvantage due to fouling of the carbon by the solids present in the uncoagulated feed.

To provide a good effluent and to utilize the sorption capacity most effectively, an approach to countercurrent contact is commonly required. This can be achieved by having the wastewater flow through a number of contactors or stages in series in one direction while the carbon moves in the opposite direction. In powdered carbon contacting systems this is the procedure used. With granular carbon the procedure is generally simplified to avoid unnecessary handling of the carbon. For most granular carbon contact systems the lead contactor in a series of adsorption columns is removed from service when the carbon it contains is exhausted (or nearly so) and, after being refilled with fresh carbon, is placed at the end of the series. Each contactor is thus advanced one position in the series by piping and valving arrangements which permit shifting of the inflow and outflow points of the series accordingly. As the number of stages increases, the piping and valving arrangement becomes more complex and costly. A compromise between the advantage of employing multiple stages to more effectively utilize carbon capacity and the cost of each additional stage must be achieved.

Principal factors which must be considered in the design of a carbon adsorption system may be summarized as:

1. type of carbon, granular or powdered;
2. contact time;
3. flow rate;
4. configuration, series or parallel;
5. number of stages;
6. mode of operation, packed-bed or expanded-bed, pumped or gravity flow; and,
7. adsorption capacity.

Table 3 gives carbon capacities obtained in field operations at several physicochemical pilot plants. In that the wastes, effluent criteria, number of contact stages, etc., varied from plant to plant, it is not surprising that some spread in the results is observed. For general planning purposes a COD capacity of 50 weight-percent is reasonable if no biologic extension of carbon capacity is taken into account. This is approximately equivalent to a requirement of 500 pounds of activated carbon per million gallons (60 grams per cubic meter) of sewage treated. However, the results obtained by Weber et al (8) with biologically-extended adsorption systems suggest that it may be possible to achieve higher effective capacities, reducing the carbon exhaustion rate to less than 200-250 pounds per million gallons (24-30 grams per cubic meter).

TABLE 3

Carbon Capacities Obtained in Physicochemical Pilot Plants

Plant	Capacities, Weight-percent	
	TOC	COD
Blue Plains (Washington)	15	41
Ewing-Lawrence (New Jersey)*	50	150
New Rochelle (New York)	20-24	60
Lebanon (Ohio)	22	50

*Biologically-extended expanded-bed operation (8)

Even for the highest capacities observed, the initial costs of carbon are such as to make regeneration and reuse of this material highly desirable. Technically and economically feasible regeneration of granular activated carbon can be accomplished by controlled heating in a multiple-hearth or rotary-kiln furnace in the presence of steam. During each regeneration cycle some carbon is lost by burning and attrition, and some by alteration of surface properties. The overall loss, expressed as percent by weight of virgin carbon required to restore the total original capacity of the batch, ranges from 5 to 10 percent. For planning purposes, carbon make-up requirements can be considered to range from 25 to 50 lbs per million gallons (3-6 grams per cubic meter) of wastewater treated, again not taking account of in-situ biological regeneration.

At present, regeneration systems for powdered carbon are being developed and tested at the pilot stage. A successful process for regeneration of the powdered form would represent a significant step toward making a system utilizing this lower cost material a technical and economic reality. The key factor will be maintaining carbon loss at a sufficiently low level during regeneration.

Design Configurations

A suggested flow sheet for physicochemical treatment of wastewaters is given in Figure 1. In this scheme, coagulant is added to the raw sewage, and flocculation takes place in a chamber which provides moderate agitation for an average detention time of 15 minutes. Clarification takes place in a sedimentation basin with an average detention time of two hours. The particular flow sheet presented here is a single-stage coagulation system.

The clarified effluent is then passed through activated car-

bon adsorption units for removal of dissolved organics. The preferred mode of operation is an expanded bed, which permits the use
of simple open-top concrete contacting basins and relatively
trouble-free operation. The use of open tanks with overflow weirs
at the surface of the contacting basin provides a means for
additional aeration of the wastewater during treatment, thus help-
ing to control anaerobic conditions in subsequent reactors. Two-
stage contacting of the activated carbon is outlined in the treatment sequence given in Figure 1. However, a larger number of
stages can be utilized if desirable for a particular application.
A typical plant layout for a design capacity of 10 million gallons
(37,850 cubic meters) per day might be based on five parallel
adsorption units of two stages each. When the granular carbon in
the first stage of one unit is spent, that unit can be taken off
stream while the spent carbon is removed and regenerated in a
furnace provided for this purpose. During the time this unit is
off-stream for regeneration, the other four units can run at 25%
higher feed rate each. Upon completion of the regeneration, the
carbon is returned to the adsorber, which then becomes the
second stage of that unit; the former second stage with partially
spent carbon becoming the first stage. Feed is then evenly
divided to the five units until another carbon bed is spent.

The water resulting from the clarification and activated carbon treatment will enhance the quality of surface waters, and
with disinfection is suitable for many reuse applications. A
final filtration may be desirable to insure a crystal clear
effluent for some uses. This post-filtration would remove any
suspended matter passing through or biologically generated in the
carbon columns.

Capital and Operating Costs

A detailed analysis of capital and operating costs has been
carried out for the physicochemical treatment sequence outlined
above, for a nominal design capacity of 10 million gallons
(37,850 cubic meters) per day. The capital cost analysis is
summarized in Table 4, and the annual operating costs analysis,
including amortization at 6% for 24 years, is summarized in
Table 5. These costs are based on a carbon utilization of 500
pounds per million gallons (60 grams per cubic meter). However,
as already noted, biological extension should significantly reduce
the carbon exhaustion rate. If the exhaustion rate were reduced
to 250 pounds per million gallons (30 grams per cubic meter), as
seems feasible from the Ewing-Lawrence pilot studies (8) the cost
of carbon tratment would drop from 8.2 cents per thousand
gallons (2.1 cents per cubic meter) to 6.8 cents per thousand
gallons (1.8 cents per cubic meter), and the combined physicochemical treatment operating costs would be less than 19 cents
per thousand gallons (5 cents per cubic meter). This assumes no
reduction in capital investment, which by use of a smaller
regeneration facility to reactivate the smaller volume of spent
carbon produced, would result in additional savings in total
operating costs as well.

TABLE 4

Estimated Capital Costs for

Direct Physicochemical Treatment of Wastewaters

Basis: 10 mgd (37,850 m^3/day)

	Total Cost	
	Expanded Bed	Packed Bed
Equipment, Piping and Instrumentation		
Pretreatment and Flocculation	$ 154,000	$ 154,000
Clarification and Sludge Handling	1,113,600	1,113,600
Adsorption System	551,000	617,600
Regeneration System	213,000	222,500
Auxiliary Facilities		
Electric Power	51,000	80,000
Fuel Handling	20,000	20,000
Buildings and Structures	475,000	493,000
Roads, Walks, Fence	170,000	170,000
Activated Carbon	288,000	288,000
Contingency	377,000	390,000
Total Fixed Capital	$ 3,412,500	$3,548,700

TABLE 5

Estimated Annual Operating Costs for

Direct Physicochemical Treatment of Wastewater

Basis: 10 mgd (37,850 m^3/day)

		Expanded	Packed
1.	Operating Labor*	$ 86,600	$ 86,600
2.	Maintenance Labor - 3% of Plant Physical Costs	53,600	55,600
3.	Maintenance Materials - 2% of Plant Physical Costs	35,800	37,100
4.	Maintenance Supplies - 15% of (2)+(3)	13,400	13,900
5.	Supervision - 15% of (1)	13,000	13,000
6.	Payroll Overhead - 15% of (1) + (2)	21,000	21,000
7.	General Overhead - 30% of (1)+(2)+(6)	48,400	49,100
8.	Insurance - 1% of Plant Physical Costs	17,900	18,600
9.	Carbon Makeup - 5% @ $.28/lb	27,500	27,500
10.	Lime Makeup - 25% @ $20/T	23,000	23,000
11.	Fuel - @ $.50/MM Btu	75,000	75,000
12.	Power - $.10/kwh	33,000	45,000
13.	Amortization - 24 years @ 6%	271,600	281,800
	Total Annual Cost	$719,800	$747,500
	Treatment Cost - ¢/1000 gal.	19.7	20.5
	Treatment Cost - ¢/m^3	5.2	5.4

*2 Shift men + 2 day men @ $4.00 per hour

Summary

Physicochemical wastewater treatment consisting of coagulation and adsorption holds significant promise as a means of economically meeting today's higher effluent standards and water reuse requirements.

The system consistently produces high levels of treatment and has a high degree of stability and reliability. Unlike biological waste treatment systems, it is highly resistant to shock loads and toxic waste constituents. Biological systems are notoriously sensitive to changes in environmental conditions. If a toxic material gains even temporary entrance to a biological plant, or a hydraulic peak occurs, not only will the efficiency of the plant drop off, but recovery may take from several days to several weeks. In a physicochemical plant serious upsets are unlikely Further, it can be expected that an immediate recovery of the plant will take place once the source of the upset is eliminated. This inherent stability of performance is also reflected in greater design and operational flexibility. Entire sections of a physicochemical plant can be cut in or out of the process stream as required, and a temporary overload can be absorbed with little effect. The major advantages of a physicochemical system over a biological system are summarized below:

1. less land area required (1/2 to 1/4 that for a biological system);
2. lower sensitivity to diurnal variations;
3. not affected by toxic substances;
4. potential for significant heavy metal removal;
5. superior removal of phosphates;
6. greater flexibility in design and operation; and,
7. superior removal or organic waste constituents.

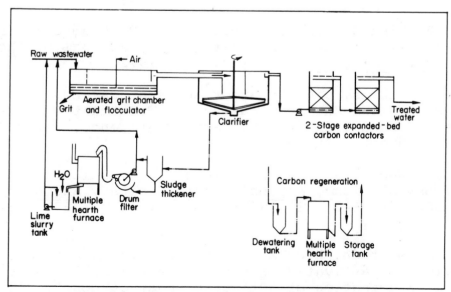

FIG.I. TYPICAL FLOWSHEET FOR TREATMENT OF WASTEWATER BY CHEMICAL CLARIFICATION AND ADSORPTION

References

(1)   J.C. Morris and W.J. Weber, Jr., "Preliminary Appraisal of Advanced Waste Treatment Processes", SEC TR W62-24, U.S. Dept. of Health, Education and Welfare, Public Health Service, R.A. Taft Sanitary Engineering Center, Cincinnati, Ohio, 1962

(2)   AWTR-1, "Summary Report - The Advanced Waste Treatment Research Program", SEC TR W62-9, U.S. Dept. of Health, Education and Welfare, Public Health Service, R.A. Taft Sanitary Engineering Center, Cincinnati, Ohio, 1962

(3)   AWTR-14, "Summary Report - The Advanced Waste Treatment Research Program", 999-WP-24, U.S. Dept. of Health, Education and Welfare, Public Health Service, R.A. Taft Sanitary Engineering Center, Cincinnati, Ohio, 1965

(4)   W.J. Weber, Jr. and J.G. Kim, "Preliminary Evaluation of the Treatment of Raw Sewage by Coagulation and Adsorption", Technical Memorandum, TM-2-65, San. and Water Resources Eng. Div., The University of Michigan, Ann Arbor, Michigan, 1965

(5)   J.L. Rizzo and R.E. Schade, "Secondary Treatment With Granular Activated Carbon", Water and Sewage Works, 116, 307, 1969

(6)   D.G. Hager and D.B. Reilly, "Clarification-Adsorption in the Treatment of Municipal Wastewaters", Journal of the Water Pollution Control Federation, 42, 5, 794, 1970

(7)   W.J. Weber, Jr., C.B. Hopkins and R. Bloom, Jr., "Physicochemical Treatment of Wastewater", Journal of the Water Pollution Control Federation, 42, 1, 83, 1970

(8)   W.J. Weber, Jr., L.D. Friedman and R. Bloom, Jr., "Biologically-Extended Physicochemical Treatment", Proceedings, Sixth Conference on Water Pollution Research, Jerusalem, 18-24 June, 1972

(9)   D.F. Bishop, T.P. O'Farrell and J.B. Stamberg, "Physical-Chemical Treatment of Municipal Wastewater", Journal of the Water Pollution Control Federation, 44, 3, 361, 1972

(10)   P.F. Atkins, D.A. Scherger and R.A. Barnes, "Ammonia Removal in a Physical-Chemical Wastewater Treatment Process", Proceedings, 27 Annual Industrial Waste Conference, Purdue University, Lafayette, Indiana, 1972

(11)   W.J. Weber, Jr. and J.C. Morris, "Kinetics of Adsorption on Carbon from Solution", Journal of the Sanitary Engineering Division, ASCE, 89, SA2, 31, 1963; Closure: Ibid., 89, SA6, 53, 1963

(12)   J.S. Mattson, H.B. Mark, Jr. and W.J. Weber, Jr., "Surface Oxides of Activated Carbon: Internal Reflectance Spectroscopic Examination of Activated Sugar Carbons", Journal of Colloid and Interface Science, 31, 1, 116, 1969

(13)   J.S. Mattson, L. Lee, H.B. Mark, Jr. and W.J. Weber, Jr., "Surface Oxides of Activated Carbon: Internal Reflectance Spectroscopic Examination of Activated Sugar Carbons", Journal of Colloid and Interface Science, 33, 2, 284, 1970

(14)   W.J. Weber, Jr., "Fluid-Carbon Columns for Sorption of Persistent Organic Pollutants", Proceedings, Third International Conference on Water Pollution Research, 1, 253, 1967

(15)   W.J. Weber, Jr., C.B. Hopkins, and R. Bloom, Jr., "A Comparison of Expanded-Bed and Packed-Bed Adsorption Systems", Report No. TWRC-2 U.S. Dept. of the Interior, Federal Water Pollution Control Administration (formerly, now EPA), Cincinnati, Ohio, 1968

(16)   C.B. Hopkins, W.J. Weber, Jr. and R. Bloom, Jr., "Granular Carbon Treatment of Raw Sewage", Report No. ORD-17050DAL05/70, Water Pollution Control Research Series, U.S. Dept. of the Interior, Federal Water Quality Administration (formerly, now EPA), Cincinnati, Ohio, 1970

Applications of New Concepts of Physical-Chemical
Wastewater Treatment
Sept.18-22, 1972

ADVANCED WASTE TREATMENT AT ALEXANDRIA, VIRGINIA

Carl W. Reh, Robert L. Hall, Thomas E. Wilson
Greeley and Hansen, Engineers
Chicago, Illinois

Introduction

Alexandria was founded in 1748, and has increased in size from an initial area of about 60 acres to its present area of more than 9,760 acres. Sewers and drains were introduced about a century after founding; another century elapsed, however, before the sewage and other liquid-carried wastes were treated prior to being discharged in the Potomac River. In 1951 the Virginia State Water Control Board ordered the City to provide sewage treatment facilities. In 1952 the Alexandria City Council created the Alexandria Sanitation Authority and charged it with the responsibility for the development of a pollution abatement program. Studies at that time indicated that considerable savings in construction and operation could be realized by the construction of a system which would not only serve the City of Alexandria, but which would also serve parts of Fairfax County. Within a year after formation, the Authority reached an agreement with Fairfax County to the effect that the Authority would undertake construction and operation of facilities which would serve a part of the County.* These facilities comprised of trunk and interceptor sewers and an 18 mgd trickling filter sewage treatment plant, were placed in operation in 1956.

In August, 1957, the Surgeon General of the U. S. Public Health Service convened the first conference of the Interstate Commission of the Potomac River Basin on the pollution of interstate water in the Washington metropolitan area. The metropolitan area is comprised of the City of Alexandria, the District of Columbia, Fairfax and Arlington Counties, Virginia; Prince William, Prince Georges and Montgomery Counties, Maryland; and adjacent Virginia

---

*Note that in Virginia, cities and counties are separate municipalities.

and Maryland communities.

The U. S. Public Health Service reported that pollution of the
Potomac River was caused by discharges of untreated and inade-
quately treated sewage and industrial wastes. The conferees,
comprising the Commonwealth of Virginia, the State of Maryland,
the District of Columbia, and the U. S. Public Health Service
concluded that the report was essentially correct and recommended
that the conference be reconvened after publication of a comprehensive master plan on pollution control by the Interstate Commission on the Potomac River Basin.

The second session of the conference was held on February 13,
1958. The conferees recommended, among other items, that "the
Virginia Communities in the Washington metropolitan area provide
treatment comparable to that provided by the District of Columbia
and Maryland political subdivision."

The Secretary of Interior reconvened the Conference on April 2,
1969, "to discuss pollutional problems, receive reports from the
State and Federal Agencies responsible for pollution control, and
recommend measures to correct pollution problems." One of the
recommendations of the conferees was that an approach be made
toward "the continuous application of the best available waste
cleansing technology to provide the highest water quality which
under no circumstances will be less than the approved water quality standards." Following approximately a month's recess, the
conferees recommended, on May 8, 1969, the discharge of phosphorus, nitrogen and 5-day biochemical oxygen demand (BOD_5) be
significantly reduced.

An additional recommendation was made that "continuous and effective effluent disinfection shall be practiced at all sewage treatment plants to meet State and Federal water quality standards."

Effluent Standards

In June, 1971, the Virginia State Water Control Board established
effluent standards on a concentration basis more rigorous than
the guidelines established at the 1969 Conference. These standards, based on a one-month average, are:

       BOD_5: not greater than 3 ppm
       Total Phosphorus (P): not greater than 0.2 ppm
       Unoxidized nitrogen (N): not greater than 1.0 ppm (during
            the period April 1 through October 31)
       Total Nitrogen (when technology is available): Not greater
            than 1 ppm

At some point in time prior to February 1, 1972, the additional
requirement that dissolved oxygen in the effluent should be
greater than 6 ppm was imposed.

The quality of effluent from the Alexandria Sanitation Authority
sewage treatment plant, constructed in 1956, could not meet these
new water quality standards. Similarly, Fairfax County's West-

gate plant, also affected by the recent rulings, could not meet
the new effluent quality standards. Further, the sewage quantities tributary to the Authority Plant and to Fairfax County Westgate Plant exceed their design capacities. Studies by the Authority and Fairfax County showed that the higher degrees of treatment and increased capacity required for both plants could best
be provided by combining the service areas and treatment facilities for Alexandria and Westgate at the Alexandria site.

Service Area

The present Alexandria Sanitation Authority system serves the
City of Alexandria and neighboring portions of Fairfax County.
The natural drainage area from which sewage is received is the
Cameron Run watershed. Fairfax County areas within the Cameron
Run watershed are now served jointly by the Alexandria Sanitation
Authority and by the County Westgate Sewage Treatment Plant. The
tributary area is shown on Figure 1. All areas within the City
of Alexandria, except the Fourmile Run area, are tributary to the
sewage treatment works. The upper Fourmile Run area sewage
drains to the Arlington County Sanitation Authority system.
Sewage from the lower Fourmile Run area is pumped to the ASA
plant via the Commonwealth interceptor. The Potomac River watershed within the City is tributary to the sewage treatment works.
The Belle Haven area, all within Fairfax County, will be pumped
from the Westgate plant to the ASA sewage treatment works.

Proposed Project

The existing trickling filter plant will be expanded to a 54-mgd
physical-chemical plant to meet the specified requirements. Layout of units will permit ultimate expansion to 81 mgd. All of
the existing plant units will be continued in service. The
sewage characteristics are given in Table 1.

TABLE 1

Sewage Characteristics

Characteristic	mg/l	1000 lbs/day - Annual Average			
		1973	1985	1992	Ultimate
Soluble BOD$_5$	30	9.0	12.1	13.4	20
Total BOD$_5$	220	66.5	89.5	98.8	148
Suspended Solids	220	66.5	89.5	98.8	148
Total Phosphorus	15	4.5	6.1	6.8	10
Total Nitrogen (as N)	42	12.7	17.1	18.9	28
Ammonia Nitrogen (as N)	28	8.4	11.2	12.4	19

Process Rationale

The current standards imposed by the Commonwealth of Virginia
posed a new challenge in that no existing plant, including South
Lake Tahoe, is capable of consistently meeting the required effluent standards. The continuing engineering challenges of space
limitations and the need to make best use of the existing facilities are also a part of this project design. The design procedure

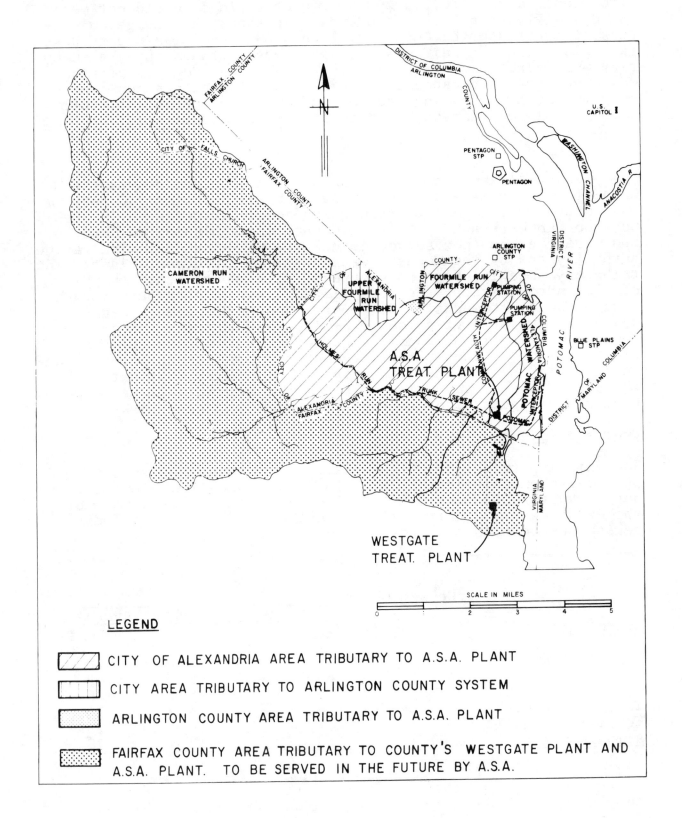

FIG. 1
Watersheds and Service Area,
Alexandria Sanitation Authority

took each standard, and critically evaluated each process or combination of processes to establish its ability to meet, under the given limitations, that standard. In the subsequent sections of this paper each of the processes selected to meet the water quality objectives of P, BOD, SS, and N removal is discussed.

Phosphorus Removal

About 75 percent of the phosphorus in the raw sewage is soluble. The existing units currently remove very little phosphorus. Laboratory tests performed on the Alexandria sewage (1) showed that about 250 mg/l of alum (as M.W.666) or 130 mg/l of ferric chloride, each with about 0.5 mg/l of anionic polyelectrolyte, would be needed to remove sufficient phosphorus to meet the standard of 0.2 mg/l effluent P. A lime system was also tested, but did not perform as well as either the alum or ferric chloride systems.

It is expected that phosphorus removal will be done in two stages. The first stage will include incidental phosphorus removal when an estimated 60 mg/l of alum is added to the primary tanks. (The major purpose of adding chemicals in this first stage is for improved efficiency of suspended solids removal.) This stage will make use of the existing and new primary sedimentation tanks. In the second stage, enough additional alum (about 180 mg/l) will be added to assure that remaining phosphorus is precipitated so that the effluent standard is met. The concentration of the phosphorus and the magnitude of the flow entering the mixing basins preceding the final sedimentation basins will be continuously monitored and the alum dose paced, accordingly.

Soluble BOD$_5$ Removal

Tests show that approximately 30 mg/l of the estimated 220 mg/l of raw sewage BOD$_5$ is in the soluble form. The existing trickling filters remove about 45 percent of the soluble BOD$_5$ applied to them, as well as much of the particulate BOD$_5$. These trickling filters are operated with recirculation at an approximately constant hydraulic loading of 27 mgd. They are capable hydraulically of handling an estimated 42 mgd.

Several basic alternative processes were considered for soluble BOD$_5$ removal. The first, addition of more trickling filters, was rejected since the filters alone would not be capable of meeting the rigid effluent standards and would occupy far too much land area.

The second alternative considered was use of an activated sludge plant. This too was rejected for reasons similar to the first alternative. A third alternative was an activated sludge plant with a tertiary physical-chemical plant, similar to the existing South Lake Tahoe plant. While this looked like it would be capable of meeting the effluent standards (excluding nitrogen), its cost and land requirements were excessive. The final alternative considered was a physical-chemical (activated carbon) plant. This alternative would require the least investment in terms of both initial cost and land. In addition, the existing trickling filters could be used, without modification, as roughing filters to reduce the soluble BOD5 load to the physical-chemical plant. This was the design chosen.

The chosen process scheme calls for granular-activated carbon columns following the existing trickling filters. This scheme provides for the removal of easily biologically degradable soluble organic matter on the filters and allows more flexibility in the operation of the carbon columns. Since the capacity of the filters will not be exceeded for several years after construction, there will be ample time to evaluate their value to the overall process. It is considered that some modification or additions to the trickling filters may be needed ultimately, but not to be part of the proposed construction. Initially, the only modification will be the elimination of the recirculation feature.

The expected mode of operation of the carbon columns is two-stage, upflow. This is similar to the mode of operation proposed and thoroughly investigated by others (2,3,4,5,6). When the granular carbon in the first stage is exhausted, the tank would be emptied, filled with fresh carbon and placed on line as the second stage, with the original second stage now becoming the first stage. The design calls for 15 minutes empty bed retention time in each stage. Laboratory column tests at Alexandria (7) have shown that this design would produce an acceptable effluent.

Suspended Solids Removal

About 85 percent of the BOD5, 25 percent of the phosphorus and 30 percent of the nitrogen in the Alexandria raw sewage are associated with the suspended solids. In addition, subsequent processes insolubilize much of the remaining BOD5 and nitrogen (in the trickling filters, for instance) and almost completely insolubilize the remaining phosphorus (in the chemical precipitation steps). It therefore becomes quite evident that almost complete removal of suspended matter will be required to meet the standards, and that a highly reliable suspended solids removal system is required.

The Alexandria design makes use of three processes and removes the suspended solids in three stages. The processes utilized are coagulation-flocculation, sedimentation and filtration.

Most of the raw suspended solids (approximately 80 percent) are expected to be collected in the primary sedimentation tanks. The existing tanks will be supplemented by new tanks for this purpose. Alum (or, possibly ferric chloride), probably in conjunction with a high molecular weight moderately anionic polyelectrolyte will be added to the raw sewage, ahead of the primary tanks. This should reduce the suspended solids to levels sufficiently low enough that they will not interfere with the following activated carbon adsorption process. Additionally, it should reduce the BOD5 loading to the trickling filters by a factor of almost two, due to the decreased particulate BOD5 loadings.

The next point where suspended solids will be removed is in the final sedimentation tanks. Here again, alum (or ferric chloride) and an anionic polyelectrolyte will be added to the wastewater just before the sedimentation tanks (just after the carbon absorbers).

In addition to the alum, a recycled stream of predominately inorganic sludge will be mixed with the incoming flow. The alum dosage will be controlled by the influent soluble phosphorus level. The effluent from these tanks should be quite free of suspended solids (less than 7 mg/l) and should provide a small, relatively constant load to the filters which follow. In addition, these clarifiers will collect the sloughings and backwashings from the proceeding carbon absorbers.

The final process to be used for suspended solids removal is filtration. Multimedia filters will follow the clarifiers. Provision for small additional chemical doses will be provided to insure a strong, easily filtered floc. The previous two stages of clarifiers should provide ample dampening of the suspended solids such that the filters may be expected to pass only a very small amount of suspended matter.

Nitrogen Removal

Almost all of the nitrogen should be in the soluble ammonia form after it passes through the biological and solids removal portions of the system. Furthermore, the effluent from the filters is expected to be quite low in organic matter and solids. Based on these properties, an ion exchange process following the filters was chosen for ammonia removal.

The ion exchange media will be the natural zeolite known as clinoptilolite, which has been used by several others (8,9,10,11,12, 13,14) or, perhaps, a synthetic version which may be developed by the time the plant is completed. The regeneration is expected to be done basically using a sodium cycle, with some calcium. How the ammonia will be removed from the regenerant solution is still under investigation. It may be means of a vacuum degasifier, resulting in an (approximately) 1-percent NH_3-water solution as a byproduct.

Most results (8 to 14) have shown that this system is capable of producing an effluent with less than 1.0 mg/l of NH_3 in it. In order to assure that the total effluent nitrogen level is maintained at less than 1.0 mg/l, enough chlorine will be provided in the subsequent disinfection step so that breakpoint chlorination can be practiced. Breakpoint chlorination has been shown to be capable of oxidizing the NH_3 to nitrogen gas (14).

Final Design and Expected Mode of Operation

Figure 2 shows a process flow diagram of the proposed plant. Note how the proposed design takes advantage of the existing process structures.

Normal operation of the plant, at design flow, is expected to be as follows: All flows will pass through all of the structures preceding the trickling filters. Chemicals will be added pre-

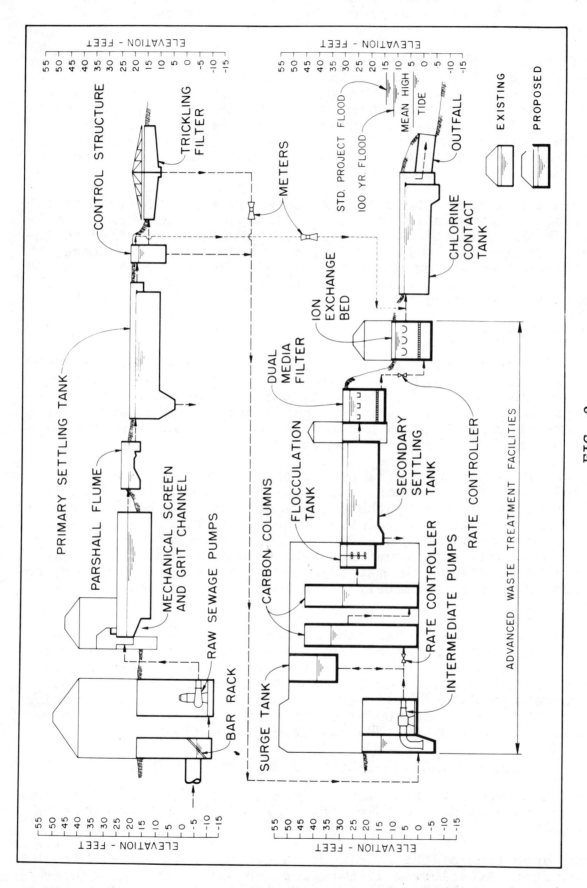

FIG. 2

Proposed Process Flow Diagram, Alexandria Sanitation Authority

ceding the primary settling tanks. Flows in excess of 36 to 42
mgd will bypass the trickling filters and will be applied directly to the carbon columns. No clarifiers are included between the
carbon columns and the trickling filters. The bypassed and trickling
filter streams will be joined immediately before the carbon columns and will be pumped, through a control structure to the activated carbon columns. There are 24 pairs of carbon "tanks," each
pair, in series, being referred to as a column. The tanks have
provision for periodic backwash (using mostly plant effluent) and
air scour to control odor problems. At average design flow the
carbon will be expanded about 10 percent, increasing to 15 to 20
percent at maximum hour conditions. Flows in excess of 95 mgd
will bypass the carbon columns to prevent washout of the beds.
Each carbon column will be paired with another carbon column and
feed into one of twelve clarifiers.

Each clarifier will consist of a mixing section, flocculation section, and a rectangular sedimentation tank. Chemicals and recycled chemical sludge will be added to the flow in the mixing section for precipitation of the remaining phosphorus. Following
each clarifier will be a multimedia filter for polishing. Surface
scour will be included for these filters. Following the filters
will be the ion exchange beds which will be designed for ammonia
removal with operation in the upflow mode. Following ion exchange, the flow will pass through the chlorine contact chambers.
Included at the head end of each contact chamber will be mechanical aerators to provide both the high degree of mixing required
for breakpoint chlorination and aeration to raise the effluent DO
to 6 mg/l or more.

While the plant is expected to be operated as just discussed, it
does have flexibility. Since this plant is a pioneering effort,
making use of processes which are all well established but which
have never been combined on such a large scale to accomplish the
functions required at Alexandria, flexibility is very important.
The carbon columns have enough free board to increase their retention time by as much as 50 percent. They also may be run as
48 parallel single-stage units instead of as 24 two-stage units.
The air scour in the carbon columns may be used continuously (at
a lower rate) to insure aerobic conditions. Chemical sludges may
be returned to the primary tanks as well as the secondary tanks.
Ferric chloride may be substituted for alum. Almost any liquid
or dry polymer may be used and enough separate polymer mixing
systems are available to feed a different polymer at each point
of application should this become desirable. Following the
trickling filters, the flow goes through twelve individually iso-
latable advanced waste treatment plants, allowing the flexibility
of full scale comparisons of various modes of operation.

Solids Handling and Disposal

The design includes provisions for combining or separating all of
the sludges. It is expected that the chemical sludges would
serve as a conditioning agent for the organic sludge. The primary
sludge will continue to be anaerobically digested. The sludges
will be dewatered by, initially, vacuum filtration followed by
incineration. (Note: Other dewatering methods may be investigated after the plant is in operation. Present construction will

not add any more vacuum filters to the existing ones.) Waste
solids from the incinerator will go to a landfill. All liquid
streams occurring after the digester will not be directly returned to the main plant flow. Instead they will be treated separately in a small "plant within a plant" and returned only after
most of the BOD, suspended solids and phosphorus have been removed from them. Details of this system are beyond the scope of
this paper.

Estimated Cost of Treatment

Table 2 presents a summary of the estimated capital costs.

TABLE 2
Estimated Capital Costs at 54 mgd (USDI Index = 127.1)*

Preliminary Treatment Units	$ 3,509,600
Advanced Treatment Units	21,451,200
Sludge Disposal Units	3,550,600
Miscellaneous	2,648,600
Total Estimated Capital Costs	$31,160,000
Estimated Cost Per Design Gallon	$0.58

The preliminary treatment units include raw sewage and other
pumps, expanded grit and scum removal facilities, expanded screen
equipment and other equipment associated with the new primary
tanks and processes which precede the activated carbon. The advanced treatment costs include all costs associated with the advanced waste treatment processes (everything which follows the
trickling filters but the chlorine tank) except those costs associated with sludge disposal. The sludge disposal costs include
dewatering, incineration and subsequent disposal of all sludge
solids. The miscellaneous costs include the chlorine contact tank,
standby power generation, roads, rail sidings, chemical storage,
and so forth.

The estimated annual costs are presented in Table 3. All chemical costs are based upon unit costs reported by Evans and Wilson
(15). Capital costs are adjusted to the same USDI index and amortization rates used by Evans and Wilson. Other costs were calculated on bases comparable to those presented by Evans and Wilson.

Table 4 shows a breakdown of the annual costs with respect to the
major design functions. It is of significant note that only
$0.076 of a total $0.362 per thousand gallons is expended for non-
advanced waste treatment purposes. This would include the costs
associated with construction of all sedimentation tanks, preliminary treatment and chlorination and approximates the cost of

*Federal waste treatment plant cost index, base year 1957-1959 =
 100

TABLE 3

Estimated Annual Costs at 54 mgd

Operation and Maintenance
Labor*	$1,125,000
Utilities	499,150
Chemicals	2,809,250
Maintenance of Structures, Grounds and Equipment (without Labor), Customer Billing, Office and Laboratory Supplies, etc	482,000
Total Annual O & M	$4,915,400
O & M Cost per 1000/gal	$0.250
Capital Costs**	$2,210,900
Capital Cost per 1000/gal	$0.112
Total Annual Cost	$7,126,300
Total Cost per 1000/gal	$0.362

* Includes both direct and indirect costs
**At 25 year, 5 percent

TABLE 4

Breakdown of Estimated Treatment Costs at 54 mgd

Advanced Waste Treatment*	Cost per 1000/gal
Activated Carbon Treatment	
O & M	$0.040
Capital	0.032
Total	$0.072
Nitrogen Removal	
O & M	$0.050
Capital	0.021
Total	$0.071
P Removal	
O & M	$0.085
Capital	0.005
Total	$0.090
Filtration	
O & M	$0.004
Capital	0.011
Total	$0.015
Solids Handling and Disposal	
O & M	$0.025
Capital	0.013
Total	$0.038
Other	
O & M	$0.046
Capital	0.030
Total	$0.076

*Excluding cost of clarifiers and flocculators

trickling filter treatment (less sludge disposal). Also included
are the costs associated with the plant in general, but not specifically the advanced waste treatment portion, such as the costs
associated with administrative and general maintenance of the
plant.

The bulk of advanced waste treatment costs are seen to be associated with the O & M costs, primarily reflecting chemical costs.
The phosphorus removal process has a particularly high chemical
cost. Therefore should the effluent standards be relaxed or more
efficient processes become available, these situations could be
taken advantage of with little loss in investment.

In order to compare the estimated costs of this plant with the
reported costs of the only full scale advanced waste treatment
plant in operation in this country, all costs were put on a basis
comparable to that used by Evans and Wilson when reporting the
South Lake Tahoe costs(15). For purposes of comparison, they
have reported an annual operating cost of $0.242 per thousand
gallons and an annual capital cost allocation of $0.142 per thou-
sand gallons plus $0.012 miscellaneous cost for the total opera-
tion of the 7.5 mgd Lake Tahoe plant.

Concluding Remarks

The Alexandria Sanitation Authority and its engineers were
charged by the Potomac River Enforcement Conference to design a
plant which would be required to produce an effluent superior to
that discharged by virtually any other waste treatment plant in
the United States. In addition, the Authority had to adapt to
site and time limitations, make the best use of the existing
trickling filter facilities and adapt to meet the more strict
effluent standards imposed by the State Water Control Board after
the process selection and much of the design was complete. In
this design, a series of processes have been put together in a
unique manner to operate as a system. Each component in this system has been carefully chosen with respect to both its function
and position to give the most reliable and flexible plant possible. The key process, the activated carbon adsorption, has been
conservatively designed and provisions have been made to allow it
to be operated in several different modes. The advanced waste
treatment portion of the plant has been designed so it can be
operated as twelve parallel, independent plants.

In conclusion, it must be said that this design represents a solution to a unique problem using what appears to be the best technology available that could be applied under the restraints imposed. This design is not to be considered a "cure all" for every-
one's pollution problems (although it is felt that many of its
features certainly merit serious consideration as components to
any municipal waste treatment plant). It is hoped that this design will result in a plant which not only will satisfy the effluent standards imposed upon it, but also one which will provide
valuable large scale information on several new waste treatment
processes.

Acknowledgments

A pioneering effort such as this plant represents, could not have been undertaken without the full cooperation, competent assistance, understanding and approval of the members of the Alexandria Sanitation Authority and in particular their Engineer-Director, Samuel W. Shafer. Mr. Shafer and his staff have provided invaluable assistance throughout the investigations and experimental phases of the work, and have advised and counseled throughout the design period.

References

1. Greeley and Hansen Report, "Laboratory Report No. 2, Phosphate Removal Studies," Alexandria Sanitation Authority, October, 1970.

2. Friedman, Weber, Bloom, and Hopkins, "Improving Granular Carbon Treatment," Water Pollution Control Research Series, U. S. Environmental Protection Agency, 17020 GDW, July, 1971.

3. Hopkins, Weber, Bloom, "A Comparison of Expanded Bed and Packed Bed Systems," Advanced Waste Treatment Research Laboratory, Federal Water Pollution Control Administration, TWRC-2, December, 1968.

4. Hopkins, Weber, Bloom, "Granular Carbon Treatment of Raw Sewage," Water Pollution Control Research Series, U. S. Environmental Protection Agency, ORD-170500A2, May, 1970.

5. G. Culp and A. Slechta, "Plant-Scale Reactivation and Reuse of Carbon in Wastewater," Water and Sewage Works, 113, 425-31, November, 1966.

6. W. J. Weber, Jr., et al., "Physicochemical Treatment of Wastewater," J. Water Pollut. Control Fed., 42, 1, 83-99, January, 1970.

7. Greeley and Hansen, "Memorandum Report to Alexandria Sanitation Authority in Response to Virginia Water Control Board letter of March 20, 1972," April 18, 1972.

8. L. L. Ames, Jr., "Zeolite Removal of Ammonium Ions from Agricultural and Other Waste Waters," Proceedings of the Thirteenth Pacific North West Industries Waste Converence, Washington State University, Pullman, Washington, April, 1967.

9. B. W. Mecer, L. L. Ames, C. J. Touhill, W. S. VanSlyke and R. B. Dean, "Ammonia Removal from Secondary Effluents by Ion Exchange," 42nd Annual Conference, Water Pollution Control Federation, October, 1969.

10. Battelle - Northwest, "Wastewater Ammonia Removal by Ion Exchange," Water Pollution Control Research Series, U. S. Environmental Protection Agency, 17010 ECZ, February, 1971.

11. D. F. G. Larkman, "Use of Clinoptilolite Zeolite for Ammonia Removal by Ion Exchange," Graver Water Conditioning Internal Report, September, 1971.

12. J. H. Koon and W. J. Kaufman, "Optimization of Ammonia Removal by Ion Exchange Using Clinoptilolite," SERL Report No. 71-5 Sanitary Engineering Research Laboratory, University of California, Berkley, September, 1971.

13. Greeley and Hansen Report, "Ion Exchange of Ammonia Nitrogen," North Shore Sanitary District, February, 1972.

14. Cassel, Pressley, Schuk, Bishop, "Physical Chemical Nitrogen Removal from Municipal Wastewater," presented at the 68th National Meeting American Institute of Chemical Engineers, March, 1971.

15. D. R. Evans and J. C. Wilson, "Capital and Operating Costs - AWT," J. Water Pollut. Control Fed. 44, 1, 1-13, January, 1972.

Applications of New Concepts of Physical-Chemical
Wastewater Treatment
Sept.18-22, 1972

PHYSICAL/CHEMICAL TREATMENT DESIGN
FOR GARLAND, TEXAS

Donald P. McDuff and W. J. Walter Chiang
Forrest and Cotton, Inc.
Consulting Engineers
Dallas—Austin, Texas

I – Introduction

The City of Garland, Texas is faced with an expansion of wastewater treatment facilities. The objectives of the plant expansion are to increase
the capacity from 10 MGD to 30 MGD and to provide treatment objectives of
reducing both the BOD and suspended solids to a concentration of less than
10 mg/1.

To meet these these objectives, the chemical/physical wastewater treatment process has been selected for the plant expansion. The major processes
of the chemical/physical plant will be equalization, chemical clarification,
filtration, activated carbon adsorption, sludge disposal and disinfection.

II – Characteristics of Wastewater

The treatability studies conducted by the Forrest and Cotton, Inc. labora-
tory(1) required six raw waste flow composited samples which were collected
by City Plant personnel and delivered to the laboratory in Austin, Texas.
Each of these wastewater samples was characterized and additionally the
City collected several samples and analyzed them for several constituents.

Average values for the constituents that directly affect the treatment
process are listed below. These parameters will be used in process design.

Parameter	Results
Total BOD$_5$ (mg/1)	266
Filtered BOD$_5$ (mg/1)	236
Total COD (mg/1)	542
Filtered COD (mg/1)	240
Suspended Solids (mg/1)	233
Alkalinity (mg/1 as CaCO$_3$)	200
pH	7.2 - 7.7
pO$_4$ (mg/1)	15.0

It should be pointed out that due to the significant industrial waste discharges into the municipal system, the Biochemical Oxygen Demand (BOD) and Chemical Oxygen Demand (COD) concentrations are higher than normal. These higher concentrations of undesirable material result in additional treatment costs. The City's industrial waste ordinance is being enforced in order to reduce the amount of BOD and/or COD that is introduced to the municipal system, or to obtain equitable compensation for treating the industrial wastes. The permit for discharge of the wastes resulting from this proposed plant expansion has the following listed criteria.

Volume:

    Average (Design Flow)   30 MGD

Quality:

    BOD (5 day)   10 mg/1, Monthly average, 24 hours daily
                             composite, and individual samples

    Total SS   10 mg/1, Monthly average, 24 hours daily
                             composite, and individual samples

    Chlorine Residual 1 mg/1, after 20 minutes contact time

The requirements of the new, amended permit are more stringent than those existing plant permits of BOD$_5$ less than 20 mg/1 and SS less than 20 mg/1.

III - Evaluation of Existing Plant Facilities

The existing Duck Creek treatment plant which was placed in service in 1962 to treat wastewater from Garland, Richardson, part of Dallas, and other areas within the Duck Creek and adjoining drainage basins. This trickling filter plant was originally constructed with a capacity of 10 MGD; however, the loading on the plant has increased through the years with the result that the plant is now overloaded. An analysis of the plant for the present quality of wastes indicates that the capacity of the existing plant should be limited to a rating of 7.5 MGD.

The existing plant will be operated at a constant flow rate of 7.5 MGD with influent wastewater being diverted from new wastewater pump station to the existing pretreatment facility.

The results of an investigation shows the expected trickling filter effluent quality to be a BOD$_5$ of 22.5 mg/1 and SS of 20 mg/1.

The solids concentration of the existing trickling filter plant sludges
will also affect the filterability and resulting efficiency of the proposed sludge dewatering process. Careful management of the sludge withdrawal from the bottom of the primary clarifier is important with sludge
density metering devices proposed for better control purposes.

A total of 15,620 lb./day (dry wt.) sludge is generated by the existing
trickling filter process. With an expected average of 3.5% solids in the
sludge, the sludge yield therefore is 7,150 cubic feet (53,500 gal.) per
day.

Anaerobic digesters are abandoned under the proposed plan.

The flow diagram for the recommended operation of the existing trickling
filter system is shown as Figure 1.

IV - Proposed Additional Treatment Facilities

The proposed additional treatment processes to the existing plant are
listed as follows (Figure 2):

>   Equalization and Aeration
>   Pretreatment
>   Chemical Clarification
>   Recarbonation
>   Filtration
>   Carbon Adsorption
>   Disinfection
>   Sludge Dewatering
>   Scum Disposal

   A. Equalization Basin. An equalization basin is employed to provide a
uniform discharge to the subsequent treatment units and maintain control
limits of variation and rate of change within a prescribed range.

The design features of the equalization basin of the Duck Creek Wastewater
Treatment Plant are listed as follows (Figure 3):

    1. Allow relatively constant flow rate into plant.
    2. Holding capacity of 8 hours at design flow.
    3. Provide uniform strength of wastewater entering plant.
    4. Aeration and mixing to prevent odor.

The wastewater is routed from the inlet diversion structure to the equalization basin wherein water is mixed and aerated. The level of the water
in the equalization basin is allowed to rise and fall as the influent to
the plant goes through daily variations. Flow from the equalization basin
to the plant is controlled at a relatively constant rate. A combination of
uniform discharge and blending of the incoming wastewater will provide a
more uniform strength and allow for a more efficient operation of the treatment plant.

The aeration provided in the equalization basin is for the purpose of
reducing odor with the objective to transfer oxygen into the wastewater
as soon as possible such that it will not become septic and produce odors.

Aeration also accelerates the microbial aerobic reaction in the equalization basin similar to an aerated lagoon with some degree of treatment being accomplished in the equalization basin. Mathematical models (2,3) for design are available for determining the relations between the influent and effluent variance as a function of detention time or basin volume.

Equalization basin design is based on an 8-hour detention time and one standard deviation, with the resulting basin effluent quality falling within a range of COD = 425±47 mg/l (i.e., COD maximum = 472 mg/l, COD minimum = 378 mg/l).

Four brush aerators (30 HP each) and four flow developers (25 HP each) will be installed in the equalization basin for aeration and mixing. Four flow developers, 25 HP each, are designed for the basin mixing. Each developer can produce 800 pounds of thrust.

The mixing velocities generated by the developers at various water depths of equalization basin are plotted in Figure 4.

Maximum particle size in suspension (in equalibrium conditions) and scouring velocity (V_c) in the equalization basin can be calculated from the following equations (4).

$$V_c = \sqrt{\frac{8\beta \ gD(S-1)}{f}}$$

Therefore:

$$D = \frac{fV_c^2}{8\beta g(s-1)}$$

Where:

D = Diameter of the particles (ft.)
V_c = Velocity of scour
β = Constant (0.05)
f = Weisback-D'Arcy friction factor, (0.03)
S = Particle specific gravity (2.5)

The relationship between the scouring velocity and the maximum diameters of sand in suspension is plotted in Figure 5 for operation at various depths of water.

Four 20 foot long floating brush aerators, 30 HP each have the function of transferring oxygen to holding wastewater and keeping the basin in aerobic condition.

The four aerators can supply a total of 9,420 lb. O_2/day. BOD will be partially removed due to the biological reaction in the basin. The removal is estimated by using the following assumptions:

    1. BOD removal is based on available oxygen which can be transferred by brush aerators.

    2. Detention time is not the limiting factor of substrate removal.

3. Available dissolved oxygen can all be utilized by microorganisms for BOD removal.

The removed BOD is converted to suspended solids. The BOD_5 and SS in the completely mixed basin can be estimated as shown in Figure 6.

All the evaluations were based on predetermined values of basin volume, fluctuation in wastewater concentration and wastewater flow. Operation conditions can be set and results can be predicted by the prepared charts and figures in this section. AT 30 MGD flow and typical day COD fluctuation pattern, effluent COD will be in the range of 425±47 mg/l.

The plan locations of the flow developers and brush aerators are very important for mixing and aeration considerations. The ideal locations are such that the flow developers can give the maximum mixing effectiveness and aerators can provide maximum oxygen transfer rate. The interference by the brush aerators with regard to the developed flow should be minimized.

The aerators and flow developers are secured in position by steel cables as this arrangement will allow the aerators and flow developers to float freely with corresponding changes in liquid levels. The anchor points of the aerators and flow developers are designed with flexibility so that the units may be relocated as required for optimization.

   B. Raw Wastewater Pump Station. The effluent from the grit basin flows by gravity to the existing raw water lift station in the trickling filter plant, and to the proposed raw water pump station. The proposed pump station is designed such that it is not necessary to operate the existing pump station; however, facilities will be provided such that it can be operated in case of emergency.

   C. Pretreatment. Pretreatment will consist of screening and degritting the wastewater.

Two mechanically cleaned bar screens with 3/4-inch openings are used to remove solids and fiber from the wastewater. Each screen is sized for normal operation at 15 MGD flow rate.

Grit will be removed from the wastewater in two aerated grit basins. Each basin has a diameter of 34 feet and a SWD of 14 feet providing a detention time of 9.14 minutes at design flow.

Each grit basin is provided with separate blower for air supply and grit pump. One additional blower and pump is piped into each system for standby purposes. The grit slurry pumped from the basins is dewatered in two cyclone hydro-degritting units with filtrate returned to the bar screen channel, and the grit disposed of by landfilling.

   D. Chemical Clarification. The wastewater is proposed to be treated with lime alone, or a combination of lime, iron salts and polyelectrolytes in each of the two chemical flash mixers. Each mixer is divided into two mixing sections into which the chemicals are added. Each mixing section has a detention time of 27.5 seconds, and the mixers produce a velocity gradient of approximately 600 fps/ft. The mixer impellers are designed to prevent rags from collecting on the blades.

The chemically dosed water flows from the flash mixers into four clari-flocculators. Features of clariflocculators are listed as follows:

Number of Units	4
Diameter in Feet	100
Sidewater Depth in Feet	14
Flocculation Well Dia. in Feet	40
Flocculating Mixer	2 per unit (G = 100 fps/ft.)
Floor Slope (Conical Bottom)	1 to 12
Sludge Rake and Turntable	85" diameter cast iron bearing race and spur gear drive rated for continuous service at 1,540,000 in.-lbs. of torque rake speed: 0.033 RPM.
Design Flow Rate	5.625 MGD
Overflow Rate	850 gals. per day per sq. ft.
Weir Loading	9800 gals. per day per ft.
Flocculation Time in Minutes	34
Settling Time in Hours	3
Total Detention Time in Hours	3 1/2

Lime is the recommended chemical for the flocculation-sedimentation process. Treatability jar test studies on flow composited samples were made to evaluate the effectiveness of treating varying dosages of lime required for removal of certain constituents by coagulation and/or precipitation.

The COD data (Figure 7) indicates that significant COD reduction (75%) can be achieved with lime dosage of 108 mg/1. Beyond this point, much smaller reductions in COD are achieved with increasing lime dosages. This phenomenon occurs because the COD is removed by the flocculation-sedimentation of solids, and removal of COD by adsorption of soluble material onto the floc. However, the flocculation-sedimentation can be readily achieved with relatively low lime dosages, whereas the second mechanism proceeds much less readily. The data also indicates that total suspended solids removal (82%) can be achieved with a lime dosage of approximately 108 mg/1 as shown in Figure 8. However, the results were obtained under laboratory conditions which was arranged to simulate the full-size clariflocculation.

Chemical dosage requirements can only be obtained more precisely from a pilot-plant study or full-size plant operation.

With regard to phosphorus reduction, the tests indicate that total-phosphorus removal, up to 70%, can also be expected with a lime dosage of approximately 108 mg/1. Figure 9 shows that higher lime dosages can increase the total-phosphorus removal significantly. Accordingly, there will be a significant increase in sludge volume and weight.

The addition of 108 mg/1 lime to the clariflocculation process will yield 3,420 lb. sludge per million gallons of wastewater treated.

Bishop, et al (5) stated that in lime precipitation of wastewater, the buffer capacities controlled the pH after lime addition by neutralizing the hydroxide ions to the calcium hydroxide. Thus, the lime dose required to overcome the buffer capacity increased with increasing alkalinity of the wastewater and the degree of clarification is affected by the original alkalinity and pH of the wastewater prior to coagulation.

Phosphorus removal increases as the pH of the water increases and, for higher wastewater alkalinities, requires more lime.

Various lime dosages were used in the treatability study performed by Forrest and Cotton, Inc. laboratory, the results of which are also plotted in Figures 10 and 11 and can be compared readily. In general, the dosages of lime, pH changes and p removed in the laboratory study compare favorably with theoretical values expected. For example, the influent of the Duck Creek plant contains approximately 200 mg/1 alkalinity (as $CaCO_3$). Lime was added at 150 mg/1 (as CaO), the resulting pH in the wastewater was 9.9. At this dosage, the treated wastewater contained 2.3 mg/1 of phosphorus as total P. Also, lime was added at 108 mg/1 (as CaO) which resulted in a pH of 8.9. At this dosage, the treated effluent contained 3.7 mg/1 of phosphorus as total P.

E. <u>Recarbonation Basin</u>. The basic purpose of recarbonation is to accomplish downward adjustment of the pH of the water by placing the water in equalibrium with respect to calcium carbonate. This avoids the deposition of calcium scale in the following unit operations for filtering and adsorption. The best filtration occurs with a pH range of 6.5 to 7.5 and optional adsorption in granular carbon beds occurs in the pH range of 5 to 9, preferable a pH below 7.

Treated effluent from the clariflocculators will be routed through the recarbonation basins for stabilization of the carbonates. Using carbon dioxide to convert the carbonates to bicarbonates, liquid carbon dioxide will be brought in by truck and stored in insulated, pressurized tanks for this purpose.

Two recarbonation basins (20' x 80' x 15' each) will be utilized and each will be divided into a mixing and reaction section. Carbon dioxide will be added in the mixing section where the detention time is set at five minutes. The water will remain in the reaction basin for 15 minutes before flowing to the ultra high rate filters.

F. <u>Ultra High Rate Filters</u>.

1. <u>General</u>. Six ultra high rate pressure filters will be used to remove the suspended solids in the effluent from the recarbonation basins. Each vessel will be 16'-6" in diameter and 16'-0" in height, and ASME pressure rated for 45 psig. The filters are equipped with an automatic air scrub and wash flush backwash system.

Water passing through the filters will flow downward under pressure through 6.5 feet of 2 to 3 mm particle size sand. The filter design capacity is 22.5 MGD.

Filtration processes are theoretically considered as two separate but sequential phonemena by Kaufman (6): 1) transport and 2) attachment. Particles must first move relatively long distances to reach the surface of the filter media and, once having reached this destination, they must become attached with sufficient energy to resist the shearing force on the moving liquid. This explains why one micron particle may be separated in 500 micron pores without the simple straining mechanism.

Based on transport phenomenon, i.e., the filtration efficiency is inversely proportional to the velocity (V) (or filter rate), media diameter (d_c) and viscosity (v), the following relationship is developed:

$$\text{Filtration Coefficient } (\lambda) = \frac{1}{v^{n_1} d_c^{n_2} V^{n_3}}$$

Where:

n_1, n_2, and n_3 are constants.

It is generally concluded that the filtration performance depends on the solids suspension condition, which is closely related to the theory of attachment.

Attachment phenomenon may be attributed to two categories: (1) electrostatic and Van-Der-Waal forces and (2) the chemical bonding of the particle to the sand surface by an intermediate material. It is well to be reminded that the same two concepts are applied to explain the coagulation of liquid suspensions.

When filtration is conceived as a combination of transport and attachment phenomenon, it is easily appreciated why even a fine filter media operated at a low loading may perform poorly, while a coarse media under heavy loading may be very efficient.

The present theory still falls short of providing explicit formulas on which the filter performance can be predicted. The design is based on a filter effluent containing less than 20 mg/l suspended solids providing that the influent solids is less than 60 mg/l.

The clariflocculator effluent should contain 50-60 mg/l suspended solids with the purpose of the ultra high-rate filters to decrease the solids level of the wastewater going to the carbon adsorption process.

COD and BOD removal are related to solids removal because a part of COD and BOD is usually present in solid form whereas in the filtration process, 20% COD and BOD are considered to be removed. Therefore, the COD and BOD in filter effluent are estimated at 80-90 mg/l and 40-50 mg/l respectively.

The ultra high-rate filtering concept in treating municipal wastewater has never been applied in flow rates in excess of 10 gpm/sq.ft. This filtration process has been successfully used at these high rates in industrial applications however.

Inasmuch as lime is selected for flocculation purposes, it is doubtful that the lime floc has the same strength as steel mill waste floc in order to stand the high shear stresses associated with the higher flow rates. However, no practical means has been perfected for determining floc strength directly and quickly which currently involves loading a pilot-type filter with a variety of coagulant doses and a hydraulic rate that would create shear stresses on the floc similar to those on the full-size plant.

The treatability study which was conducted by Forrest and Cotton, Inc. included the in-depth single media chemical pretreated filter study. The following are the results obtained from the pilot filter study:

Media Size   =   2-3 mm Sand
Media Depth   =   5 Feet

Filter Rate (GPM/sq.ft.)	SS Removal %
3	78.0
6	68.5
10	59.5

The above data show that the SS removal efficiency decreases as the filter rate increases. The required SS removal is 67% (influent 60 mg/l, effluent 20 mg/l) in the proposed filtration process. Therefore, the filter rate for the design should be normally 6-8 gpm/sq.ft.

High filter rate (V) and large media particle size (d_c) have the adverse affect on filtration coefficient (λ). The transport mechanism is playing an unimportant role in filtration process, and likewise a successful filtration will mostly rely on the attachment mechanism which should be obtained from more study with related chemical pretreatment.

   G. Carbon Adsorption Basins. Treated wastewater from the ultra high-rate filters and the biological trickling filter plant are combined and routed through the carbon adsorption basins. Practice and theory of carbon adsorption process had been illustrated by Culp (7) and Weber (8).

Ten common wall concrete basins are provided for the carbon adsorption process with each basin having a surface area of 950 square feet and a 10 foot carbon depth containing 9,500 cubic feet of granular activated carbon. Nine basins can adequately treat the 30 MGD design rate while the remaining basin is off line for backwashing or carbon reactivation. The wastewater is in contact with the carbon for a minimum of thirty minutes, and the up-flow rate is 2.5 gpm/sq.ft. at design flow.

The direction of flow is upward through the carbon with the bed experiencing an expansion of 2 to 3 percent at design flow. Each bed of carbon is washed by increasing the influent flow and expanding the bed to 40 percent in conjunction with an air scrub system to assist in cleaning and to control odor. The effluent washwater is returned to the raw wastewater pump station for recycling via the process wastewater reservoir.

Exhausted carbon is removed from the bottom of each basin and reactivated in a multiple hearth furnace. The furnace is natural gas fired, has a 14'-6" inside diameter with seven hearths, and is capable of regenerating

80,000 pounds of carbon per day. An afterburner and scrubber system is supplied with the furnace for odor and air pollution control.

The adsorption process is to be used to treat 30 MGD of wastewater which has previously undergone the pretreatment either in the ultra high rate filter or biological trickling filter process.

The blended influent COD to the adsorption basins is calculated to be 75 mg/l, whereas the plant effluent standard is BOD_5 = 10 mg/l which corresponds to COD of 15 mg/l. Therefore, 80 percent of the COD is required to be removed in the carbon adsorption process.

The design includes the following:

1. The minimum rate of carbon applied to the basins is 970 lb./carbon/ MG.

2. A design rate of makeup carbon which will provide 20% excess capacity.

3. The quantity of carbon required in the basin at any time to accomplish the desired removal of COD is 2,120,000 lb. carbon.

4. The required depth of the carbon basins is 10 feet.

5. According to the treatability study, thirty minutes of superficial carbon contact is required to treat filtered wastewater with laboratory results showing the effluent quality with COD less than 15 mg/l and BOD_5 less than 10 mg/l.

The carbon beds are moving bed up-flow design with exhausted carbon removed from the bottom by specially designed tube devices. The exhausted carbon is then conveyed to multi-hearth furnace for reactivation. Makeup carbon is then applied on top of the bed to maintain the proper bed depth.

H. Sterilization Basins. Chlorine is proposed for effluent sterilization. Two sterilization basins will receive the effluent from the carbon adsorption system with each basin sized for a detention time of thirty minutes at design flow.

I. Sludge Dewatering and Disposal. The sludge removed from each clariflocculator in the proposed plant and from the biological plant are combined in two sludge equalizing and holding tanks. The tanks are sized for an 18 hour detention time; however, sludge will not normally be held for more than 2 hours in the tanks.

The combined sludge will then be chemically conditioned and pumped into four filter presses. The presses are capable of dewatering 130,000 pounds of sludge (on a dry weight basis) per day to a sludge cake which is 40 percent solids by weight.

The recommended chemical treatment is the low lime process. An estimate of 3,420 lb./MG of sludge will be produced from the clariflocculators. The sludge from the trickling filter plant is estimated at 15,620 lb./day.

Additional amounts of lime, ferric salt and polymer are added for sludge dewatering. Total sludge is estimated at 115,960 lb./day (58 tons/day - dry weight).

The sludge cakes are trucked to a sanitary landfill site along with the grit and screenings for ultimate disposal.

The proposed future sludge disposal is at a regional incineration plant, which is under study at this time. The 40 percent solids dry cake contains enough heat value for self-supporting incineration.

J. <u>Process Wastewater Reservoir</u>. Process wastewaters are as follows:

1. Ultra High-Rate Filter Washwater = 5% of water treated = 1.12 MGD.
2. Carbon Adsorption Bed Washwater = 3% of water treated = 0.9 MGD.
3. Plant service water = 0.5 MGD.

Total process wastewater requirements is 2.52 MGD.

The process wastewater will flow into a reservoir with a volume of 21,300 cubic feet (or 159,000 gallons), which can hold approximately 1.5 hours of process wastewater. However, the purpose of the reservoir is to act as a holding basin to provide for a more uniform return flow to the raw wastewater pump station.

K. <u>Scum Disposal</u>. Scum collected at the surface of each clariflocculator is pumped into two grease units where the grease is concentrated, mixed and heated prior to feeding into two grease burners provided with air pollution control devices, each having a capacity of 300 pounds of grease per hour.

V - Nutrient Removal

A. <u>Phosphorus Removal</u>. According to the treatability study, 70% of the total-P can be removed at pH 8.9 by adding 108 mg/1 of lime in the chemical clariflocculation process. Further phosphorus removal will take place in the following ultra-high rate filters. A higher percentage of phosphorus removal can be achieved by increasing lime dosages, or by adding metal salt or polyelectrolytes.

B. <u>Nitrogen Removal</u>. Nitrogen removal in wastewater is equally important as the phosphorus removal to the receiving water entrophication problem control. It should be mentioned that the chemical/physical treatment plant does not include nitrogen removal facilities.

Due to the mild winter weather in Garland, Texas and the use of lime clarification process, the ammonia stripping may be superior to other methods in future nitrogen removal considerations.

VI - Summary

A. <u>Performance and Cost</u>

1. The diagram showing the unit operations of the existing plant along with the proposed plant is shown in Figure 12.

2. The Garland Duck Creek plant will produce an effluent containing concentrations of BOD and SS equal to or less than 10 mg/1. The process will also accomplish a degree of phosphate removal which will be in the range of 75-80%. Based on laboratory study results and the experience of other studies, the anticipated cumulative reduction of these parameters by the various treatment units is illustrated in Figure 13.

3. Capital and operating costs have been estimated for the treatment plant. The preliminary estimated capital cost is $10,600,000 for the proposed facilities and controls. For minimizing the operation of the chemical/physical plant, the interrelationship of the treatment performance of the various units must be considered. For example, the degree of treatment accomplished by the chemical clarifiers will have an appreciable effect on the operation of the recarbonation unit, the filters and the carbon beds. The operating costs have been estimated to range from $0.13 to $0.15 per thousand gallons of wastewater treated.

4. The process control system for the Duck Creek Wastewater Treatment Plant is designed for automatic, computer controlled, optimized plant operation. The entire operation of the plant is to be centrally controlled. A computer will be used to condition the incoming data such that the plant operators will be warned of abnormal occurrences. Incoming data will be conditioned and recorded in a digital presentation for observation by the operating personnel.

B. Conclusions

1. The chemical/physical process will provide the City of Garland with a plant capable of accomplishing superior treatment.

2. The proposed chemical/physical plant will provide the City with a unit operations arrangement featuring flexibility of operation in order that the cost of operation can be optimized in the production of a superior effluent.

3. The chemical/physical plant will produce an effluent that will not only enhance the quality of the receiving stream but will also be available for water reuse considerations.

)

References:

1. Forrest and Cotton, Inc., "Engineering Design Report on Garland Wastewater Treatment Facilities Duck Creek Plant Expansion," January 1971.

2. P. V. Danckwerts and E. S. Sellers, "The Effect of Hold-up and Mixing on a Stream of Fluctuating Composition," The Industrial Chemist, 27, 395, (1951).

3. A. T. Wallace, "Analysis of Equalization Basins," Proceedings A.S.C.E., Journal of the Sanitary Engineering Division, 94, No. SA6, 1161, (1968).

4. W. W. Eckenfelder, Jr., "Industrial Water Pollution Control," p. 31, McGraw-Hill Book Company (1966).

5. J.B. Stamberg, D. F. Bishop, H. P. Warner, and S. H. Griggs, "Lime Precipitation in Municipal Wastewater," U.S. Department of the Interior FWPCA Publication Nov. (1969).

6. W. J. Kaufman, "Report Theoretical Concept of Filtration," Proceedings of the Symposium of Water Filtration, California State Department of Health, May (1969).

7. R. L. Culp and G. L. Culp, "Advance Wastewater Treatment," Van Norstrand Rienhold Company, New York 1971.

8. W. J. Weber, C. B. Hopkins, and R. Bloom, "Granular Carbon Treatment of Raw Sewage," EPA Contract No. 14-12-459, May 1970.

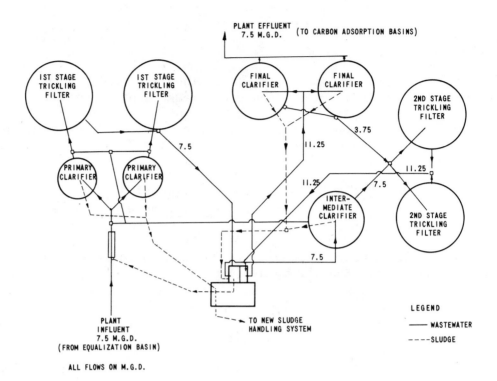

FIG. 1
Recommended Flow Diagram for the Existing
Duck Creek Wastewater Treatment Plant

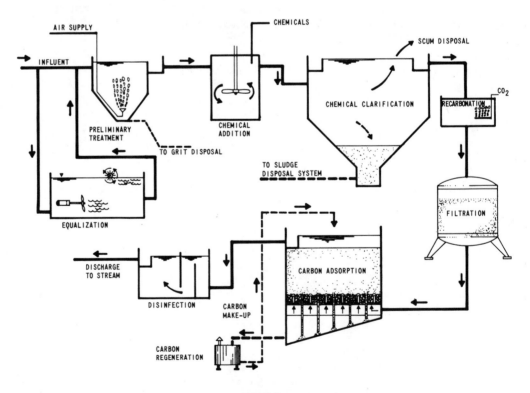

FIG. 2
Flow Diagram of Chemical - Physical Treatment

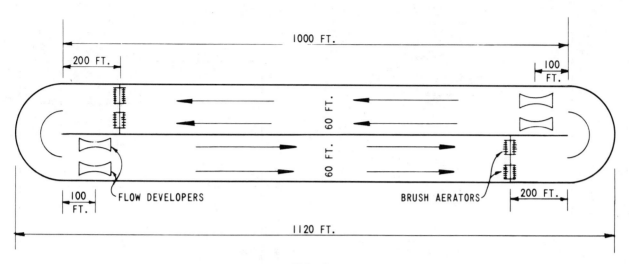

FIG. 3
Equalization Basin

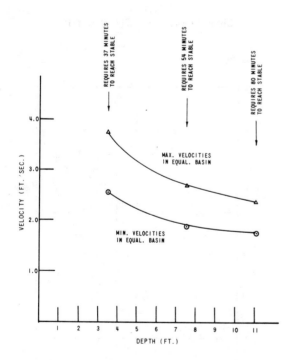

FIG. 4
Velocities of Water for Equalization Basin

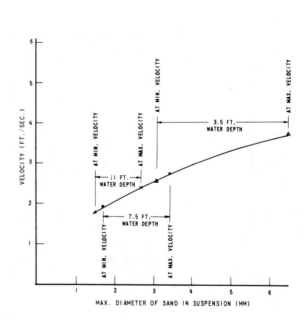

FIG. 5
Relationships Between Scouring Velocity
and Maximum Diameter of Sand in Suspension

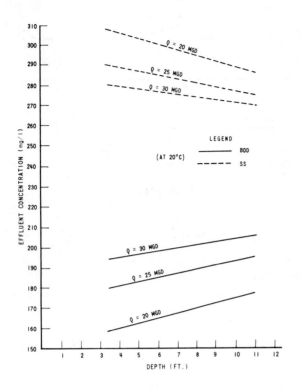

FIG. 6
Equalization Basin Depth vs Effluent Quality

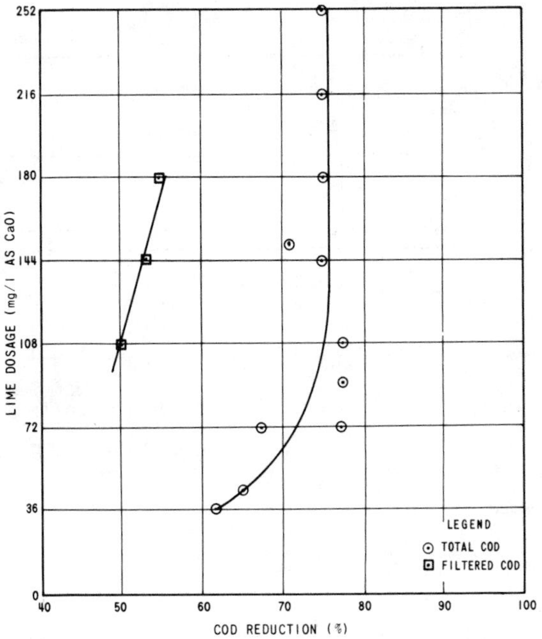

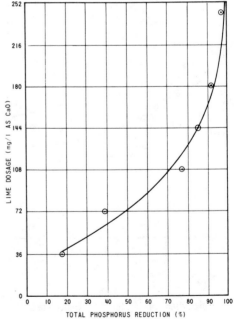

FIG. 7
Lime Dosage vs COD Reduction

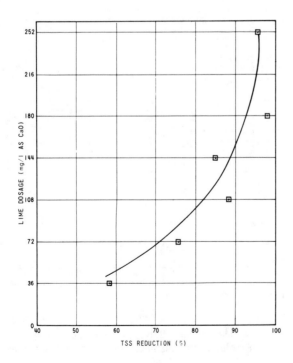

FIG. 8
Lime Dosage vs TSS Reduction

FIG. 9
Lime Dosage vs Total Phosphorus Reduction

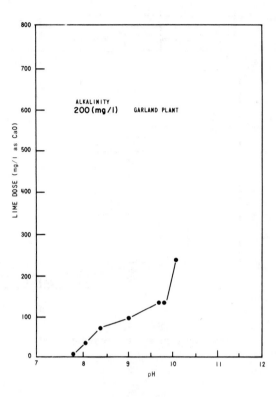

FIG. 10
Lime Dosage and pH

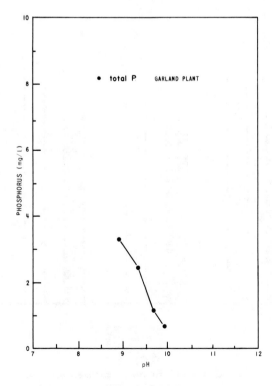

FIG. 11
Lime Precipitation and Phosphorus Removal

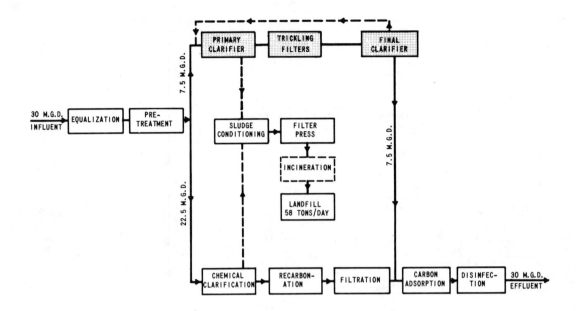

FIG. 12
Flow Diagram of the Garland Plant

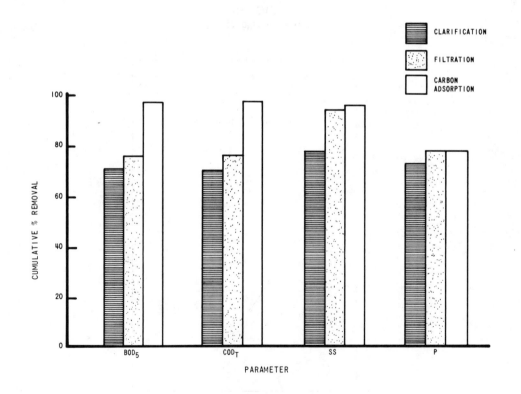

FIG. 13
Estimated Removal Efficiencies

ULTRAHIGH RATE FILTRATION OF MUNICIPAL WASTEWATER

Joseph F. Malina, Jr., P.E.
The University of Texas at Austin
William R. Hancuff, Jr.
Environmental Protection Agency

Introduction

Filtration with sand or other granular media has been used for the removal of suspended solids from wastewater in tertiary treatment systems following biological processes or chemical coagulation and sedimentation. However, the use of ultrahigh rate in-depth multilayered filtration also has great potential for application to the treatment of raw wastewater. Investigations by Tchobanoglous (1) on multilayer filtration indicated that with the proper selection of material and loading conditions the entire depth of the filter can be employed in the filtration process, thus effecting a more efficient use of the filter. Frank (2) and Donovan (3) have demonstrated the feasibility of using high rate in-depth dual media filtration for the efficient removal of suspended solids at concentrations normally encountered in municipal wastewater.

The removal of suspended or colloidal material from water by filtration is accomplished by one of a number of mechanisms which may be generally classified as straining or transport-attachment. The multiplicity of filtration studies and the variety of conditions specific to each study have addressed the large number of potential filtration mechanisms (4,5,6,7,8,9). However, the physical and chemical complexities of the filtration process have defied for years the effective theoretical characterization. Numerous mathematical models have been developed to describe filter performance under controlled laboratory conditions, but none of these models are applicable to the design of filters.

The highly unpredictable nature of municipal wastewater complicates the development of a valid design model further. The direct adaptation of water filtration models was impossible. However, the models do present the relative significance of the numerous variables and provide a starting point for the design of a filter.

The objective of this paper is the evaluation of a multilayered filtration system for the in-depth removal of suspended solids from untreated municipal wastewater at ultrahigh flow rates. Pilot-scale multimedia filters were operated using municipal wastewater.

Filtration System

A schematic diagram of the filtration system is presented in Figure 1. The head loss in the filters was monitored by means of manometers connected to the filters. The filter design embodied two important considerations: a) the determination of reasonable flow conditions which would permit simple sampling techniques; and b) the versatility necessary to cope with unpredictable operations conditions. The columns were constructed of 1/8-inch steel plate bent into a channel resulting in a cross-sectional area of 0.1 square feet. The sides of the channel were flared at the open face to permit fastening of a 1/2-inch transparent plexiglas face plate which was held in place by C clamps and sealed by a 1/8-inch rubber gasket material coated with silicone lubricant. The top and bottom of the channel were also flared to permit attachment of the removable cover plate and the flow dispersing chamber. The medium in the dispersing chamber consisted of: a) a 3/4-inch pipe inlet deflected by a 1.5-inch diameter cone; b) a 3-inch layer of marbles; c) a 1-inch layer of 1/4-inch gravel; d) a 1-inch layer of 1/8-inch gravel; e) a 2-inch layer of 2 mm garnet. Sample taps consisting of 1/8-inch nipples were secured in the rear wall with epoxy at a spacing of 1.0-inch on center. Serum caps covered the ports, and permitted the use of No. 18 sampling needles to monitor the chamber pressure. The filters were operated at flow rates of 10, 20, and 30 gallons per minute per square feet (gpm/sq ft). A detailed discussion of the experimental equipment and the characterization of the filter media are presented elsewhere (10).

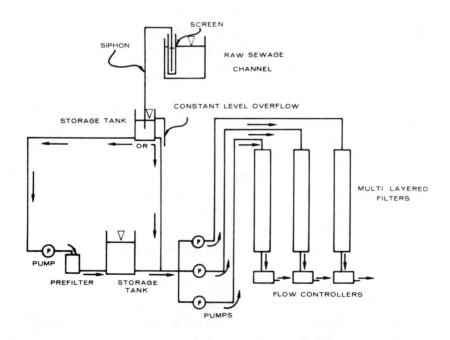

FIG. 1

Flow Diagram for Wastewater Filtration

The characteristics of the filter media used are presented in Table 1, and the three filter designs are summarized in Table 2. The nonuniformity coefficient was defined as the ratio of the 60 percentile size seive to the ten percentile size.

TABLE 1

Characteristics of Filter Media

Material	Sphericity	Mean Diameter (MM)	Specific Gravity
PVC Pellets	0.90	4.4	1.20
Anthracite Coal	0.73	1.85	1.68
Silica Sand	0.96	0.77	2.65
Garnet Sand	0.78	0.49	4.08

TABLE 2

Filter Design

Media	Depth (inches)	Non-Uniformity Coefficient		
PVC	8	1.07	1.07	1.07
Coal	5	1.07	1.15	1.45
Sand	3	1.07	1.22	1.39
Garnet	2	1.13	1.13	1.13
Media Inter-mixing		None	Intermediate	Intense

The experimental work was performed at the Govalle Wastewater Treatment Plant in Austin, Texas. The wastewater used in these studies was passed through bar screens, degritted and skimmed for grease removal. The underflow sludge collected in the skimming tank was returned into the primary effluent channel from which the supply to be filtered was drawn. A screen box constructed of wood and standard 14 x 18 mesh window screen was placed within the channel.

A prefilter was used in order to prolong the length of the runs of the multilayered filters. The prefilter was designed based on the requirement of supply to three filters at a maximum loading of 30 gpm/sq ft. The prefilter was constructed from 3/8-inch plexiglas with dimensions of 30 inches high by 5-5/8 inches

square. The calculated inside cross-sectional area was 0.22 square feet. Four
spray nozzles located approximately eight inches above the medium were used
for surface wash. Provision for air wash also was included. The underdrain
system consisted of a 3/4-inch pipe with perforations in the horizontal plane
placed in four inches of one to one and one-half inch stones covered by two
inches of 1/8 inch gravel. The medium used in the prefilter consisted of one
foot of four by five mesh PVC pellets. The inlet at the top of the filter was provided with a three-inch diameter disc located approximately one inch below the
entrance port to distribute the influent. A pressure gage and a pressure relief
system were built into the prefilter.

Experimental Results

The equipment was designed specifically for this study and was constructed and
hydraulically tested in the laboratory before location at the treatment plant. The
mean BOD and suspended solids concentration in the wastewater during these
studies were 150 mg/l and 155 mg/l, respectively. The volatile content of the
suspended solids was 86.5 percent. Experimental determinations included the
essential variables with emphasis placed on trubidity and suspended solids.

The relationship between total head loss and time for the multilayered filters is
presented in Figure 2(a) for flow rates of 10, 20, and 30 gpm/sq ft. These
linear results indicate in-depth removal of suspended solids and no surface mat
was observed.

The in-depth removal potential of PVC pellets was evaluated in more detail to
determine the headloss in the filter and the penetration of suspended solids
into the medium. A typical incremental headloss with depth curve for increasing time is illustrated in Figure 2(b). These curves are similar to those observed
in water filtration, however the time scale is only 60 minutes for filtration of
wastewater compared to 60 hours commonly encountered for water treatment.

The total removal of suspended solids throughout the depth of the filter nor the
internal removal function could be determined from the headloss curves. Therefore, the prefilter efficiency, the dependence on varying influent concentrations
and removal efficiency with depth were evaluated. The results of loading the
filter at 40 gpm/sq ft under a wide variety of influent suspended solids concentrations are presented in Figure 3. The data indicate that as the influent concentration increases, the efficiency of removal increases. The depth-efficiency
relationship for the filtration of wastewater is illustrated in Figure 4. These
curves simulate the mathematical functions developed for water filtration.

Three multimedia filters described above were operated in parallel at an average
influent suspended solids concentration of 150 mg/l. The effects of hydraulic
loading and degree of media intermixing are illustrated in Figure 5. These data
indicate that the non-intermixing filter was more efficient in suspended solids
removal. The filter efficiencies at a 30 gpm/sq ft loading were 63.5 percent
for non-intermixing, 56 percent for intermediate intermixing, and 58.5 for
intense intermixing. The data indicate a decrease of efficiency with increase
in loading; however, the rate of decrease is relatively small. A three-fold

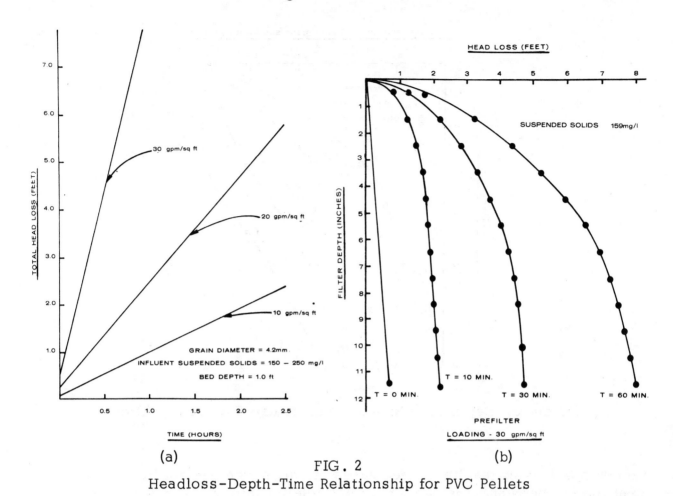

FIG. 2

Headloss-Depth-Time Relationship for PVC Pellets

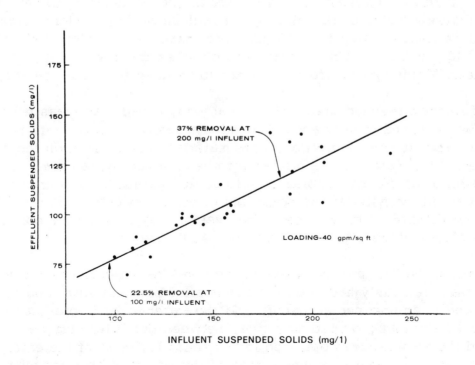

FIG. 3

Suspended Solids Removal by PVC Pellets

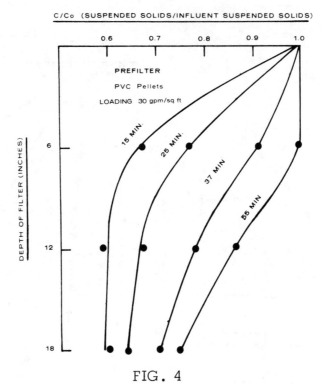

FIG. 4

Efficiency-Depth-Time Relationship for Raw Sewage Filtration

increase in loading resulted in decreases of only seven percent from 70.5 percent to 63.5 percent. The depth headloss relationship is presented in Figure 6. These data indicate that the headloss in the non-intermixed filter was less than in the other filters. The headloss curves for the non-intermixed filter are presented in Figure 7 and illustrate the rapid headloss buildup under average suspended solids loading conditions. A nine foot headloss developed after 20 minutes at 30 gpm/sq ft. This observation obviates the need for a pressure filter system, if this type of treatment were to be used in a full-scale system.

Once the filter had reached the terminal headloss, a backwash was necessary. The filter was cleaned rapidly and a two-minute backwash was sufficient. However, after backwashing, large flocculent material accumulated on the surface of the filter. This material was too large to be removed hydraulically and would present a significant problem in full-scale operation. A four-foot depth of water overlaid the filter at the beginning of the backwash cycle resulting in a substantial dilution of the sludge. The average maximum solids concentration in the sludge ranged between 1,000 and 4,000 mg/l.

In backwashing the PVC prefilter the surface spray resulted in destruction of the large floc that previously had been a problem after backwashing, and the air wash resulted in an unexpected benefit. If the water level in prefilter were dropped to a point that provided an air gap between the filter surface and the outlet, and the air wash was operated for 30 seconds, most of the entrapped suspended solids moved above the surface of the filter medium and within five minutes settled into a sludge blanket containing three to five percent solids.

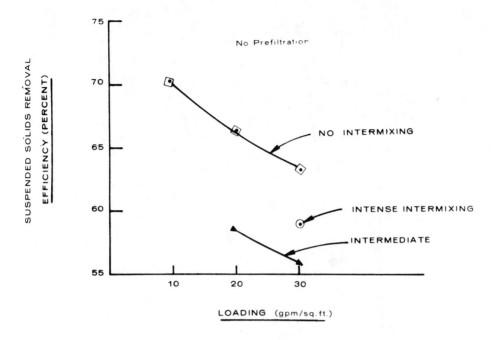

FIG. 5

Efficiency of Multilayered Filtration of Raw Sewage

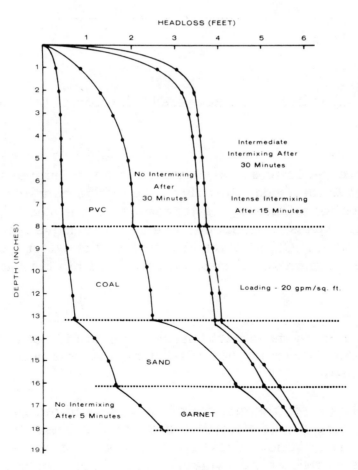

FIG. 6

Headloss for Multimedia Filters at Terminal Headloss

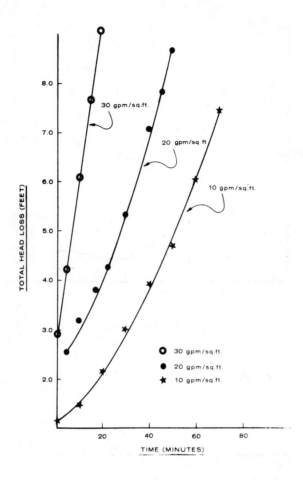

FIG. 7

Total Headloss for Multilayered Filtration of Wastewater

A device similar to the Patterson siphon used in many European water filtration plants could remove the sludge directly to the sludge handling facilities, and eliminate the need for thickening. After removal of the sludge, the filter was backwashed to an expansion of about 50 percent to remove the residual solids entrapped in the filter. Chlorination of the backwash water was included primarily to eliminate the growth of microorganisms within the filter.

Design Model

The various theoretically valid mathematical models did not provide a suitable design relationship. Therefore, the filter design models must be based on an empirical relationship.

Sakthivadivel (11) showed the validity of the hydraulic radius theory relating the hydraulic conductivity as a function of porosity. The relationship of head-loss and the volume of solids removed was independent of the concentration of suspended solids as well as the hydraulic rate. This phenomenon appeared to be applicable to wastewater filtration.

The experimental data observed for the prefilter corroborate this relationship and are presented in Figure 8. The slope of the headloss buildup with solids removed is constant regardless of loading; however, there is a difference in the extrapolated intercept. The equation which describes this function is:

$$HL = ae^{b\int ECQdt} \tag{1}$$

in which

HL = headloss (ft)
E = efficiency
C = influent suspended solids concentration (pounds/gal)
Q = hydraulic loading (gpm/sq ft)
dt = increment of time

The calculated constraints for these equations are presented in Table 3 and illustrate the mathematical uniformity of slope under the various influent conditions. These constants increased significantly at extremely low influent suspended solids concentration, for example 50 mg/l. However, at higher than average influent concentrations, for example 180 mg/l, the exponent constant decreased.

The efficiency of the filter varied exponentially with time in the same manner as the headloss function:

$$E = a'e^{b't} \tag{2}$$

The constants of this function also varied with the depth of medium. The variation of constants for various depth of a PVC filter is presented in Table 4.

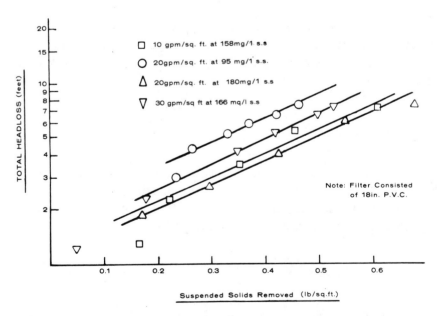

FIG. 8
Suspended Solids Removal

TABLE 3

Constants of Empirical Equation: $HL = ae^{b\int EQCdt}$

Q(gpm/sq ft)	C (mg/l)	a	b
10	158	1.16	3.20
20	95	1.87	3.23
20	180	1.11	3.02
30	166	1.36	3.20

TABLE 4

Constants for Empirical Equation $E = a'e^{b't}$

Filter Depth (inches)	a'	b'
6	0.85	-2.72
12	0.61	-1.27
18	0.46	-0.46

Note: Constants for filtration time in hours.

The effects of depth and time on the suspended solids removal efficiency for PVC under constant conditions are presented in Figure 9 . These data indicate that in a filter of sufficient depth operating under relatively constant influent conditions, the efficiency is independent of time and the equation can be approximated as:

$$HL = ae^{b\bar{E}cqt} \tag{3}$$

The above equation can also be written for the multilayer filter, but with the constants a and b chosen for the entire filter as a unit. This approach was shown to be empirically valid and is illustrated in Figure 10. The constants determined graphically for this function resulted in:

$$HL = 3.25\,e^{1.90\int EQCdt} \tag{4}$$

The design of a filter of this type must include several conditions and constraints: a) each layer of the multilayered filter should be sufficiently deep to keep from

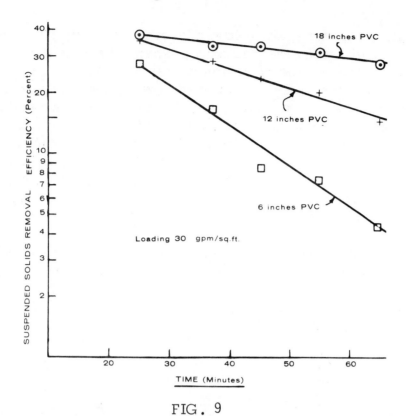

FIG. 9

Suspended Solids Removal Efficiency for Various Depths of PVC Pellets

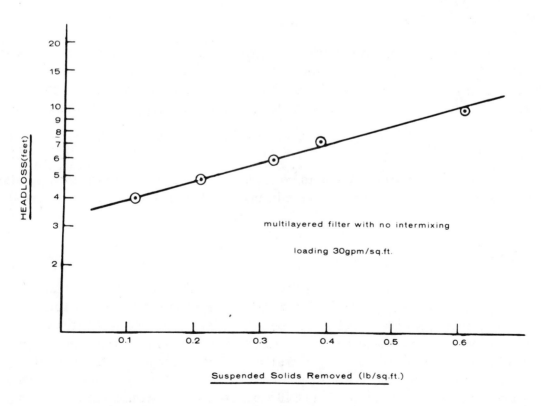

FIG. 10

Headloss and Suspended Solids Removal for Multi-Layered Filter

early breakthrough to the immediately lower layer; b) the filtration system should have the capacity to withstand the maximum suspended solids and hydraulic loadings; the design flows in the operation of this filtration process must be the maximum hourly or maximum three to four hour flow; c) the rate of headloss build-up combined with the maximum pumping capacity must not be exceeded by the time requirements of the backwash. For example, if the rate of headloss buildup is 100 ft/hr with maximum pump capacity of 100 feet and a typical backwash down time of 15 minutes, five filters would be required with one constantly being backwashed. Therefore, the time of run is confined to one hour, which also would mean that the design conditions would have to be for the maximum hourly flow and/or suspended solids loading.

Using the constants developed for the case investigated, the required design parameters are calculated by estimating the most serious conditions expected. As an example assume:

HL   = $4.0 e^{1.2\ EQCt}$
Maximum SS   = 450 mg/l
Hydraulic Loading   = 30 gpm/sq ft
Efficiency   = 70 percent
Pump Capacity   = 100 psi (231 feet)

Therefore:

$231 = 4.0 e^{1.2(0.70)\ (6.75\ lb/hr\ sq\ ft)t}$
$4.05 = 1.2\ (0.70)\ (6.75)\ t$
$t = 0.71$ hr or 42 min.

The area needed for the filtration facility must include the loading and backwash requirements. The area required is based on the flowrate divided by the loading, and the backwash requirement is included as a percentage of this calculated loading area. Therefore, the requirement for a 14-minute backwash cycle is:

$$14/42 = 0.33$$

Therefore, 33 percent additional area is required to maintain constant backwash of the system. In a four filter treatment plant, one filter would be being back-washed continuously.

Process Potential

The multilayered filter can remove from 60 to 65 percent of the influent suspended solids and provides the same degree of treatment as primary clarification. It should also be noted that the results of this study are for a moderately weak wastewater. However, the data indicate that the efficiency of filtration increased with increased suspended solids concentration. The multimedia filter would require considerably less land than a primary clarifier. For example, a clarifier designed to operate at 1.0 gpm/sq ft (1440 gal/day-sq ft) would require from 10 to 30 times as much land as multimedia filters, depending on the hydraulic load-ing.

The land value purposely was not included in evaluating the economics to illustrate the competitiveness of the process. A comparison of the cost of a 20 MGD facility is presented in Table 5. The capital cost is $192,000 for filtration and greater than $207,000 for primary sedimentation.

The results of the studies on backwashing indicate that an air wash coupled with a very low rate hydraulic wash for a short period would remove most of the entrapped material in the form of a sludge blanket containing two to five percent solids.

Conclusions

(1) A new concept in municipal wastewater treatment is introduced which provided a design model for the ultrahigh rate in-depth multilayered filtration of untreated municipal wastewater.

(2) The use of PVC pellets as a fourth medium overlaying coal, sand and garnet in the multimedia filter was the major contribution to the success of the process. The large grain size was responsible for prevention of surface mat formation which made the process unfeasible.

TABLE 5

Cost Comparison of Primary Sedimentation with Multilayered Filtration

Primary Sedimentation	
Capitol Cost	$207,000 (12)
	315,000 (13)
Filtration	
Disc Screen	15,000
Tanks and Controls	65,000
Media	17,000
Piping	80,000
Pumps, Motor and Base	
5 @ $3,000	15,000
	$192,000

Note: Design for 20 MGD facility 1970 ENR Construction Cost Index

(3)   The suspended solids removal efficiency at the ultrahigh loading rate of 30 gpm/sq ft ranged between 60 and 65 percent, and the efficiency increased as the influent concentration increased.

(4)   A non-intermixed media filter was more efficient in the removal of suspended solids and rate of headloss increase than a moderately and intensely intermixed media.

(5)   A backwash technique that produces a two to four percent sludge was possible using an air backwash followed by a short settling period. After sludge removal, the normal backwash with surface wash was employed.

(6)   An empirical relationship for filter design was developed, which can be applied after observing the total headloss and filter efficiency with time.

(7)   The cost of the filtration process is economically competitive with primary clarification, excluding the cost of land. A conservative estimate of the difference of land requirements for a 20 MGD facility would be 12,000 square feet for clarification and 400 square feet for filtration, representing a savings of approximately 97 percent.

Bibliography

1.   G. Tchobanoglous and R. Eliassen, J. SED Proc. ASCE, 96, 243 (1970).

2.   V. F. Frank and J. P. Gravenstrater, J. Water Pollution Control Federation, 41, 292 (1969).

3.   E. J. Donovan, Jr., Water Quality Improvement by Physical and Chemical Processes (E. F. Gloyna and W. W. Eckenfelder, Jr., eds.), Center for Research in Water Resources, The University of Texas, Austin (1970).

4.   A. Deb, J. SED Proc. ASCE, 95, 399 (1969).

5.   J. P. Herzig, D. M. Leclerc and P. LeGoff, Flow Through Porous Media, Amer. Chem. Soc., Washington, D. C. (1970).

6.   C. R. Ison and K. J. Ives, Chem. Eng. Sci., 24, 717 (1969).

7.   T. Iwasaki, J. Am. Water Works Assoc., 29, 1591 (1937).

8.   S. S. Mohanka, J. SED Proc. ASCE, 95, 1079 (1969).

9.   R. Sakthivadivel, HEL 15-5, Hydraulic Engineering Laboratory, University of California, Berkeley (1966).

10. W. R. Hancuff, Jr. and J. F. Malina, Jr., Technical Report CRWR, EHE, Center for Research in Water Resources, The University of Texas, Austin, 115 (1972).

11. R. Sakthivadivel, HEL 15-7, Hydraulic Engineering Laboratory, University of California, Berkeley (1969).

12. U.S. Environmental Protection Agency, Technology Transfer, Washington, D.C. (1971).

13. U.S. Federal Water Pollution Control Administration, Rept. No. TWRC-6, Robert A. Taft Water Research Center, Cinn. Ohio (1968).

INTERRELATIONSHIPS BETWEEN BIOLOGICAL TREATMENT
AND PHYSICAL-CHEMICAL TREATMENT

W. Wesley Eckenfelder, Jr.
Vanderbilt University

In the past few years there has been an increased emphasis on achieving
effluent qualities superior to those attained by conventional biological
treatment systems. There is little doubt that this trend will continue in the
future.

In order to maintain a high and consistent effluent quality at least
cost it is necessary to optimize the biological process.

This paper will discuss alternatives presently available for optimizing
the effluent quality from biological wastewater treatment systems.

Water quality requirements in the future will consider, in addition to
BOD and suspended solids, non-degradable organics (COD or TOC), nitrogen and
phosphorous, and dissolved inorganic solids. These latter substances are
removed by tertiary treatment processes or in some cases by a physical-chemical
process sequence which replaces or acts in combination with the biological
process.

If tertiary treatment is to be considered, it is essential to optimize
the biological process in order to effect optimal performance and economics
from the tertiary system.

Optimal performance from the biological process relates itself to maintaining a favorable sludge age (or F/M) and imposing constraints on the
influent wastewater variability to avoid upsets to the process.

It has been found that effluent quality from the activated sludge process
can be related to sludge age (1)(2) or F/M (Figure 1). When the sludge age
is too low, filamentous and/or dispersed growth results yielding poor settling
properties and a high suspended solids carryover from the final settling
tank. This in turn increases the total BOD discharged from the plant. For
example, in a kraft pulp and paper mill wastewater treatment plant, the
soluble and total BOD removal was 94 and 88 percent removal respectively at an
F/M of 1.0 (based on active mass). The BOD contributed by the carryover

suspended solids was 0.2 mg BOD/mg SS. Increasing the loading (F/M) to 2.0
reduced the soluble and total BOD removals to 90.5% and 76% respectively, and
increased the BOD contribution of the suspended solids to 0.5 mg BOD/mg SS.
While there was only a small decrease in soluble BOD removal, the effluent
deteriorated markedly due to increased carryover of suspended solids with a
high active fraction.

When the sludge age becomes too high (or the F/M too low) the biological
floc is oxidized and dispersed.

Chudoba (3) has shown that the biological oxidation of degradable organic
compounds yield refractory organics as a by-product equal to 0.5-1.3 percent
of the original COD. By contrast, other investigators have reported residual
COD values of 5-15% of the original COD. It would appear that the residual
COD after treatment bears a relationship to the F/M employed in the process.
As a result, while the effluent BOD will remain relatively constant with
increasing initial concentration of biodegradable substrate, the effluent COD
will increase due to the increased bio-resistant products of the oxidation.
This phenomena results in a changing BOD_5/COD ratio through treatment. Raw
biodegradable wastewaters will have a BOD_5/COD ratio of 0.5-0.7. Activated
process effluents will decrease to 0.03-0.2. Chudoba showed that afer very
long periods of aeration, the ratio further reduced to 0.007-0.04.

The COD of the effluent will therefore be composed of bio-resistant
materials present in the wastewater, refractory metabolic by-products and
residual compounds resulting from cell lysis and auto-oxidation. Chudoba (4)
showed that the residual COD of an activated sludge effluent increased with
sludge age as a result of the release of refractory organic compounds to
solution. While most of the residual COD and BOD is in suspension, the soluble
COD increased from about 20 mg/l at a sludge age of 10 days and an F/M of 0.2
to 40 mg/l at a sludge age of 50 days and an F/M of 0.082. The soluble BOD
remained constant over this period. Chudoba's results are summarized in
Figure 2. Filtration of the biological effluent will effect a substantial
reduction in COD and BOD due to removal of the finely suspended organics. In
extended aeration plants operating at high sludge ages, the effluent suspended
solids has been found to vary from 30-65 mg/l under normal operating conditions. Coagulation preceding filtration is necessary to remove the finely
dispersed organics as indicated by the difference between the coarse filtered
and soluble COD. At high sludge ages, the auto-oxidation of the cell mass will
release nitrogen and phosphorous back to solution. The nitrogen and phosphorous discharged in the effluent will therefore depend upon the BOD/N and
the BOD/P ratio in the wastewater and the sludge age in the process (see
Table 1).

Phosphorous Removal

Phosphorous can be removed in the primary treatment or in a tertiary step
by precipitation with lime, alum or ferric chloride. Phosphorous can also be
removed by the addition of chemicals (aluminum or iron salts) near the outlet
of the aeration basin or at a point between the aeration section and the final
clarifier. A total phosphorous residual of 0.19 mg/l PO_4 with a 94.8 percent BOD removal has been reported by precipitation with alum in the activated
sludge process. While the quantity of excess sludge from the biological
process will increase due to the additional chemical sludge, there is very
little increase in excess sludge volume due to the increased density of the
combined sludge. The combined sludge has been reported to be more readily

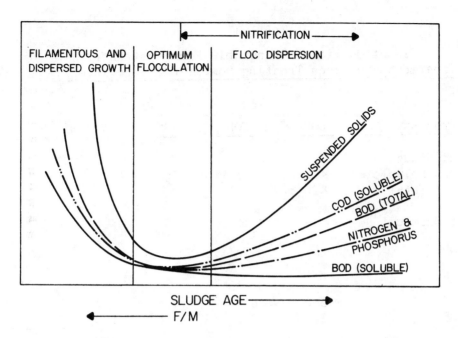

FIG. 1
Effluent Characteristics as Related to
Sludge Age and F/M

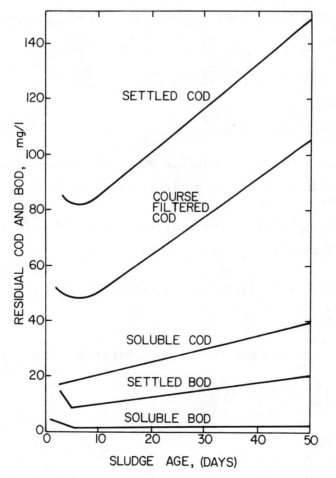

FIG. 2
Relationship Between COD, BOD, and Sludge Age in the
Activated Sludge Process (Chudoba[4])

TABLE 1

Effluent Qualities Attainable From Wastewater
Treatment Processes Treating Domestic Sewage in mg/1

Process	BOD	COD	SS	N	P
Activated Sludge					
Soluble	<10	a	<20	b	c
Total	<20				
Physical-Chemical[d]	5	13	5	4.6	0.15
Tertiary[e]	<1	10	0	-	0.06
Tertiary[f]					
inf	5.8	8.0[h]	4.4	17.4[g]	13.5
eff	3.6	2.2	1.6	1.5	1.4

[a] $COD_{inf} - (BOD_u \ removal/0.92) + COD_B$

COD_B is nondegradable organics generated in process and is a function of sludge age.

[b] $N_{inf} - 0.123x\Delta X_v/0.77 + 0.07 \frac{(1-x)}{0.77} \Delta X_v$

[c] $P_{inf} - 0.026x\Delta X_v/0.77 + 0.01 \frac{(1-x)}{0.77} \Delta X_v$

in which ΔX_v is the excess volatile suspended solids and x is the biodegradable fraction of the volatile suspended solids.

[d] Blue Plains, Washington - lime clarification, filtration, ion exchange, carbon adsorption (11).

[e] Lake Tahoe - chemical clarification, filtration, carbon adsorption, chlorination (12).

[f] Strong base anion exchange following activated sludge (13).

[g] NO_3-N

[h] Permanganate value.

dewaterable.

When considering phosphorous removal in the biological process it is necessary to have:

(a)   a low sludge age to minimize phosphorous feedback to the system
(b)   proper selection of the chemical addition point and dosage level
(c)   efficient solids - liquids separation

In the process the soluble phosphorous is converted to an insoluble form by complexing with metallic ions in addition to microbial phosphorous assimilation. The precipitated phosphorous is incorporated in the sludge mass and removed in the clarification step. Goodman (5) has shown that a chemical dosage in the range of 1.75-2.25:1 on a molar basis of Al:P is sufficient to remove 90 percent of the total phosphorous entering the system.

Sludge age is significant because high sludge ages result in floc dispersion and loss of phosphates as fine particulates as well as feedback of phosphorous to solution through endogenous metabolism. The relationship between sludge age and phosphorous removal is shown in Figure 3.

Improved overall phosphorous removal in the system can be achieved by the addition of synthetic organic polyelectrolytes for particle agglomeration prior to the final clarifier or modifying the final clarifier to a sludge blanker type of device.

Nitrogen Removal

Nitrogen may be present in a wastewater as organic nitrogen, ammonia nitrogen and nitrite and nitrate nitrogen. During the course of biological treatment, organic nitrogen is broken down to ammonia and depending on the process conditions ammonia may be progressively oxidized to nitrite and nitrate. Some organic nitrogen forms are refractory and pass through the process. During the BOD removal phase of the process nitrogen is assimilated into bacterial protoplasm. A portion of this nitrogen is released back to solution through the endogenous metabolism of the biomass. Nitrogen is essential in effective biological treatment for the synthesis of cell mass. Based on empirical formula for biomass of $C_5H_7NO_2$, the nitrogen content is 12.3 lbs. of N per 100 lbs. of dry organic weight. The weight of sludge produced is proportional to the amount of organics removed. The cell mass undergoes auto-oxidation in the process releasing nitrogen back to solution where it may be utilized by other organisms for synthesis, oxidized to nitrate or discharged in the effluent. The degree of auto-oxidation depends on the sludge age or the length of time the sludge is under aeration. The non-degradable residual biomass will contain about 7 percent nitrogen by wt (VSS). The nitrogen discharged in the effluent will therefore depend on the BOD/N ratio in the wastewater and the sludge age in the process. The nitrogen present in the excess sludge can be computed from the relationship:

$$N(lbs) = 0.123x\Delta X_v/0.77 + 0.07 \frac{(1-x)}{0.77} \Delta X_v \qquad (1)$$

Figure 4 shows the relationship between the effluent nitrogen content and sludge age for various initial wastewater nitrogen levels. Figure 4 is developed for a wastewater with an initial BOD of 237 mg/l. Equation (1) would also define the minimum nitrogen requirement for effective biological treatment.

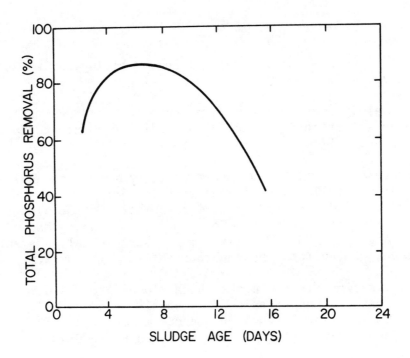

FIG. 3
Phosphorous Removal as Related to Sludge Age (Goodman [8])

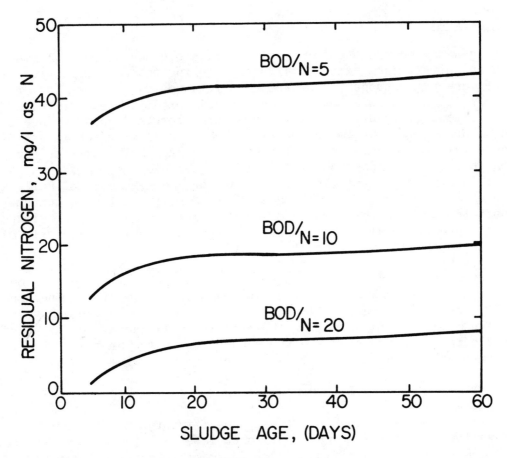

FIG. 4
Effluent Nitrogen As Related to Influent Characteristics
and Sludge Age

Biological removal of nitrogen is attained by first oxidizing ammonia nitrogen to nitrate followed by denitrification to nitrogen gas. This can be achieved by one of several modes of process operation. Both BOD removal and nitrification can be achieved in a single aeration basin providing that the sludge age be sufficient for the critical temperature of operation and the dissolved oxygen level be sufficiently high to insure unrestricted nitrification. Denitrification is achieved in a subsequent step either through nitrate reduction , through endogenous respiration in a mixed basin (6), the addition of methanol as a carbon source to increase the denitrification rate in the basin (7), or through upflow denitrification filters with the addition of methanol. If nitrification is achieved in the biological process, nitrogen removed by columnar denitrification is feasible. In this process, a carbon source such as methanol is added to the feed to the column. The methanol theoretical dosage has been estimated by McCarty (7) as 2.5 mg/1 methanol/mg/1 NO_3-N (13). (Additional methanol as a dosage of 0.87 mg/1/mg/1 of dissolved oxygen is required if oxygen is present in the feed.) In excess of 90 percent reduction of NO_3-N has been achieved in columns with retention periods as low as 5-15 minutes at 20°C. Alternatively, the BOD removal and nitrification steps can be two staged in which case the first stage is optimized by (F/M) for BOD removal and the second stage optimized (by sludge age) for nitrification. Two staging the process can result in a substantial reduction in total aeration requirements, although two varieties of final clarifiers and return sludge pumping systems are required. The third stage can be one of the denitrification stages described above or a modification there of. The advantage of the two stage process relates to improved overall effluent quality and a greater degree of process control.

Recently, a modified process operation in Vienna (8) has achieved nitrification and denitrification in a single basin at an overall loading (F/M) to the process of 0.11. Reported data is shown in Table 2. The mode of operation is such that the dissolved oxygen is alternatively 1-2 mg/1 through the aerator and auoxic or zero prior to the aerator. Under these conditions, denitrification occurs concurrently with BOD removal in the basin thereby achieving a high rate of denitrification.

TABLE 2

Nitrogen Removal at Blumental, Vienna

	Raw Sewage mg/1	Effluent mg/1	Percent Removal
BOD	257	13	95
COD	475	50	90
TOC	153	14	91
TKN	13.8	0.4	
NH_3-N	21.4	3.8	
NO_2-N	0.2	0	
NO_3-N	0.3	0	
Total N	35.7	4.2	88

Operating Conditions:
    Temp.   18°C
    Detn.   8.5 hrs.
    F/M   0.11

Variation in Effluent Quality

While it is possible to estimate the average effluent quality from a secondary treatment facility as design input for tertiary treatment design, variation in effluent quality can exert a profound effect on tertiary performance. Variations in effluent suspended solids from secondary clarification will result from hydraulic upsets and from load variations which markedly alter the biological sludge characteristics resulting in sludge bulking. Organic upsets, slugs and spills in the case of industrial wastewaters can result in a marked increase in the BOD and COD entering the tertiary treatment system.

There are a number of control measures that can be taken in the design of a secondary wastewater treatment facility to stabilize the quality of the effluent. When industrial wastes are to be treated and there is a chance of severe spills, a monitoring station with a diversion should be considered. For example, a biological treatment facility is designed for a maximum TOC of 3000 mg/l after equalization. This represents a 98% probability during plant operations. Higher TOC levels are usually due to product loss, severe spills, etc. The monitors divert the flow to a holding basin and sets an alarm for the plant personnel to investigate and correct the source of pollution. After correction, the wastewater is again diverted to the equalization basin. The high strength waste in the holding basin is then metered at a constant rate to the equalization basin.

Several investigators have shown that sludge characteristics, to a large measure, are related to the F/M. The F/M should be kept as constant as possible in order to maintain a stable process operation and a uniform quality effluent. This is best achieved by employing one or more completely mixed basins. Variations in organic loading, pH, etc., are equalized throughout the aeration basin. It is also desirable to maintain a high mixed liquor solids level (if the sludge age gets excessively high, however, there will be deterioration of the effluent). If organics which are toxic in high concentration but degradable in low concentration are present in the wastewater, such as phenol, the minimum retention period in the basin is that to reduce the concentration to a level less than the toxic threshold.

It has been found that sludge bulking and resulting effluent deterioration can be controlled in many cases by maintaining the activated sludge under anerobic conditions in the final clarifier, prior to return to the aeration tank.

Biological Process Modifications

In recent years, a number of biological process modifications have been developed to enhance the treatment efficiency and to provide a more uniform effluent. A few of these will be mentioned here.

High purity oxygen has recently been employed in which dissolved oxygen levels in the order of 10 mg/l are maintained in the aeration basin. Under these conditions, a more dense sludge is attained and filamentous growths substantially eliminated even at high F/M values. A rationale for this phenomena can be made. It is postulated that under the moderately turbulent conditions (0.2-0.3 HP/1000 gal) usually employed in the activated sludge process, only a portion of the floc is aerobic. Increasing the biological activity by establishing fully aerobic conditions require either increased dissolved oxygen levels or an increased power level.

The hypothesis described above could also offer an explanation for the increased sludge density and lack of filamentous growths when using high purity oxygen. A biomass functioning under facultative and anaerobic conditions will probably have a lower density than a fully aerobic floc. This is due to the anerobic by-products produced in the floc. The bacteria normally encountered in the biological floc should have a higher growth rate than the filamentous organisms.

To support this hypothesis, Zahradka (9) further showed that at high carbohydrate loadings, filamentous growths abundantly developed under low turbulence conditions, while high turbulence and resulting small floc size showed no filamentous growths at the same organic loadings. This supports the contention that under low oxygen tensions when only a portion of the floc is aerobic, the bacterial growth rate on a mass basis is relatively low. The filamentous growths, on the other hand, with a large surface area/volume ratio, will exhibit unrestricted growth. Under these conditions, low dissolved oxygen concentrations will favor filamentous growth, providing environmental and loading conditions are conducive to the growth of the filamentous organisms.

A recent modification of the activated sludge process is the DuPont Pact process in which powdered activated carbon is added to the aeration basin. Carbon dosages are in the range of 150 mg/l. Pilot plant studies have indicated that carbon addition improves BOD and COD removal (some of the refractory organics are removed in the carbon). Improved process stability and sludge settling characteristics are reported.

An interesting biological-chemical process variation has been reported by Humenick and Kaufman (10). This process employs high rate activated sludge followed by lime clarification. A portion of the lime sludge is recycled to the biological stage resulting in a more concentrated sludge in the biological process and resulting higher organic loadings (400-500 lbs BOD/day/1000 ft^3). Removals of BOD and COD in excess of 90 and 80 percent respectively are reported with phosphorous removal in excess of 95 percent.

References

1. Ford, D. L. and Eckenfelder, W. W., Journal of the Water Pollution Control Federation, 39, 11, 1950, 1967.

2. Eckenfelder, W. W., "The Role of Secondary Treatment in Advanced Waste Treatment Schemes", Water and Wastes Engineering. (In press)

3. Chudoba, J., Scientific Papers of the Institute of Chemical Technology, Prague F12 (1967) Technology of Water.

4. Chudoba, J., Scientific Papers of the Institute of Chemical Technology, Prague F16 (1971), Technology of Water.

5. Goodman, B., Notes on Activated Sludge, 3rd Edition, Smith and Loveless, 1971.

6. Wuhmann, K., "Removal in Sewage Treatment Processes". Verhandl. Intern. Verein, Limnol. XV, 580-596 (1964).

7. McCarty, P.L., et al., "Biological Denitrification of Wastewaters by the Addition of Organic Materials". 24th Ind. Waste Conf., Purdue University, 1969.

8. Von Der Emde, W., Private Report.

9. Zahradka, R., Advances in Water Quality Improvement, Prague Conference Proceedings, Pergamon Press, 1970.

10. Humenick, M. J. and Kaufman, W. J., "An Integrated Biological-Chemical Process for Municipal Wastewater Treatment," Proceedings 5th International Conference on Water Pollution Research, San Francisco, California, 1970.

11. Kugelman, I. J. and Cohen, J. M., "Chemical-Physical Processes," Advanced Waste Treatment Symposium, Cleveland, Ohio, March, 1971.

12. Culp, R. and Culp G., Advanced Wastewater Treatment, Van Nostrand Rienhold Publishing Co., New York, N.Y., 1971.

13. Gregory, J. and Phond, R. V., "Wastewater Treatment by Ion Exchange," Water Research (In press).

TERTIARY TREATMENT - THE CORNER STONE OF WATER QUALITY
PROTECTION AND WATER RESOURCES OPTIMISATION

Dr. Gerrie J. Stander
Water Research Commission
Pretoria, South Africa

Introduction

The average annual economic growth rate for the Republic of South Africa
for the years 1946-1970 was 5.2 per cent as compared with an annual increase
water consumption of 7 per cent. This difference of nearly 40 per cent between
the two respective growth rates clearly demonstrates the absolute dependence
of the nation's economic growth on the optimisation of the country's water
resources. It is, however, unfortunately true that if the current growing
imbalance between demand and available supplies is permitted to continue, the
country's water resources will be completely exhausted by the turn of the
century.

The aforementioned situation is rendered even more critical by pollution
problems. For South Africa, widespread pollution will have a disastrous effect
in that it will reduce still further the already limited quantity of usable
water available. This effect, together with the growing imbalance between water
supply and demand, will culminate in a complete stagnation of development. The
essence of water pollution control settles in the concept of clean-water resources
in accordance with the requirements of the user whether he be an industrialist,
a domestic consumer, a farmer, a fisherman, a holidaymaker, a bather or a
nature lover. By striving for the achievement of this objective, we shall not
only succeed in forcing every drop of use out of our limited water resources,
but will help to maintain a positive gap between demand and supply of clean
water.

The Challenge

The Prime Minister's Water Plan Commission (1) appraised the country's
water household exhaustively and its report to the Government contains some
most pertinent recommendations which pose far reaching challenges to scientists
and engineers engaged in the many fields of water and wastewater technology to
generate the knowledge required for the achievement of the Water Plan Commission's far-sighted measures to ensure and maintain a credit balance in the
country's water budget, not only until the turn of the century but for many

more years to come.

In order to place into perspective the particular challenges involved in the fields of wastewater reclamation and pollution control the undermentioned data indicate quantitatively the estimated contribution which, in the light of current technological advances, several practical measures could make in maintaining a positive water economy. The figures to be quoted may sound like a drop in the ocean for the United States, Canada, United Kingdom and Europe with their mighty rivers and lakes, but for South Africa with an estimated population of some 50 million people by the turn of the century the problem is real and critical.

(i) The current assured yield of surface and underground water resources is estimated at 27.5 Km^3 per year and by the year 2000 the fresh-water requirements are expected to be 29.5 Km^3 per year. The expected deficit in the water balance is therefore 2.0 Km^3 per year (2.0 Km^3 per year is equivalent to 1,200 million imperial gallons per day).

(ii) Measures to increase the assured yield:
  (a) Control of Evaporation.................... 1.7 Km^3 per year
  (b) Desalination of Sea-water................. 0.5 Km^3 per year
  (c) Interconnection of river systems......... 2.3 Km^3 per year

  Total   4.5 Km^3 per year
  The assured yield could be increased to.. 32.5 Km^3 per year

(iii) Measures to curtail fresh-water allocations:
  (a) Improvement of irrigation farming......... 1.5 Km^3 per year
  (b) Augmentation of industrial and domestic
       water supply through domestic and
       industrial wastewater reclamation and
       pollution control........................ 7.2 Km^3 per year

  Total   8.7 Km^3 per year

It is obvious from the aforementioned statistics that provided research and development accelerate the current advances in water and wastewater technology, the fresh-water requirements of 29.5 Km^3 per year could be reduced by augmentation with reclaimed water and by improvement in irrigation techniques to 20.8 Km^3 per year and with the anticipated increase in the assured yield the water credit could be 11.2 Km^3. It is significant to note that of all the above measures proposed by the Commission, wastewater reclamation and pollution control measures are expected to contribute almost 65 per cent of this water credit.

Water Quality

The tremendous increase in the production of synthetic chemicals and new products has introduced into the water environment sophisticated toxic, biocumulative, carcinogenic and subtle pollutants. Furthermore, the ever-increasing proportions of conventionally purified domestic and industrial wastewaters, containing these pollutants and viruses, reaching rivers, lakes, and impoundments must undoubtedly effect marked changes in the quality of the receiving waters.

It is, therefore, not surprising that in countries blessed with abundant fresh water resources there is a growing scarcity of usable water and an increasing destruction of the usefulness of water resources. The various sectors of water users in particular and the public generally have become critically aware of water quality and the effects of pollution on public health, aesthetics and the amenities of the water environment. As a consequence water quality criteria, goals and standards have been developed for a wide range of water uses. As in other countries the experience with the application of new water criteria in South Africa has necessitated a critical investigation of:

(i) The capability of biological wastewater purification facilities to produce a finished water conforming to prescribed quality requirements for return of purified effluents to the water environment or for industrial, domestic, agricultural or recreational use.

(ii) The development of wastewater reclamation processes for:
    (a) The production of water which conforms to the quality requirements of specific uses.
    (b) The augmentation of biological wastewater purification facilities for upgrading the quality of effluents discharged to the natural water environment to conform to the requirements for public health, for aquatic life, for prevention of eutrophication and for raw water intakes to conventional water purification facilities producing public and industrial water supplies.

(iii) The efficiency of conventional water purification plants and their upgrading to conform to quality changes in the natural water environment serving as raw water intakes.

(iv) Reliable and accurate analytical tools to detect, measure and interpret changes in the quality of river, lake and marine waters with a view to developing realistic water criteria for effective water pollution control and for the progressive refinement and upgrading of wastewater installations in accordance with these changes.

Wastewater Technology and Water Purification

Wastewater technology had it's beginnings more than a century ago in the exploitation of the diluting and self-purification capacity of the water environment for the disposal of sewage and industrial wastes. The modern wastewater biological purification plant constitutes effectively the harnessing of these unit processes under controlled engineering operation; the physical and self-purification capacities of the water environment are not only accelerated manifold in the unit processes of the plant, but the variables can also be controlled. Moreover, the significance of the dilution capacity of the receiving body of water is minimised. With the increasing proportions of wastewater being discharged into rivers and lakes, it has become obvious that there is an over-confidence in the natural capability of rivers and lakes to maintain the quality of the water at a level suitable for all recognised uses. Furthermore, experience with biological wastewater purification plants, even with their capacity of accelerating biological purification processes, has exposed notable limitations in the degradation of organic substances and the destruction of water-borne disease organisms. This experience has also provided confirmatory evidence that dilution and self-purification processes in rivers and lakes are equally limited in their capability of degrading micropollutants, carcinogenic compounds, toxic and complex organic substances, in the destruction of viruses, pathogens and parasites, and in the removal of bio-cumulative substance, phosphates and nitrogen. Scientists and engineers

are very much alive to the limitations of secondary biological treatment processes as a pollution control measure, and have during the past two decades dili-
gently applied themselves to the development of advanced wastewater unit processes (2) with the object of:

(i) purifying industrial wastewaters for internal recycling and
reclaiming biologically purified sewage effluent for augmenting industrial water
demand.

(ii) Introducing changes in manufacturing process to eliminate or minimise
undesirable contamination of water and wastewater circuits.

(iii) Improving biological purification processes to remove phosphates,
ammonia, nitrates, to destroy or remove viruses, pathogens and parasites, and
to degrade toxic, carcinogenic and bio-cumulative substances - in essence the
objective is the production of purified sewage effluent conforming to quality
requirements for the water intakes to various advanced wastewater reclamation
systems.

(iv) The refinement of advanced physical and chemical unit processes
and the development of advanced wastewater reclamation systems to supplement
biological wastewater treatment systems in order to produce a product water
conforming to the quality requirements of various uses of water or for return
to the water environment.

Pollution and its control have inseparably linked water purification
and wastewater technology. There is convincing evidence that intakes to
water purification plants drawing from polluted sources must contain wastewater
which has escaped the factors of self-purification, dilution and time lapse.
In fact the reservations expressed regarding wastewater reclamation for
domestic re-use effectively incriminate secondary purified sewage effluents
which reach our rivers and therefore the conventional water purification plants
which treat these river waters. In fact, many simple water purification
systems in some of the world's most developed countries are effectively forcing
(2) poorly reclaimed water down the public's throat with concomitant risk to
health; the increasing presence of micro-pollutants (3) responsivle for taste
and odours in water has become a matter of grave concern to water supply
undertakings - in fact this problem rather than viruses and carcinogens is
currently a much stronger driving force behind the inclusion of advanced wastewater process units in conventional water purification plants. This is even
more the objective in the design of wastewater purification plants for the
return of effluents to the natural water environment. As a consequence many
municipal water treatment authorities are extending the conventional water purification facilities by the installation of wastewater unit processes to produce a finished water which is safe and aesthetically pure. The implication
of this action is obvious; in course of time conventional water purification
plants will be nothing less than wastewater reclamation systems. This would
indeed be a sad situation for South Africa for, as long as rivers and lakes
are exploited to purify the residues in biologically treated wastewater effluents,
the backdoor to increased pollution will be kept open - the ultimate target
must rather be the final purification of wastewater by advanced processes before
discharge to the river, thus keeping our clean rivers clean.

Planning of Wastewater Reclamation

The above elaboration largely dispels the many misconceptions regarding advanced wastewater reclamation. A proper understanding of the achievements of advanced wastewater treatment and it's key role in water resources optimisation in South Africa calls for the immediate exploitation at national level of the experiences and lessons of Chanute (4), Lake Tahoe (5), and Windhoek (6,7,8). The latter two projects have unquestionably produced some outstanding hardware technology to set the course which the decision makers in South Africa is following with fixity of purpose.

In South Africa the problem of water pollution arises from the use of water for domestic purposes, industry, mining, power generation and irrigation farming. We know that between 70 and 100% of the water intake of urban and industrialised areas become effluents whereas in irrigation farming mineralised return flows constitute between 10 and 20% of the water intake. The net result is that the water utilisation, water supply and waste disposal cycles tend to become a closed system (Fig. 1) (9) and public authorities are faced with a dilemma of planning the use and augmentation of available supplies and, simultaneously, protecting their usefulness; indeed, a delicate balance of factors is involved, which may differ from region to region and almost from user to user, requiring critical path planning and research to optimise water resources exploitation. Further socio-economic development will depend entirely on the country's ability to expend available resources by efficient use and re-use within the water utilisation and supply cycles. It is obvious from Figure 1 that current conventional purification of sewage and industrial effluents yields secondary purified effluents which, if discharged to rivers would not adequately protect and preserve the quality of the natural raw water intakes to conventional water purification plants situated on such rivers. The costs of purifyins such polluted natural waters will soar and will be no different from reclaiming water from secondary effluents. The ultimate result will be that water supply systems based on rivers receiving increasing proportions of secondary purified wastewaters will grind themselves to standstill as conventional water purification schemes and the costs of supplying a safe water to the public will become a prohibitive price to pay for failing to keep the natural waters in rivers clean - positive cost benefit factors of water supply schemes will be maintained only if reclamation is practiced.

Because of the aforementioned situation, it stands to reason that for effective pollution control and water conservation, the management of wastewaters and pollution control must be planned and implemented as an integral part of the Country's water household. This involves costs to the user, who more often than not objects to requirements for the return of purified effluents to rivers or disposal into the sea. Therefore, by subsidising fresh water schemes directly or indirectly by too low water rates, water pollution control becomes difficult and complicated. Obviously therefore by charging the user for the actual costs of water supplied to him and by subsidising wastewater reclamation and water pollution prevention schemes, arguments against water pollution control measures on economic grounds will have to be strongly motivated.

The aforementioned concepts constitute the main driving force behind the economic incentives embodied in the Water Plan Commission's (1) recommendation for the development of municipal and industrial water schemes. This recommendation indeed poses great challenges for the development of the necessary water and wastewater technology and the relevant cost benefit criteria on which wastewater reclamation could be harnessed as the effective control of pollution.

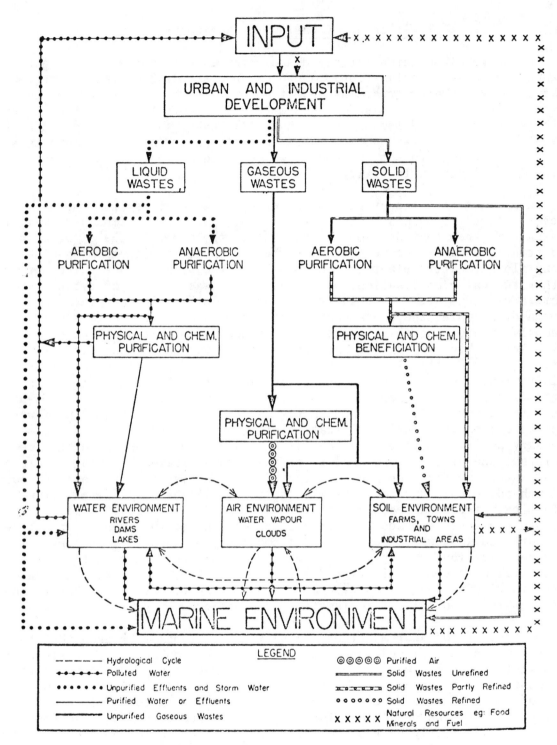

FIG. 1

UTILIZATION OF THE WATER CYCLE

It is becoming increasingly obvious that efforts to achieve effective water pollution control by the application of advanced wastewater technology are resulting in the not so surprising discovery that reclaimed water is a valuable by-product.

From Figure 2 it will be seen that by the judicious selection and combination of biological and advanced wastewater unit purification processes it is

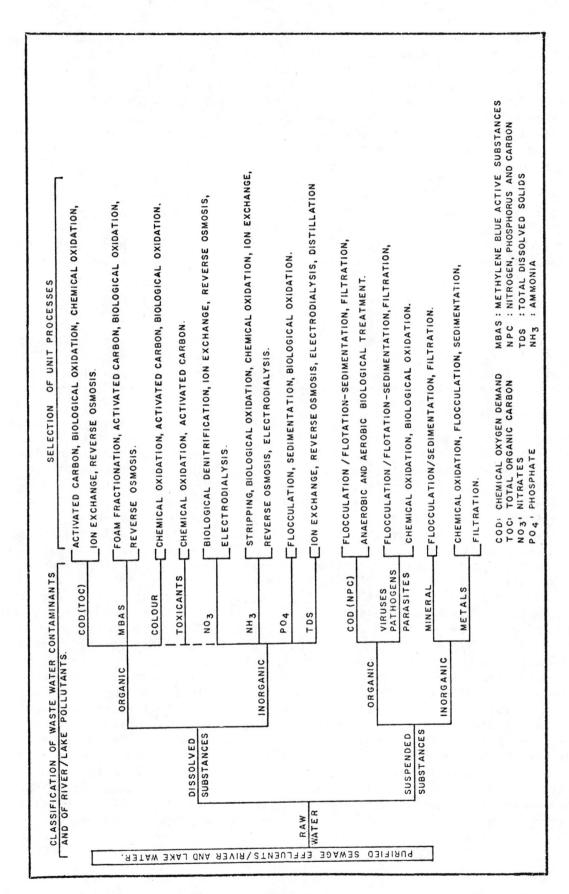

FIG. 2

CLASSIFICATION OF WASTEWATER CONTAMINANTS AND SELECTION OF PROCESS UNITS

possible to produce water of virtually any desired quality. Whereas the cost benefit factors may not as yet be favorable for all situations, as will be reported at this symposium, experience in South Africa has shown that sufficient hardware technology is available to produce waters which match the quality requirements of various uses at costs which are competitive with the price of current freshwater supplies. Furthermore, the removal of nutrients which enrich the water environment can also be effectively and economically achieved. In fact studies of a seriously eutrophied water source have shown that the cost increase of conventional water purification to produce a water of potable quality is of the same order as the costs of advanced wastewater reclamation; the benefits which will accrue from cleaning up the catchment of this water supply by investing the increase in costs in the extension of biological purification plants by the addition of wastewater reclamation units are obvious, not only with respect to domestic and industrial water supply schemes but also with respect to the many other important amenities which have developed around this water environment.

Conclusion

For South Africa the route which should be followed in the optimisation of its water resources is very clear. Water supply, water pollution control, wastewater reclamation for re-use or return to rivers and the upgrading of conventional water purification plants constitute an integral part of the country's water household.

Whereas research and development should be sharpened and accelerated in a number of critical areas in order to generate the new knowledge and expertise required to resolve the many technological problems which will be encountered in the critical planning of the Republic's water resources, current achievement in technological innovation have proved to be a key factor in ensuring a credit balance in South Africa's water budget by the turn of the century. It cannot, however, be emphasized too strongly that this technological innovation can reach it's full potential only if water supply and water pollution control authorities and decision makers practice a farsighted policy in the implementation, at national level, of available hardware wastewater technology. One thing is sure, the judicious application of available expertise will accelerate the development of new knowledge and the refinements so vitally needed for sensible and effective pollution control and beneficial use of our water resources. Any activity by the water supply and water pollution control authorities and decision makers which stimulates the utilisation and exploitation of new wastewater technology is without doubt their most effective contribution to the efforts of scientists and engineers to keep abreast of problems of water supply and pollution control. This incentive, together with farsighted planning and implementation of wastewater reclamation schemes and pollution control measures as major national water schemes within their own right, will constitute the key routes along which a substantial credit in the Country's water budget could be ensured by the turn of the century. Failure to do this and to accelerate further research and development in water and wastewater technology as a tool to ensure more 'kilometers of water use per kiloliter of water' will create irreversible situations in the country's socio-economic progress.

References

1. Report of the Commission of Enquiry into Water Matters. Republic of South Africa, R.P. 34/1970.

2. Stander, G. J.: "Re-Use of Wastewater for Industrial and Household Purposes." Paper delivered at the Ninth International Water Supply Congress, New York, 11-14 September, 1972.

3. Gomella, C.: "Prevention of Micro Pollutants." Paper delivered at the Ninth International Water Supply Congress, New York, 11-14 September, 1972.

4. Metzler, D. F.: "Emergency use of reclaimed Water for Potable Supply at Chanute, Kan." Journal of the American Water Works Association, Vol. 50, 1958, pp. 1021.

5. Sebastian, Frank P.: "Wastewater Reclamation and Re-Use." Water and Wastes Engineering, July, 1970, pp. 46-77.

6. Van Vuuren, L. R. J., Henzen, M. R., Stander, G. J. and Clayton, A. J.: "The full-scale reclamation of purified sewage effluent for the augmentation of the domestic supplies of the City of Windhoek." Published in: "Advances in water pollution research," Proceedings of the 5th Water Pollution Research Conference, San Francisco-Hawaii, 26th July - 5th August, 1970.

7. Stander, G. J. and Van Vuuren, L. R. J.:"The reclamation of potable water from wastewater." Journal Water Pollution Control Federation, Vol. 41, No. 3, Part I, March, 1969, pp. 355-367.

8. Stander, G. J., Van Vuuren, L. R. J., and Dalton, G. L.: "Current Status of research on wastewater reclamation in South Africa." Journal of the Institute of Water Pollution Control, Vol. 70, No. 2, pp. 213.

9. Stander, G. J. et al: "Occurrence and Development of Water Resources in the Republic of South Africa." International Conference on Water for Peace, Washington, D. C., May, 1967, Doc/40.

10. Henzen, M. R., Stander, G. J. and Van Vuuren, L. R. J.: "The current status of technological development in water reclamation." Paper to be presented at Workshop Session of the 6th International Conference of the International Association on Water Pollution Research, Jerusalem, June, 1972.

THE EFFECTIVE UTILISATION OF PHYSICAL-
CHEMICALLY TREATED EFFLUENTS

O.O. Hart* and G.J. Stander**
* National Institute for Water Research of the South African
  Council for Scientific and Industrial Research, and
** Water Research Commission.

Introduction

The population explosion, increased industrialisation, agricultural development and recreational activities all led to a corresponding increase in the volume of effluents and diversity of pollutants discharged into the natural environment. By seepage these pollutants have also found their way into underground water resources. The result is that both surface and underground water resources which serve as major water supply for mankind are today threatened by pollution.

Pollution occurring in water supplies must obviously be removed to render such water fit for beneficial use. Although the modern conventional water purification plant with its various water treatment processes provides a number of safety barriers against such pollution, these processes have definite limitations regarding the efficient removal of pollutants, particularly of organic nature.

The solution to this problem lies in the abatement of pollution at source, but since such abatement measures cannot ensure complete removal of all con- taminants which may be deleterious to man's health, it is essential to provide additional safety barriers in subsequent water purification systems. Research in the field of wastewater treatment which has been directed primarily at the abatement of pollution at source has led to the development of processes which are capable of effective removal of a wide range of pollutants. The inte- gration of these processes with conventional water purification systems in con- junction with their incorporation into wastewater treatment facilities, will greatly alleviate the overall problem of minimising pollution of natural water resources and the production of high quality potable water.

This paper presents some data to demonstrate that this objective can be achieved by the effective utilization of physical-chemical treatment processes.

Reuse Application

Research in the field of advanced wastewater treatment has been directed mainly towards the improvement in wastewater quality as a prerequisite for the protection of the aquatic environment.   The resultant technology has progressed to a stage where it is now possible to design systems capable of producing any desired quality of water, ranging from high quality potable water to that required for recreational, agricultural and industrial purposes. Although the application of the processes and techniques developed in waste-water technology has not yet attained widespread recognition, the following examples clearly demonstrate the efficacy of the processes and systems in achieving the desired results.

Recreational reuse

A good example of the application of advanced wastewater technology is to be found in the Santee Recreation Project (1).   The Santee Lakes were designed to use reclaimed sewage for fish propagation and for recreational purposes such as swimming and boating.   The project has demonstrated the feasibility and social acceptability of using water reclaimed from sewage as a supply for recreational lakes.   The study also demonstrated that nutrients can be controlled to create a balanced food chain that supports several species of fish.

In this project the tertiary treated effluent from an activated sludge plant is retained in a maturation pond for 15 days, providing some removal of nitrogen, bacteria and viruses, and then chlorinated and allowed to percolate through 3,6 to 4,5 m alluvial deposits for a distance in excess of 120 m. Phosphorus and nitrogen are removed to an average level of 8,0 mg/l each. Effluent from the percolation area is again chlorinated to a minimum chlorine residual of 0,4 mg/l free or 0,7 mg/l combined chlorine for swimming purposes. The quality of the secondary treated effluent is thereby upgraded considerably.

Industrial reuse

In industry, process selection is based both on economic considerations and on specific water quality requirements.   In the pulp and paper industry these requirements relate primarily to minimal concentrations of suspended solids, iron and manganese.   In addition phosphate concentration levels must be sufficiently low to prevent microbiological growth.

In the case of the South African Pulp and Paper Industries' plant at Springs, the judicious selection of advanced wastewater treatment processes provided an excellent quality process water with relatively low capital expenditure.   The resultant annual saving in water purchased was comparable to the initial capital costs of the treatment plant (2).   Flotation with a hydraulic retention of less than 30 minutes is applied using aluminium sulphate and a polyelectrolyte as flocculating agents in conjunction with an aeration vessel with high speed dispersers operating at 10 psig.   Sodium hydroxide is added for stabilisation purposes and chlorine to prevent algal growth.

Environmental protection

A good example of successful protection of the environment against pollution is to be found in the South Lake Tahoe Water Reclamation Project (3,4,5).   By the application of advanced wastewater treatment techniques,

the South Tahoe Public Utility District not only preserved the pristine
beauty and crystal clarity of Lake Tahoe, but in so doing created another
beautiful reservoir.   The treated water, although of drinking water quality,
is pumped 43 km to Indian Creek reservoir where, apart from boating and
swimming, a rainbow trout fishery is supported.

The treatment facilities at South Lake Tahoe Water Reclamation Plant
consists of chemical clarification, i.e. lime to pH 11 plus polymer addition,
ammonia stripping, two-stage recarbonation to pH 7,5, settling, mixed-media
sand filtration and finally activated carbon filtration through moving bed
counter-current columns.   The effluent, low in dissolved organics, is
odourless and sparkling clear.   The chlorine demand is low, which permits
the effective use of chlorine for disinfection of the reclaimed water.

Potable reuse

The ultimate acceptance of water reclamation pivots on the ability of
advanced wastewater treatment techniques to consistantly produce a quality
water suitable for unrestricted reuse.   In this regard the Windhoek reclamation project, which has been in operation since October 1968, and has been
producing approximately 14 per cent of the city's domectic water requirements,
is unrivalled in the field of water reclamation (6,7).   In Windhoek conventional secondary treatment is followed by maturation ponds, primarily for
nitrogen, bacteria and virus reduction, pH correction with carbon dioxide,
aluminium sulphate flotation, flocculation with lime, followed by sedimentation, rapid gravity sand filtration, breakpoint chlorination and allowing
for a contact period of $1\frac{1}{2}$ hours at a free residual chlorine concentration of
0,5 mg/l, activated carbon treatment with fixed bed granular carbon columns
and finally post chlorination to maintain a free residual chlorine concentration
of 0,5 mg/l.

Reuse Incentives

The preceeding examples serve to indicate the scope of physical-chemical
technology to produce water of any required quality.

The isolated cases quoted - and there are only a few more - illustrate
how the demand for any quality water can be met by selection of a number of
unit processes that today can be taken from the shelf.   The implementation
of each of these treatment plants were, however, due to an incentive in one
or other form.

The community of Santee was confronted with the problem of either aban-
doning its existing sewage treatment works and joining the San Diego Metropolitan System or of incurring higher costs for providing advanced treatment
facilities but defraying the latter partly by judicious utilization of the
reclaimed water.   In deciding on the latter they not only reclaimed sewage
effluent but created a pleasant appearing oasis with its adjacent grassed
areas in an otherwise unattractive and arid land.

In the case of the South African Pulp and Paper Industries, the production of pulp and paper reached static levels due to limitations in water
supply.   The installation of a 27 Ml/d reclamation plant resulted not only
in a 30 per cent increase in the pulp production but also in a saving of
R156 000 per annum in the costs of water.

In order to protect Lake Tahoe from nutrient enrichment by sewage, the State of California issued an edict in 1963, directing the South Lake Tahoe Public Utility District to export all sewage effluent by 1965. Various export routes were investigated and in each case the effluent had to be of drinking water quality. The District had no alternative but to instigate research which culminated in the 28,4 Ml/d advanced wastewater treatment plant. The benefits derived from this action is the creation of the crystal clear Indian Creek reservoir.

Far-sightedness on the part of the Windhoek Municipality resulted in the realisation of the Windhoek reclamation plant during a critical period of water shortage. As early as 1961 it was realised that the demands of the rapidly growing city would exceed the assured supplies from the available water resources by 1969. Under these circumstances the city was forced to investigate the possibility of reclaiming the water from its own wastewaters for unrestricted reuse. The success of this scheme need no further elaboration other than mentioning that for the first two years about 14 per cent of the City's needs were derived from this source.

It can be postulated that the general application of physical-chemically treated effluents will become established once the demand for high quality water exceeds the available supply. It can also be accepted that the effective utilization of physical-chemically treated effluents not only results in the primary goal being achieved, whether it be the need for a high quality water supply or because of an economic incentive, but that the effort is nearly always rewarded with an added advantage in one or other form.

Factors Influencing Physical-Chemical Treatment

Various factors may create a demand for high quality water which will result in the implementation of advanced wastewater treatment processes. Cases where the application of advanced techniques can provide immediate benefit or relief can be classified under potable water supply, industrial water supply, ecology enhancement and environmental protection.

The utilization of advanced treatment techniques to provide a potable water supply need no elaboration. An urgent demand develops and must be complied with. The application of advanced techniques for the benefit of industry is again dependent upon an economic incentive.

The factors that mostly influence the application of advanced techniques for ecology enhancement and environmental protection is the removal of phosphorus and nitrogen. It has been recognised for decades that these biostimulants produce adverse effects such as intensification of eutrophication and unbalancing of food chains. Only 37 - 46 per cent of the phosphate contained in raw sewage is removed by conventional biological treatment; removal of higher percentages demands advanced treatment. It has been estimated (8) that the amount of phosphorus and nitrogen contributed by each person every day is 2 grams and 9 grams respectively. About 21 per cent of the phosphorus in sewage is contributed by synthetic detergents. A massive and expensive national corrective effort will therefore have to be initiated to minimize the disposal of phosphorus and nitrogen. Sweden has set an example worth following. Here all sewage works will be required to remove phosphate from their effluents by 1975. The additional treatment facilities necessary will be subsidised by the Swedish Government (9).

The requirement for phosphorus removal results in overall improvement of effluent quality because the precipitation reaction that removes phosphorus significantly enhances overall removal of organics. Hence physical-chemical treatment processes should become standardised and normal components of waste-water treatment systems of the future.

Those factors which are more difficult to correlate with immediate response in terms of aesthetic or economic benefits are the long-term ecological and physiological aspects. The aforementioned factors have an influence on our immediate requirements whereas the latter should rather be seen in terms of an insurance policy. While vast amounts are spent on wildlife protection, very little is done from a national point of view to protect our natural water systems.

It is impossible to predict with certainty the long-range ecological effects of maintaining the so-called minimum adequate levels of water quality. Although we cannot achieve pristine water quality in an industrial age, we can steer away from the doubtful concept of 'minimum' water quality which was laid down to suit secondary or tertiary treated effluents. With so many physical-chemical treatment processes available, we can design our waste treatment plant to suit optimal water quality requirements as related to ecology and thereby not only preserve and protect but even enhance the natural environment.

When considering the long-term physiological effects that might be expected from continuous exposure to organic chemicals in our water supplies, we are even more insecure as regards safe water quality criteria.

A survey of the available literature on organic chemical pollution of fresh water showed that 496 organic chemicals have been reported to be found or are suspected to be in fresh water (10). Industrial sources were responsible for the largest number and variety of structural types of organic chemical pollutants. Reported agricultural sources of pollutants were all pesticides and domestic sources were all detergents. Pesticides were shown to be the most acutely toxic organic chemicals in water and only methyl mercuric chloride was found to be more toxic.

From the health point of view, chemical quality criteria of water should be based on concentration of chemicals which, when exceeded is likely to produce physiologic, toxicologic, histopathologic, carcinogenic, teratogenic, mutagenic and any other undesirable effects. The maximum allowable concentration of chemicals that might produce the above effects obviously cannot be obtained in a practical manner.

The geneticist, Dr James V. Neel of the University of Michigan, Department of Human Genetics, attempted to answer the question - " 'Is chemical mutagenesis a threat to man'? He pointed out that 'such is our ignorance of chemical mutagenesis in man or any other mammal that it is impossible to find hard data to sustain even a brief presentation'. The thesis of his analysis was that 'while there seems no immediate danger of massive genetic damage from trace chemicals, i.e. no genetic catastrophe, it does seem quite possible that current exposures to trace chemicals have increased human mutation rates'. How much he does not know" (11).

Whereas it is impossible to determine for each chemical the maximum concentration that would have no ill effects on humans, and while the dilution afforded by our natural water systems are no longer effective, the only alternative seems to be the general application of advanced wastewater treatment techniques to achieve the degree of purification required for each application.

Existing Technology

That the technology exists to achieve any desired degree of water renovation can hardly be questioned.   A number of advanced wastewater treatment processes are sufficiently well-developed to permit application on a much wider scale with confidence.   Furthermore, advances in physical-chemical treatment during the last decade have significantly reduced the cost of such treatment.   Physical-chemical treatment can, therefore, now be applied as an alternative or in conjunction with conventional treatment, especially for situations where significant phosphorus removal is required.

In 1968, The American City Magazine disclosed that 26 States already had required one or more municipalities, subdivisions, or institutions to install advanced wastewater treatment facilities.   Sixteen additional States indicated that they were considering advanced wastewater treatment and only five States reported that they saw no necessity for advanced wastewater treatment (12).   'To meet the spectrum of needs, almost 100 different processes and process variations for treatment and disposal of waterborne wastes have been considered.   Some 85 of these processes are under active study at this time, at almost 150 different locations throughout the United States. These studies are aimed at determining the efficacy and the cost of the various unit processes which may make up the advanced waste treatment systems of the future'.

All of these processes are by no means cut and dry.   Some technical problems do still exist but by continuing research the prospects are good for overcoming these problems.   However, to restore and protect the quality of our natural water systems, we are obliged to utilize the present proven technology to the best advantage.   It is thereby not advocated that all secondary effluents and industrial wastewaters be treated to potable water quality but rather to a standard comparable with the potential use thereof.

All the effort and money put into research will be well awarded if physical-chemically treated effluents are effectively utilized.   This can only happen if appropriate targets are set for each application.   If the target is augmentation of potable water supplies then we have no problems because the necessary criteria exists.   Except for the isolated case of Lake Tahoe where the criteria for discharge into a natural system had been expected to comply with potable water quality, the general minimum quality criteria applicable to effluent discharges should be upgraded to optimal quality with respect to the local ecology and domestic raw water requirements.   Thereby we would protect as fully as possible the aquatic life regime that are natural to the area and simultaneously assure a 'natural' raw water for man's own need.

Conclusions

1. Conventional water and wastewater treatment plants are no longer capable of removing the ever increasing complexity of pollutants from the water environment.   If pollution is allowed to continue indefinitely we will eventually be faced with an ecological catastrophy.

2. Advanced wastewater treatment technology has advanced to the point where
    any degree of treatment can be obtained by the judicious selection of suit-
    able combinations of a large number of processes. These unit processes
    can today be taken from the shelf and harnessed as effective tools against
    pollution.

3. As a result of the ever-increasing demand for high quality water on the
    one hand and the complexity of industrial wastes on the other hand,
    physical-chemical treatment methods may well prove to be the only way to
    provide adequate treatment. It appears that in the near future the
    basic technology for municipal treatment will consist of physical-chemical
    treatment alone or in conjunction with biological systems.

4. Wastewater treatment technology has advanced to the point of being able
    of yielding a potable water supply. Equally important is the public
    acceptance of reclaimed water for unrestricted reuse.

5. Physical-chemical treatment methods offer protection against the pollu-
    tion hazard of water used for domestic, recreational and industrial pur-
    poses. The ever-increasing number of organic chemicals in our natural
    water supplies necessitates the effective use of physical-chemical treat-
    ment methods.

6. Although the technology is available, pollution will continue because it
    appears cheaper to pollute than to reclaim. If the decision rests with
    the industrialist or local authority, economics will have the final say.
    However, if it is our earnest desire to protect the water environment
    the only solution seems to be the upgrading of effluent quality criteria.

References

1. United States Department of the Interior, Federal Water Pollution
    Control Administration, Santee Recreation Project Final Report (Water
    Pollution Control Series, WP - 20 - 7) (1967).

2. L.R.J. van Vuuren, J.W. Funke and L. Smith, 6th International Conference
    on Water Pollution Research, Jerusalem, 18-24 June, 1972.

3. A.F. Sleghta and G.L. Culp, J.Wat. Pollut. Control Fed. 39, p. 787 (1967).

4. R. Culp, Wat & Wastes Eng. 6, p. 36 (1969).

5. H.E. Moyer, Publ. Wks, N.Y. 99, p. 87 (1968).

6. L.R.J. van Vuuren, M.R. Henzen, G.J. Stander and A.J. Clayton, 5th
    International Water Pollution Research Conference, San Francisco, 26
    July - 5 Aug., 1970, pp. 1-32/1 - 1-32/9.

7. G.J. Stander and L.R.J. van Vuuren, Water Quality Improvement by Physical
    and Chemical Processes, p.31, University of Texas Press, Austin and
    London (1970).

8. H.A. Painter and M. Viney, J. biochem. microbiol. Technol. and Engng. 1,
    p. 143 (1959).

9. A.J. O'Sullivan, Water Pollution as a World Problem, the Legal,
    Scientific and Political Aspects, p.143, Europa Publications, London (1970).

10. T.R.A. Davis, A.W. Burg, J.L. Neumeyer, K.M. Butters and B.D. Wadler,
     Water Quality Criteria Data Book, vol. 1 - Organic Chemical Pollution
     of Freshwater.   (Water Pollution Control Research Series - 18010DPV
     12/70) (1970).

11. A. Wolman, J. Am. Wat. Wks Ass. 62, p. 746 (1970).

12. D.G. Stephan and R.B. Schaffer, J.Wat. Pollut. Control Fed. 42, p. 399
     (1970).

13. D.F. Metzler and H.B. Russelmann, J. Am. Wat. Wks Ass. 60, p. 95
     (1968).

14. E.F. Barth, J. Wat. Pollut. Control Fed. 43, p. 2189 (1971).

15. P.H. McGauhey and E.J. Middlebrooks, Wat. Sewage Wks, 119, p. 76 (1972).

16. P.D. Haney, J. Am. Wat. Wks Ass. 61, p. 73 (1969).

17. J.E. Vogt, J. Am. Wat. Wks Ass. 64; p. 113 (1972).

18. R.L. Culp and G.L. Culp, Advanced Wastewater Treatment, Van Rostrand
     Reinhold Company, New York, (1971)

Acknowledgement

The permission of the Director of the National Institute for Water Research
to read this paper is acknowledged.

Applications of New Concepts of Physical-Chemical
Wastewater Treatment
Sept.18-22, 1972

UPGRADING EXISTING WASTEWATER TREATMENT PLANTS

John M. Smith, Arthur N. Masse and Walter A. Feige
Environmental Protection Agency
National Environmental Research Center
Advanced Waste Treatment Research Laboratory
Cincinnati, Ohio 45268

INTRODUCTION

It is estimated that an investment of 18.1 billion dollars will be required
for the construction of municipal wastewater treatment facilities in the
United States to meet the projected 1976 Federal State Water Quality Standards
(1). Approximately one fourth of this amount will be used for upgrading the
performance of the existing wastewater treatment facilities.

The distribution of these facilities according to population is presented in
Figure 1. This figure indicates that 32% of the population receiving treatment is served by primary-intermediate treatment facilities, 31% by the activated sludge process, and 21% by the trickling filter process.

FIG. 1   FIG. 2

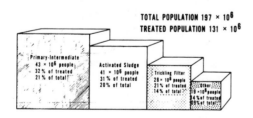

Treatment Classification by Population
EPA Survey 1968

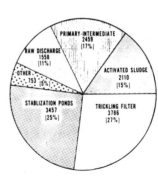

Total Facilities 14,123
EPA Survey 1968

Another way of looking at the in-ground municipal treatment plant investment
is to examine the total number of existing plants and their distribution according to type. This is shown in Figure 2. Comparison of Figures 1 and 2
indicates that there are more trickling filter plants than activated sludge
plants, but that more people are served by the activated sludge process. Simi-

lar comparisons can be made for the remaining treatment types.

The application of todays technology in upgrading plant performance includes
(a) techniques that can be used to maintain the original treatment plant efficiency under increasing organic and/or hydraulic loading, (b) the addition of
processes that can be used to increase overall plant removal efficiencies and
(c) process additions or modifications for specific contaminant removal. Physically, these upgrading procedures may be applied ahead of the plant; as modifications of the treatment process itself; or as effluent polishing techniques.

Regardless of the techniques employed, cost effective treatment plant upgrading requires efficient use of available tankage and equipment, along with implementation of the best possible operating and maintenance programs. Initial
investigations of potential upgrading situations must include a thorough examination of existing plant equipment and past performance history as well as a
complete understanding of existing plant deficiencies and future treatment
requirements. This information can then be used to formulate alternate courses
of action, and finally to select the most effective solution.

PRE-PLANT CONSIDERATIONS

Infiltration and Flow Reduction

Historically, the construction of municipal wastewater treatment plants has
been a "catch up" phenomenon. Under our current system of plant monitoring
there has been little if any motivation to improve treatment plant efficiency
until the plant is severely overloaded, many times to the point of routinely
by-passing untreated or poorly treated wastewater.

A first step then, in upgrading many facilities, is to examine techniques to
 alleviate or reduce hydraulic overload. The first approach here is not at
 the plant. Control of groundwater infiltration, the reduction of extraneous
 surface sources into sanitary sewers, and the reduction of household water
 usage should all be thoroughly examined before considering any in-plant changes.

Nationally, there are nearly 3 billion feet of public sewer, the majority of
which are constructed below the prevailing groundwater tables and therefore
are subject to infiltration. Defective sewer pipe, faulty pipe joints and
poor manhole construction are the principle causes of infiltration. In the
United States infiltration flows average 15% of the total sewage flow, and
during prolonged rainy periods amount to as much as 30% of the total flow.
The entrance of extraneous surface water from such sources as roof leaders,
manhole covers, cellar and foundation drains and other illegal connections can
add another 20% to the total sewage flow (2).

While the adoption of tighter (200 gal/day/inch diameter/mile) infiltration
specifications, better inspection during construction, improved construction
methods and materials, and legal control of house laterals inspection, can
virtually eliminate excessive infiltration for new systems, the reduction of
excess flows in older systems is a perplexing and costly proposition. The use
of closed circuit TV, dye testing and smoke testing have proven quite valuable
in locating faulty joints and illegal connections. Studies have shown that
TV inspection of 8" diameter sewer lines can be accomplished for about $0.20-
0.30 per lineal foot, including labor, equipment, and supplies. Complete
sealing of all joints can be accomplished at a speed of about 300 ft/day using
a three man crew at a total cost including inspection of $1.70 per lineal

foot. This is equivalent to a joint cost of $5.79. In one study (2), sewer sealing by this method reduced the total sewage flow by 40%. Effective over-all excess flow control will require reduction of extraneous surface sources in addition to adequate infiltration control. Control of infiltration and extraneous surface sources can be complimented by reduction of water usage in the home.

A recent EPA sponsored study (3) has shown that both water-saving devices and recycling systems for non-potable use in the home can significantly reduce the per capita sewage contribution from the average household. The installa-tion of shallow-trap and dual flush toilets resulted in a toilet water usage savings of 20%, while flow restricting shower heads decreased the shower us-age by as much as 35%. Washwater recycle systems have decreased total house-hold water usage from 24 to 35%.

Flow Equalization

Equalization of the diurnal variation in incoming sewage flows to a treatment plant can relieve hydraulic overload, and if properly designed, can signifi-cantly dampen the variation in mass flow of contaminants into the plant. It has long been recognized that operation under these quasi-steady-state condi-tions is necessary and desirable for optimum operation of both biological and chemical-physical plants.

One way of determining the required equalization volume for a given plant is to plot an inflow-mass diagram of the hourly fluctuations in sewage volume for a typical day. This variation is shown in Figure 3 along with the associ-ated inflow mass diagram. The ordinate of this diagram is obtained by accu-mulating the hourly flows for a given plant and converting them into volumes. The slope of line "A" in Figure 3 represents the average "outflow" pumping rate. The vertical distance between lines "B" and "D", which are drawn para-llel to line "A" at the maximum and minimum points of the inflow mass diagram, represent the minimum required equalization volume for constant outflow. For most diurnal variations, this volume amounts to 12-15% of the average daily flow. Either a flow-through or side-line basin may be employed for flow equalization. The volume requirements are the same in either case. In the flow-through tank, all the flow passes through the equalization basin and much better mixing or mass-flow dampening is possible. In the side-line tank only the amount of flow over the average is retained. This scheme minimizes pumping requirements at the expense of less effective mixing. A flow diagram for a treatment plant employing flow equalization using a side-line tank is shown in Figure 4, and a summary of operating data for this plant is shown in Table 1.

Flow equalization tanks should be designed as completely-mixed basins, using either diffused air or mechanical surface aerators. For systems using sur-face aeration equipment, the mixing requirements will vary from 0.02 to 0.04 HP/1000 gallons of storage volume for a typical municipal wastewater having a suspended solids concentration of 200 mg/l. Aeration to prevent septicity must also be provided. This equipment should be sized to supply 10-15 mg $O_2/l/hr$.

FIG. 3

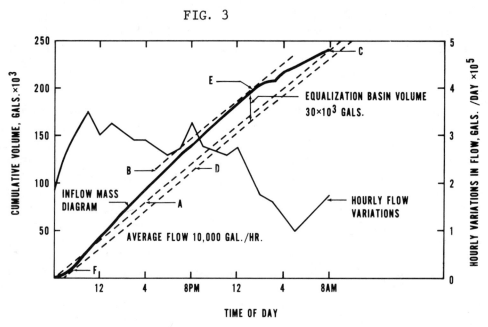

Equalization Requirements

FIG. 4

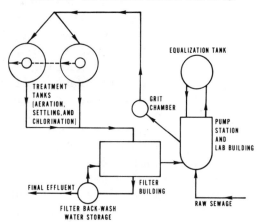

Walled-Lake Novi Wastewater
Treatment Plant 2.1 mgd

TABLE 1
Performance Data - Walled Lake Novi
Plant

PERFORMANCE DATA WALLED LAKE-NOVI PLANT				
MONTH	**REMOVAL %**		**EFF. mg/l**	
	BOD	SS	BOD	SS
SEPT. 71	91.6	96.5	6	4.1
OCT. 71	98.6	98.2	2	3.4
NOV. 71	98.6	93.7	2.3	18.7
DEC. 71	98.8	92.3	* 2.3	19.0
JAN. 72	98.2	96.1	3.5	9.0

IN-PLANT MODIFICATIONS AND ADDITIONS

Clarification

Improvement in primary and secondary clarification can be accomplished either
by structural modifications of the existing basin, by improved operation and
sludge management practices, or by the addition of chemicals to improve coagulation and settling. Improper basin inlet design, high weir overflow rates,
short circuiting and lack of surface skimming devices are the principle causes
of sub-standard clarifier operation. Many of these deficiencies can be reme-
died by inexpensive structural changes to the existing clarifier such as the
addition of scum baffles, strategically placed distribution plates, or ad-

ditional weirs. West (4) has reported that increasing the slude recirculation capacity, measuring effluent turbidity, and using a sludge blanket finder to control sludge wasting decreased the effluent BOD from 20 to 10 mg/l and the effluent suspended solids from 35 to 13 mg/l at a 3.5 mgd activated sludge plant in Sioux Falls, South Dakota. West (4) reports further that in St. Louis Missouri the use of a sludge blanket finder, a turbidimeter, increased air supply and <u>reduced</u> sludge recirculation, decreased effluent BOD from 40 to 9 mg/l and effluent suspended solids from 92 to 16 mg/l in a conventional activated sludge plant.

The most recent innovation in clarifier design is the tube settler concept. A discussion of this subject by Hansen and Culp (5) and later papers by Culp, Hansen and Richardson (6), and Hansen, Culp and Stukenberg (7) have described the theoretical and practicable aspects of both the horizontal and steeply inclined tube settlers. Tube settlers have applicability for upgrading both primary and secondary clarifiers. Culp et.al. (8) has described the performance of several plant scale tube settler installations. Results of these studies indicate that overflow rates of up to 5,000 gpd/ft^2 can be maintained in primary clarifiers without sacrificing removal efficiency. Tube settlers in this application however, do not improve the basin efficiency much beyond the 40-60% removal range. Conley (9) has recommended the use of 1.0 gpm/sq. ft. as a maximum overflow rate and 35 lbs./sq.ft./day as a maximum solids loading for the design of secondary clarifiers with tube settlers. Since the flocculating and settling characteristics of sludge vary from plant to plant, each case should be evaluated separately for suitable design criteria.

Fouling due to attachment and growth of biological slime on the sides of the tubes is sometimes a problem. Some form of cleaning device (water jet or air) is required so that the solids build-up can be removed occasionally. The performance of clarifiers using tube settlers at various installations is summarized in Table 2. Little information is available at the present time to establish cost information on tube settlers. However, an estimating cost figure of 12 to 20 dollars/sq.ft. for tube settlers with an installation cost of 5 to 15 dollars/sq.ft. has been recommended by the manufacturer (9).

TABLE 2
Performance of Tube Settlers (10)

Plant Location	Type	Size mgd	Tube Location	Existing Facility SOR[1] gpm/sq.ft.	Eff.SS mg/l	Operational Data Using Tube Settlers SOR[1] gpm/sq.ft.	Eff.SS mg/l
Philomath, Oregon	Trickling Filter	0.15	Secondary Clarifier	0.6	60-70	3.3-4.6	60-70
Philomath, Oregon	Trickling Filter	0.15	Primary Clarifier	0.84	40-45[2]	2.1-3.3	34-41[2]
Hopewell Township, Pennsylvania	Activated Sludge	0.13	Secondary Clarifier	0.34	60-70	2-3	27
Miami, Florida	Activated Sludge	1.0	Secondary Clarifier	1.3	500	1.7	33

[1]SOR - Surface Overflow Rate
[2]Percent removal rather than concentration

Chemical Addition

The use of chemicals to improve treatment efficiency, or for specific contaminant removal is now standard practice in many locations (11)(12). Chemicals commonly used are the salts of iron and aluminum, lime, and synthetic organic polyelectrolytes. The metal salts (iron and aluminum) react with the alkalinity and soluble orthophosphate in the wastewater to precipitate the metal hydroxides or phosphates. Of equal importance in upgrading applications, is the ability of these hydroxides to destabilize colloidal particles that would otherwise remain in suspension. The destabilized colloids then flocculate and settle readily, thus improving the solids removal efficiency during clarification.

Iron and aluminum salts have been widely used in many locations to precipitate phosphorus from both activated sludge and trickling filter plants. The most common locations within the plant for adding these minerals are (1) the primary clarifier, (2) directly into the aerator of an activated sludge plant, and (3) before the final clarifier. Although minimum metal dosage is dictated by incoming phosphorus levels and desired removal efficiencies, this amount is normally sufficient for improved flocculation of suspended solids. Results of metal salt addition on suspended solids and BOD removal for some selected locations are shown in Table 3.

TABLE 3
Effect of Metal Salt Addition on
Suspended Solids and BOD Removal Efficiency

Location	Type of Plant	Location of Mineral Addition	Metal and Dosage As Shown	EFF.(COD) or BOD Before Metal Add. mg/l	Total Phos. Removal	EFF. SS Before Metal Add mg/l	EFF. (COD) or BOD After Metal Add. mg/l	SS After Metal Add mg/l	(COD) or BOD Removal After Metal Add mg/l	SS Removal After Metal Add.	Reference
Richardson, Texas	Trickling Filter std. rate	Before Final Settlers	150 mg/l $Al_2(SO_4)_3$	20	93	20	<7	10	-	-	(13)
Chapel Hill, North Carolina	Trickling Filter high rate	Before Final Settling	AL/P Mole Dosage 1.7/1	40	78	47	19	30	-	-	(13)
Texas City, Texas	Activated Sludge	Before Primary Settling	19 mg/l $FeSO_4$ as Fe	-	87	-	(85)	-	-	-	(14)
Wyoming, Michigan	Trickling Filter	Before Primary Before Final	15 mg/l Fe^{+3} 3-4 mg/l Fe^{+3}	-	80	-	-	75(1)	60-87	-	(12)
Lake Odessa, Michigan	Trickling Filter	Before Primary	15-25 mg/l Fe^{+2}	89(2)	82	78(2)	82(2)	62(2)	-	-	(14)
District of Columbia (Pilot Plant)	Activated Sludge Step Aeration	In Aerator	AL/P 2/1 wt. Ratio	-	-	-	12	33	89(3)	68(3)	(15)
Manassas, Va.	Activated Sludge high rate	In Aerator	Sodium Alum. 14.8 mg/l Al^{+3}	-	61	-	-	41	(74)	69	(16)
	Three Sludge System - 2 Stage Activated Sludge Denitrification	In First Stage Aerator	Sodium Alum. 14.8 mg/l Al^{+3}	-	78	-	-	29	(91)	80	
Pomona, California	Conventional Activated Sludge	In Aerator	Alum Al/P Molar Ratio 1.9/1	(35.7)	74	9.8	(44.6)	29.7			(16)
	Conventional Activated Sludge	In Aerator	Iron Fe/P Molar Ratio 2.2/1	(35.7)	86	9.8	(38.6)	18.2			
			Fe/P 1.5/1	(35.7)	79	9.8	(37.2)	18.0			
			Fe/P 3/1	(35.7)	77	9.8	56.7	54.4			
Penn. State	Conventional Activated Sludge	Aerator Effluent	Al/P wt. Ratio 3/1	13	86	26	9	22			(17)

(1) removal across primary clarifier
(2) % removal
(3) removal across secondary treatment

In general, the addition of minerals in the dosage range shown above will produce 80-90% total phosphorus removal. If added to the plant ahead of the primary, the addition of an anionic polymer is often recommended to minimize the carry-over of insolubilized phosphorus to the subsequent treatment stages. The addition of minerals to the aerator or aerator effluent of an activated sludge plant will generally result in more efficient coagulant utilization than when added to the primary because, during biological treatment, the complex phosphorus forms are hydrolyzed to the ortho form which is more easily precipitated. The addition of metal salts to the primary tends to dilute the resulting primary sludge. If added to the aerator, the metal has a weighting effect which results in a thickening of normal activated sludge.

The addition of metal salts to the aerator of an activated sludge plant produces moderate increases in suspended solids removal efficiency in some instances, but on many occasions results in significant increases in suspended solids carry-over. Mulbarger (16) states that this is due more to an incorrect ratio of volatile solids/aluminum added than to a strict overdose of chemicals. Results at Pomona, California indicate that increased turbidity is due to lowering the pH beyond the optimum limits for alum flocculation. The addition of 2 mg/1 of polyelectrolyte prior to final clarification reduced the effluent turbidity to its previous value.

The addition of polyelectrolytes alone, in the 0.25-1.0 mg/1 range has been shown to improve suspended solids removal of primary clarifiers from 37.7% to 64.7%, and BOD removal from 31.0% to 46.7% (10). The effectiveness of lime addition in increasing the efficiency of primary clarifiers at several locations is shown in Table 4.

TABLE 4
Effect of Lime Addition on
Primary Clarifier Performance

Location	Lime Added mg/l CaO	Percent Removal Before Lime Addition		Percent Removal After Lime Addition		Remarks
		BOD	SS	BOD	SS	
Duluth, Minnesota	75	50	70	60	75	—
	125	55	70	75	90	—
Rochester, New York	140	—	—	50	80-90	Jar tests
Lebanon, Ohio	145	—	—	66	74	Pilot plant

The addition of lime to primary clarifiers can nearly double the mass of primary sludge to be handled, depending on the alkalinity of the incoming wastewater, and will raise the primary effluent pH to the 9-11 range, depending on the lime dosage. The addition of lime at this location will decrease the organic loading to the subsequent biologically stage in many instances to where complete nitrification will occur. If this happens, the nitric acid and carbon dioxide produced by the nitrification and carbon oxidation reactions are adequate to maintain the pH within the aerator at a near optimum level for nitrification (12). A principal consideration in selecting lime addition to the primary clarifier as an upgrading technique is whether or not the existing sludge handling facilities are adequate and if not, whether alternate sludge handling schemes are possible.

EFFLUENT POLISHING

The use of effluent polishing techniques is recognized as one of the most cost effective methods of upgrading existing plants to obtain increased organic and suspended solids removal. Overall plant performance can be improved from the 70-85% efficiency range to the 95-99% range depending on the process employed. Four unit processes are considered here for effluent polishing. They are: 1) filtration, 2) microscreening 3) granular carbon adsorption, and 4) reverse osmosis.

Filtration

Simple filtration of secondary effluent provides a positive method of suspended solids control, and as such, is the most widely used and the most efficient single unit process for upgrading treatment plant performance today.

J.M. SMITH, A.N. MASSE and W.A. FIEGE

Contemporary filtration systems can be broadly classified as either deep-bed or surface filters. The most popular trend recently in deep-bed filter design is the use of the dual or tri-media filter. Here, the use of two or more layers of different media having increasing specific gravity with increasing bed depth allows gradation of the filter bed from coarse to fine in the direction of flow. This allows more efficient utilization of the total bed depth for solids storage than conventionally graded, single medium filters.

The approach taken in the development of surface filters on the other hand, is to allow filtration to take place on or near the top of relatively shallow single medium filters, and to optimize removal of the accumulated solids. In addition to the standard pressure and gravity surface filters, several innovative techniques for providing a continuous clean filter surface have been developed. The moving bed filter developed by Johns-Manville Products, the radial flow filter developed by the Dravo Company, the radial flow-external wash filter developed by the Hydromation Corporation, and the Hardinge traveling bed filters are typical examples.

The performance of each of the above filters in polishing secondary effluent will depend on such factors as surface loading rate, temperature, floc size and strength, degree of biological or chemical flocculation, media depth, grain size, solids loading, run length, and method of filter operation. Because of the numerous variables involved, and the ease of obtaining reliable design data from small pilot filters, it is recommended that final filter selection be based on pilot plant results where possible. Operating results from typical filter installations are shown in Table 5.

TABLE 5
Filtration Performance

Filter Type	Feed Type	Media Size mm	Filter Depth ft.	Hydraulic Loading gpm/sq.ft.	SS Removal %	BOD Removal %	Effluent SS mg/l	Effluent BOD mg/l	Reference
Deep-Bed									
Gravity Downflow	T.F. Eff.	1.0-2.0	-	6	70	55	5-7	-	(18)
Gravity Downflow	T.F. Eff.	0.9-1.7	2-3	3	67	58	-	2.5	(19)
Pressure Upflow	T.F. Eff.	0.9-1.7	5	3	85	74	5.0	2.5	(19)
Pressure Upflow	A.S. Eff.	0.9-1.7	5	3	77	-	-	-	(19)
Pressure Upflow	A.S. Eff.	1.0-2.0	5	2.2	50	62	7.0	6.4	(20)
Pressure Upflow	A.S. Eff.	1.0-2.0	5	4.0	67	73	4.9	6.4	(20)
Pressure Upflow	A.S. Eff.	1.0-2.0	5	4.9	56	65	5.7	7.1	(20)
Mixed Media	E.A. Eff.	0.25-2.0	2.5	5.0	74	88	4.6	2.5	(21)
Mixed Media	A.S. Eff.			2.0	73	74	3.8	6.0	(20)
Mixed Media	A.S. Eff.			4.0	73	85	4.3	3.9	(20)
Mixed Media	A.S. Eff.	0.25-2.0	2.5	2.5					(22)
Surface Filters									
Moving Bed	T.F. Eff. (1)	0.6-0.8	4.2	2	47	71	-	-	(23)
Moving Bed	T.F. Eff. (2)	0.6-0.8	4.2	2	67	80	-	-	(23)
Gravity Downflow	A.S. Eff.	-	-	2.2	55	64	7.2	7.3	(20)
Gravity Downflow	A.S. Eff.	-	-	4.0	69	70	4.8	7.4	(20)
Gravity Downflow	A.S. Eff.	-	-	8.0	48	64	6.1	6.7	(20)
Gravity Downflow	A.S. Eff.	0.9-1.7	2.0	1.6-4.0	72-91	52-70	-	-	(19)
Gravity Downflow	A.S. Eff.	0.95	1.0	2.0	46	57	-	-	(24)
Gravity Downflow	A.S. Eff.	0.58	-	2.0-6.0	70	80	-	-	(25)
Gravity Downflow	C.S. Eff.	0.45	1.0	5.3	62	78	5	4	(26)

T.F. - Trickling Filter   E.A. - Extended Aeration
A.S. - Activated Sludge   C.S. - Contact Stabilization
(1) 100 mg/l alum & 0.2-0.75 mg/l anionic polymer added
(2) 200 mg/l alum & 0.2-0.75 mg/l anionic polymer added

As the table indicates, all of the filters mentioned produce an effluent with suspended solids and BOD_5 generally less than 7 mg/l. As a rule, the deep-bed filters are better suited to treating strong biological floc, will yield longer run lengths, and are less sensitive to solids loading than are the surface filters. The surface filters are better adapted to removing the more fragile chemical flocs, yield shorter run lengths, and require less backwash water than the deep-bed filters. A key element in efficient effluent polishing is to match correctly the type of filter used with the flocculant nature of the solids to be removed, and also to design the filter for solids-loading and run lengths that are compatible with normal operating schedules.

Microscreening

The microstrainer is another surface filtration device that has found increasing utility for polishing secondary effluents. The system consists of a specially woven stainless steel fabric mounted on the periphery of a partially submerged horizontal revolving drum. Influent enters through the upstream end of the drum and flows radially outwards through the fabric leaving the intercepted suspended solids behind. Microstrainers are available in sizes ranging from 5'0" diameter x 1'0" wide having a 0.06 to 0.6 mgd capacity to 10' 0" diameter x 10'0" wide with a capacity of 4.0 to 12.0 mgd. Filtration efficiency depends primarily on fabric size and the character of the solids being removed. For wastewater polishing applications, these microstrainers are available with automatic controls to increase drum speed and backwash pressure to accomodate variations in flow, and to a lesser extent, variations in solids loading. One of the chief advantages of using a microstrainer for polishing secondary effluent is its low head requirement of 1 to 1-1/2 feet. Pumping secondary effluent prior to microstraining tends to shear the biological floc and decrease solids removal efficiency.

Microstrainers are washed continuously at 20-50 PSIG and require 4-6% of the filter throughput. Continuous backwash is advantageous for upgrading small plants since it eliminates sur-charging of the upstream units which must be considered when using conventional intermittently backwashed filters. Operational data from various microstrainer installations are presented in Table 6 (10).

TABLE 6
Microstrainer Performance

Location	Plant Size mgd	Feed Type	Fabric Opening microns	SS Removal percent	Effluent SS mg/l	BOD Removal percent	Effluent BOD mg/l	Backwash % of flow
Brampton, Ontario	0.1	A.S.[1] Effluent	23	57	-	54	-	-
Lebanon, Ohio	Pilot	A.S. Effluent	23	89	1.9	81	-	5.3
	Pilot	A.S. Effluent	35	73	7.3	61	-	5.0
Chicago, Illinois	3.0	A.S. Effluent	23	71	3.0	74	3.0	3.0
Lutton, England	3.6	Effluent from A.S. and T.F.[2]	35	55	7.3	30	-	3.0
Bracknell, England	7.2	T.F. Effluent	35	66	5.7	32	8.4	-

[1]A.S. - Activated Sludge

[2]T.F. - Trickling Filter

Carbon Adsorption

The necessity to upgrade secondary effluent quality beyond the levels that can be obtained by implementation of the process modifications previously discussed, or by the application of tertiary filtration, will require a substantial investment in additional plant equipment as well as a 30-50% increase in operating cost. It is doubtful that this level of expenditure can be justified except as part of a major plant expansion where the expected lifetime of the new facility is sufficiently long to permit reasonable amortization of the high capital investment required. The situation is especially difficult for small plants in the 0.5 mgd range that serve rapidly growing urban areas.

The effectiveness of granular activated carbon for upgrading the treatment efficiency of larger plants is well established. Operating experiences at Pomona and South Lake Tahoe, California; Nassau County, New York and Colorado Springs, Colorado have left little doubt regarding process efficiency, operating cost and reliability of these systems. Operating results and principle design parameters for three of these locations are shown in Table 7 (10).

TABLE 7
Tertiary Granular Carbon Adsorption
Design Parameters and Operating Results

	Pomona		Lake Tahoe		Nassau County	
Operating Data						
Capacity	200 gpm		1,800 gpm		400 gpm	
Source of Waste	Domestic		Domestic		Domestic	
Secondary Treatment	Standard Activated Sludge		Standard Activated Sludge		High-rate Activated Sludge	
Pre-treatment	Chlorination		Coagulation & Filtration		Coagulation & Filtration	
Carbon Type	16 x 40 mesh		8 x 30 mesh		8 x 30 mesh	
Column Configuration	4-Stage Downflow		2-Upflow in Parallel		4-Stage Downflow	
Column Dimensions	6' dia. x 9' deep		12' dia. x 14' deep		8' dia. x 6' deep	
Nominal Contact Time	36 minutes		13 minutes		24 minutes	
Loading Rate	7 gpm/sq.ft.		8 gpm/sq.ft.		7.5 gpm/sq.ft.	
Carbon Column Performance	Influent	Effluent	Influent	Effluent	Influent	Effluent
COD, mg/l	47	10	20-30	2-10	-	5
BOD, mg/l	-	-	5-20	2-5	-	-
Color, Pt-Co Units	30	3	20-50	5	-	-
Carbon Dosage	350 lbs/million gallon		250 lbs/million gallon		500 lbs/million gallon	

The capital cost of tertiary granular carbon systems will vary widely depending on the particular system design and pretreatment provided. Direct application of secondary effluent to downflow carbon adsorption columns as practiced at Pomona, California will result in a smaller capital investment than the tertiary system used at Lake Tahoe, but may increase operating costs due to more frequent column backwashing.

Operating cost will depend primarily on the organic loading and associated carbon dosages. Table 8 compares the actual capital and operating costs for the 7.5 mgd conventional activated sludge plant at South Lake Tahoe with the corresponding tertiary carbon adsorption costs at that location (27). Other investigators have estimated capital and operating costs for tertiary carbon adsorption systems that are considerably higher than those shown. The above data was used to estimate operating costs for carbon dosages of 350 #/mg and 500 #/mg. These higher dosages would be anticipated from most activated sludge effluents applied directly to adsorption columns.

The following tabulation of costs illustrates that at the 7.5 mgd scale the use of granular carbon adsorption with regeneration can increase total plant operating cost by 60% and require a capital expenditure as great as 30% of the original plant investment. Culp and Culp (27) estimate that the total costs for tertiary carbon adsorption for a 2.5 mgd plant would be 50% greater

than for the 7.5 mgd Tahoe system. Little operating information is available
on carbon adsorption operating costs below this level.

TABLE 8
Tertiary Granular Carbon Adsorption Costs

	Operating Cost $/mg	Capital Cost $/mg	Total Cost $/mg
Primary and Activated Sludge + Organic Sludge Handling and Chlorination 7.5 mgd	103	67.50	170.50
Granular Carbon Adsorption and Regeneration 7.5 mgd			
250#/mg	30	21.5	51.5
350#/mg	42*	21.5	63.5
500#/mg	60*	21.5	81.5

* estimated based on South Lake Tahoe Data

Recently some investigators have advocated adding powdered activated carbon
directly into the aerator of an activated sludge plant to upgrade the organic
removal efficiency. The DuPont Company has developed a PACT (powdered activated carbon treatment) process for this purpose. Preliminary data obtained in
a parallel 0.45 mgd study showed that the application of 308 mg/l of powdered
activated carbon into a completely mixed aerator of a conventional activated
sludge plant reduced the soluble effluent BOD from 20 to 11 mg/l, for an overall BOD removal of 96% (28). In another test, the flow to one side of a parallel activated sludge plant receiving 295 mg/l of powdered carbon was more
than doubled without sacrificing soluble BOD removal efficiency. Total BOD removal decreased from 98 to 96% with suspended solids increasing from 58 to 89
mg/l (28). These experiences indicate that powdered activated carbon addition
to activated sludge plants will produce excellent soluble BOD removal, but
these plants must be followed by filtration, to produce high suspended solids
removals.

The operating cost for the addition of 300 mg/l of powdered activated carbon
on a once through basis would be about $220/mg which is prohibitively high for
continuous use. Regeneration of this carbon could bring the cost down to
about $70/mg which is competitive with granular carbon adsorption systems. Al-
though a great deal of work has been completed on powdered carbon regeneration
schemes (29,30,31,32), they have not been successfully used for this application. In any case, it is doubtful that powdered carbon adsorption should be
used to compete with biological oxidation in the aerator of an activated sludge
plant except for the removal of "hard" or refractory BOD substances. The most
practical application would be to use the powdered carbon as an operational
tool to improve treatment efficiency during times of "biological upset" due
to toxic materials, or during short periods of extremely heavy organic loads.

Reverse Osmosis

The most efficient effluent polishing process available today is reverse osmosis. When applied to sand filtered secondary effluent, it is capable of producing water suitable for nearly any reuse application. Typical effluent
quality for reverse osmosis treatment of secondary effluent using cellulose
acetate mambranes at an applied pressure of 400-600 PSIG is shown in Table 9.

TABLE 9
Reverse Osmosis Performance
Treating Secondary Effluent

Parameter	Effluent Quality mg/l unless noted	Removal %	Reference
Suspended Solids	0.0	100%	(33)
Dissolved COD	1.5	>96.1	(34)
Calcium	0.3	>99.6	(34)
Magnesium	0.1	>99.5	(34)
Sulfate	1.9	>99.5	(34)
Phosphate	0.1	>99.7	(34)
TDS	54	95.2	(34)
Ammonia Nitrogen	1.0	95.1	(34)
Chloride	8.6	94.6	(34)
Sodium	10.6	94.0	(34)
Potassium	1.1	94.7	(34)
BOD_5	<1.0	92.0	(35)
Color	3.0	97.0	(35)
Zn^{++}(1)	0.0	100.0	(35)
Pb^{++}	0.0	100.0	(35)
Cr^{++}	0.0	100.0	(35)
ABS	0.0	100.0	(35)
Bacteria (2)	700	99.8+	(35)
Turbidity	0.0	100.0	(35)

(1) ppb
(2) Colonies/100 ml, MF technique

Notwithstanding the high quality water than can be produced by reverse osmosis, it is unlikely that this process will find wide utility in upgrading treatment plant effluents except for special reuse applications, where the value of the reclaimed water will help defray total treatment costs, or in instances where partial or total demineralization is required.

Pending Federal legislation for "zero discharge" of point sources of wastewater by 1981 (36), the adoption of similar standards by the State of Vermont (37) and the recently announced plan to desalt portions of the Colorado River (38), are all examples of potential applications of a high removal efficiency process such as reverse osmosis. It must be remembered however, that reverse osmosis is a separation process that produces a brine stream having approximately 1/10 of the volume and 10 times the strength of the original waste. Treatment of this brine must also be included in any successful application of this process.

Over seven years of experimental results in testing the three principle reverse osmosis configurations on variously treated municipal wastewater streams have shown that module productivity declines rapidly due to organic fouling, and that routine membrane cleaning is required for continuous operation on any wastewater stream. Optimization studies have shown that filtration of chemically clarified primary effluent or secondary effluent is the most cost effective pretreatment for the reverse osmosis process. Only the spiral-wound and tubular configurations have been found suitable for processing organically laden wastewaters.

Experience to date has shown that the total operating costs exclusive of brine disposal will vary from $350/mg for treating secondary effluent to $450/mg for treating primary effluents. Approximately 48% of the operating cost is for membrane replacement, 22% for capital amortization, 18% for power, and 12% for operation and maintenance (34). Brine treatment and disposal will increase the above total operating cost by another 12-18%.

REFERENCES

1. "Water in the News," Soap and Detergent Association, New York, N.Y. (June 1972).

2. "Control of Infiltration and Inflow into Sewer Systems," EPA Publication, 11022-EFF (December 1970).

3. S. Cohen, "Demonstration of Waste Flow Reduction from Households," Report for EPA, 68-01-0041, (February 11, 1972).

4. A. F. West, "Case Histories of Plant Improvement by Operations Control," Nutrient Removal and Advanced Waste Treatment Symposium, FWPCA, OBR, Cincinnati, Ohio (1969).

5. S.P. Hansen and G. L. Culp, Jour. Amer. Water Wks. Assoc, Vol. 59, p. 1134 (1967).

6. G. L. Culp, S. P. Hansen, and G. H. Richardson, Jour. Amer. Water Wks. Assoc. Vol. 60, p. 681 (1968).

7. S. P. Hansen, G. L. Culp and J. R. Stukenberg, Jour. Water Poll. Cont. Fed. Vol. 4, p. 1421 (1969).

8. G. L. Culp, K. Y. Hsiung, and W. R. Conley, Jour. Sanit. Engr. Div., ASCE, 95, Vol. 5, p. 829.

9. W. R. Conley and A. F. Slechta, Presented at the 43rd Annual Wat. Poll. Cont. Fed. Conference, Boston, Mass. (October 1970).

10. "Process Design Manual for Upgrading Existing Waste Water Treatment Plants," EPA, Contract 14-12-933 (October 1971).

11. "Process Design Manual for Phosphorus Removal," EPA, Contract 14-12-936 (October 1971).

12. D. S. Parker, "Process Development for Nitrogen Removal at the Central Contra Costa Water Reclamation Plant, Contra Costa County, California," Paper presented at the EPA, AWT Design Seminar held at the University of California, Riverside, California (March 24, 1972).

13. R. C. Brenner, "Advances in Treatment of Domestic Wastes," Paper presented at the National EPA Training Course held in Athens, Georgia, (October 18-22, 1971).

14. J. J. Convery, "The Use of Physical-Chemical Treatment Techniques for the Removal of Phosphorus from Municipal Waste Water," Presented at FWQA, AWT Seminar in San Francisco, (October 28-29, 1970).

15. A. B. Hais, et.al; "Alum Addition to Activated Sludge with Tertiary Solids Removal," EPA, Advanced Waste Treatment Research Laboratory, National Environmental Research Center, Cincinnati, Internal Report (March 1971).

16. M. C. Mulbarger and D. G. Shifflett, Chem. Engr. Prog., Vol. 67, No. 107 (1970).

17. L. S. Directo, R. P. Miele, A. N. Masse, "Phosphate Removal by Mineral Addition to Secondary and Tertiary Treatment Systems," 27th Purdue Industrial Waste Conference, (May 2-4, 1972).

18. Private Communication with Peter Kaye, Municipal Sales Manager, Dravo Corporation, Pittsburgh, Pennsylvania, June 2, 1971.

19. J. J. Convery, Solids Removal Processes, Nutrient Removal and Advanced Waste Treatment Symposium, FWPCA, Cincinnati, Ohio (April 29-30, 1969).

20. D. R. Zeny, "Hanover Tertiary Plant Studies," April 9-June 30 Quarterly Report WPRD Grant 92-01-68 (Unpublished).

21. G. L. Culp and S. Hansen, Water & Sewage Wks., 114, No. 2, p. 46 (1967).

22. Private Communication with Mr. Michael Strachow, Johnson & Anderson, Consulting Engineers, Pontiac, Michigan, (March 1972).

23. G. R. Bell, D. V. Libby and D. T. Lordi, "Phosphorus Removal Using Chemical Coagulation and a Continuous Countercurrent Filtration Process," FWQA, No. 17010-EDO (June 1970).

24. F. B. Laverty, R. Stone and L. A. Meyerson, Jour. Sanit. Engr. Div., ASCE, 87, 6, 1 (November 1961).

25. B. Lynam, G. Ettelt and T. J. McAloon, Jour. Water Poll. Cont. Fed., Vol. 41 p. 247, (February 1969).

26. Performance Data Contained in Hydroclear Corporation Catalogue, Avon Lake Ohio as tested by the Clark County Utilities Department, Springfield, Ohio, May 1969.

27. R. L. Culp and G. L. Culp, "Advanced Wastewater Treatment," Van Nostrand-Reinhold Company, New York (1971).

28. DuPont PACT Process "Advertised Pilot Plant Results," Jackson Laboratory, P. O. Box 525, Wilmington, Delaware, 19899.

29. R. Bloom, Jr., et al; Env. Sci. Technology, Vol. 3, p. 214 (March 1969).

30. E. L. Berg, R. Villiers, A. N. Masse and L. Winslow, Chem. Engr. Prog. Symp. Series, Vol. 67, No. 107, p. 154, (1970).

31. D. S. Davies and R. A. Kaplan, Chem. Engr. Prog. Symp. Series, 60, 12, p. 46, (1964).

32. FWPCA Contract 14-12-400 (GATX), Infilco Products Co.

33. J. M. Smith, A. N. Masse and R. P. Miele, "Removation of Municipal Wastewater by Reverse Osmosis," FWQA Report ORD 17040 --5/70.

34. "Water Renovation of Municipal Effluents by Reverse Osmosis," Gulf Oil Corporation, Report for EPA Contract 14-12-831, (February 1972).

35. J. Wilford and F. Perkins, "Test of G. A. Reverse Osmosis Unit in New
     Jersey," p. 25, (January 1966).

36. United States Senate Bill (S2770).

37. "Regulations Governing Water Clarification and Control of Quality,"
     Public Act 252, Vermont Water Resources Board, (1969 Session).

38. Wall Street Journal (June 22, 1972).

SOME PROBLEMS ASSOCIATED WITH THE TREATMENT OF SEWAGE

BY NON-BIOLOGICAL PROCESSES

R. W. Bayley, E. V. Thomas, and P. F. Cooper
Water Pollution Research Laboratory of the Department of the Environment,
Elder Way, Stevenage, Herts, England.

Introduction

Conventional systems for treating sewage which employ physical and biological processes have been developed over a period of many years and, when carefully operated, plants employing these processes can consistently produce final effluents containing no more than 20 mg BOD/l and 30 mg suspended solids/l; if a polishing process is employed final effluents containing as little as 5 mg BOD/l can be obtained. However, recognition of problems resulting from eutrophication hastened by discharge of nutrients in waste waters, and the growing need to use surface waters as sources for domestic water supply, are almost certain to lead to further constraints being imposed on the quality of treated effluents in some areas. It is already clear that additional processes involving the use of chemical coagulants may be required to control the rate of discharge of phosphates, and, should it be necessary to remove the last traces of organic material from polished sewage effluent, processes which are at present not familiar in waste treatment will be required. As there appear to be several disadvantages and limitations inherent in biological systems, including sensitivity to toxic substances, low rates of reaction when sewage temperatures fall below $10^{\circ}C$, and the production of surplus "secondary" sludges which may be extremely difficult to dewater, it is appropriate to examine the possibility of developing a purely non-biological treatment system.

Though chemical engineering technology has brought many separation processes to a high level of performance in the chemical industry, few technically feasible and economically acceptable patterns of treatment of sewage, which employ purely non-biological processes have been designed, and of these systems only two have apparently been studied in detail with the aid of pilot-scale equipment.

In the first system the waste liquor is passed across the surface of a semi-permeable membrane; the liquor may be whole or settled sewage or a treated effluent containing only a fraction of the organic material present

in the raw waste. The membrane may be cast from a specially prepared
solution or "dope" or alternatively it may be formed by depositing a layer
of prepared particles known to have semi-permeable properties. Under
carefully controlled laboratory conditions high-quality treated effluents
can be obtained using either of these processes, but the rate of flow of
product water tends to fall to an unacceptably low figure and current
research is largely concerned with developing economical techniques for
overcoming this problem (1-4).

Much of the current research into membrane systems has been carried out
in small-scale equipment and concentrate or reject liquors have merely been
discarded, but reject liquor from a full-scale sewage-treatment plant
employing a membrane separation process would have many of the characteristics of a highly concentrated waste liquor - as yet no process has been
developed specifically for treating such liquors.

In the second system most of the suspended matter in raw sewage is
removed with the aid of a chemical coagulant and organic material in
solution is subsequently removed by adsorption. Polyelectrolytes, lime,
alum, and ferric salts have been used in the first stage and though sedimentation has been employed in the majority of the experimental pilot plants
for which details are available, separation might be effected by flotation
or possibly a combination of sedimentation followed by flotation.

Some of the capabilities of processes employing adsorption on granular
activated carbon have been amply demonstrated in several pilot plants in
the USA which have been supplied with raw sewage containing 80-120 mg
organic carbon/l; these experimental plants have consistently produced
final effluents containing no more than 5-20 mg organic carbon/l (5-9).
These encouraging results suggested that a satisfactory process suitable
for use in full-scale plants might be developed in the near future;
consequently much of the current interest in physico-chemical systems is
concerned with this type of treatment. However, several important
questions remain unresolved and the present work describes some of these
problems and shows how they may have a major influence on the future
development and possible application of novel processes in the British
Isles.

Sewage Treatment in the British Isles

Some aspects of river management and water supply, commonplace in
the British Isles, are seldom encountered in many countries; these peculiar
conditions have resulted in quite stringent standards being imposed on the
quality of treated sewage effluents discharged to the upper reaches of
some streams and rivers.

The daily consumption of water per person in the UK for purely domestic
purposes is generally between 0.15 and 0.25 m^3/d; consequently domestic
sewage may contain up to 400 or 500 mg organic carbon/l and have a 5-day BOD
between 400 and 700 mg/l, though less concentrated sewage is sometimes
encountered as a result of infiltration and dilution with weak industrial
wastes. As a large proportion of the urban population in England is situated
in areas where the flow in many rivers and streams falls to values
insufficient to afford adequate dilution during parts of the year,
concentrations of BOD, suspended solids, and ammonia (as N) in final
effluents may be limited to between 5 and 10 mg/l; to comply with such

standards 98 to 99 per cent of the BOD initially present must be removed.

In eastern and south-eastern England treated effluent forms a significant proportion of the flow in many rivers which are, and in some instances have for many years been, sources of raw water. To reduce the risk of pollution of such rivers by failure of conventional biological sewage-treatment plants elaborate systems have been developed for monitoring and controlling discharges of toxic industrial waste waters which might interfere with treatment processes.

The considerations set out above indicate some of the criteria on which any proposed alternative to conventional sewage-treatment processes might be assessed in the British Isles; the process would be required to treat "strong sewage" and though a high quality effluent might not be required for discharge to an estuary or the sea very stringent standards would be imposed at some inland sites. In contrast with some parts of North America and the mainland of Europe eutrophication has not presented any serious widespread problems in the British Isles (10); consequently it seems unlikely that much advantage would be gained in the near future by widespread adoption of waste-treatment plants designed to remove phosphorus. However, as mentioned above, many plants are already required to produce well nitrified effluents and, if the level of nitrate in some rivers approaches a limit beyond which abstracted water could not be distributed without hazard to health, it would no doubt be necessary to employ denitrification systems either at water-treatment plants or during sewage treatment.

Experimental Work

Towards the end of 1970 a decision was taken to construct at the Water Pollution Research Laboratory a pilot plant for examining the potentiality of non-biological treatment and in October 1971 a small laboratory-scale plant was assembled so that information could be obtained which might prove useful when designing the larger plant. The relative merits of aluminium and ferric salts and lime were examined in a series of preliminary tests. Having regard to the quantities required to bring about significant and generally similar removals of slowly settling material during quiescent settling of crude sewage it was apparent that lime seemed to offer the greatest economy in use; consequently a process using lime as coagulant has been studied.

Fresh whole sewage has been taken twice every week from a sewer carrying domestic sewage from a nearby residential area; freshly prepared lime slurry has been added to a sample of approximately 0.04 m^3 sewage and after 0.5 h quiescent settling the alkaline settled liquor has been recarbonated in two stages, firstly to a pH value of about 9.3 and after a further 0.5 h quiescent settling to neutrality. Neutral liquor has been pumped at a constant rate of 0.01 m^3/d from a small reservoir to a sand filter discharging to four beds of granular carbon arranged in series. Detention times and other information concerning the equipment are shown in Fig. 1 and a summary of results is shown in Table 1. Concentrations of organic carbon in raw sludge, lime treated settled sewage, and final effluent are shown in Fig. 2.

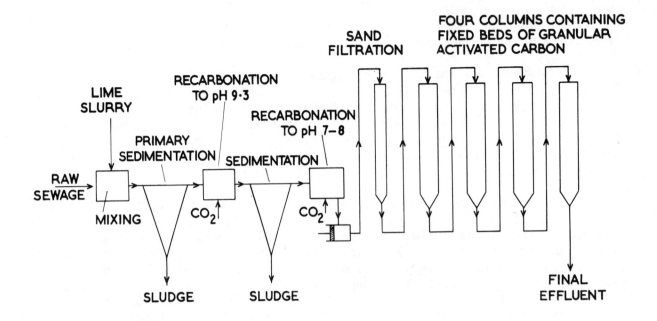

Fig. 1.   Flow diagram representing laboratory-scale equipment
                 Detention time in each carbon contactor, 0.48 h
                 Diameter of carbon columns, 30 mm
                 Each bed contained approximately 210 ml of 'Filtrasorb 400'
                      supplied by Chemviron Ltd, Grays Thurrock, Essex
                 The sand filter and the four columns containing activated
carbon were supplied with liquor flowing at a rate of 0.01 m^3/d;
the preceding stages of the process were operated manually in
batches

TABLE 1

Summary of Results Obtained During 182 Days' Operation of a Laboratory-Scale
Non-Biological Treatment Plant

Sample	Organic carbon	COD	BOD
	(mg/l)		
Raw sewage	333	1028	546
Settled lime-treated sewage	124	–	–
Recarbonated settled lime-treated sewage	107	275	–
Liquor leaving			
1st carbon contactor	76	220	–
2nd " "	56	166	–
3rd " "	52	150	–
4th " "	40	110	50

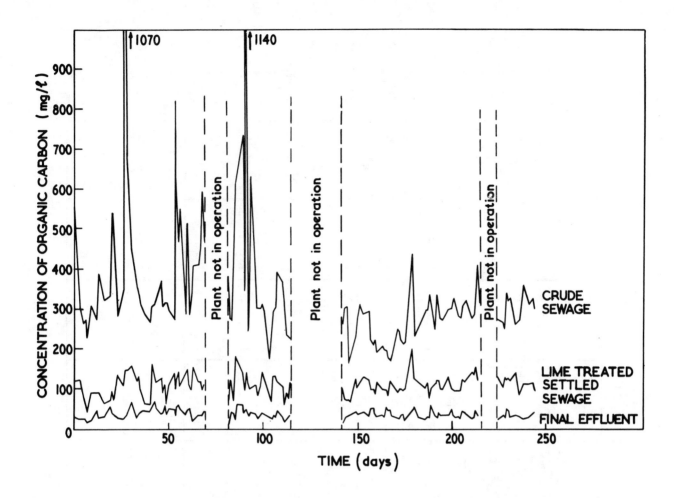

Fig. 2.   Concentration of organic carbon in raw sewage, lime-treated
                settled sewage, and effluent from a series of four columns contain-
                ing granular activated carbon

     As well as providing information on the removal of BOD, suspended solids,
organic carbon, and phosphorus from normal domestic sewage the laboratory-
scale equipment was used in two tests in which two frequently encountered
constituents of industrial discharge, known to create difficulties in conventional treatment if present in excess, i.e. nickel and cyanide, were added
to raw sewage so that the ability of systems to remove them could be examined.
Concentrations of phosphorus, nickel, and cyanide in raw waste and liquors
sampled after various stages of treatment are shown in Tables 2, 3, and 4.

TABLE 2

Removal of Phosphorus from Lime-Treated Crude Sewage During Quiescent
Settling and After Subsequent Two-Stage Recarbonation

Type of compound		1 Raw sewage	2 Lime treated settled sewage	3 As in column 2 after recarbonation
Orthophosphate,	soluble	6.6	0.025	0.42
" "	insoluble	5.8	0.50	0.08
" "	total	12.4	0.52	0.50
Polyphosphates,	soluble	1.1	0.04	0.09
"	insoluble	4.5	0.20	0.07
"	total	6.6	0.24	0.16
Organic phosphorus,	soluble	0.39	0.30	0.17
" "	insoluble	2.80	0.03	0.19
" "	total	3.2	0.33	0.36
Total phosphorus,	soluble	8.1	0.36	0.68
" "	insoluble	13.1	0.73	0.34
" "	sum	21.2	1.09	1.02

TABLE 3

Two sets of results showing removal of nickel from sewage during treatment
with lime (pH 11.5), recarbonation, and contact with
granular activated carbon

Sample	A		B	
	Concentration of nickel (mg/l)	Percentage removal	Concentration of nickel (mg/l)	Percentage removal
Whole raw sewage	10.0		10.0	
Filtered raw sewage	6.8		7.2	
Settled lime-treated sewage	1.5	85.0	1.4	86.0
Filtered " "	0.8		1.0	
Recarbonated liquor	1.2	88.0	1.2	88.0
Filtered "	1.2		1.1	

Effluent from fourth
  column containing
  activated carbon
	0.18	98.2	0.2	98.0

Table 4

Removal of Cyanide from Lime-Treated Crude Sewage During Quiescent Settling,
After Recarbonation and Contact with Granular Activated Carbon

Sample	Concentration of cyanide (mg CN/1)	Removal (per cent)
Raw sewage	82	-
Settled lime treated sewage	77	6.1
Recarbonated settled sewage	73	11.0
Final effluent from activated carbon contactors	49	40.3

Nitrogen Removal

Before discussing the results mentioned above some consideration will be given to the question of nitrogen removal. No provision was made for removing ammonia in the simple laboratory equipment but it had been supposed that an ammonia stripping tower, or some alternative non-biological process for removing ammonia, would be included in the larger pilot plant but these provisional plans have had to be modified in the light of results obtained from a survey in which concentrations of urea were determined in sewage sampled at the inlet works of ten treatment plants in the British Isles. Samples taken during periods of maximum flow contained over 20 mg urea/1 (as N) though levels of 5 mg/1 were common during times of minimum flow; urea was also detectable in settled sewage, concentrations of 15 mg/1 being found at one works.

Examination of fresh urine and faeces (11,12) has indicated that adults excrete a total of 19 to 23 g of nitrogen daily and between 68 and 84 per cent of this is in the form of urea. During passage through a sewerage system sufficient urease is present to hydrolyse urea to ammonia but conversion is seldom complete; tests carried out by Painter (13) showed that the rate of hydrolysis in a large sample of domestic sewage was constant and required over nineteen hours to reach completion at a temperature in the range from 11° to 14°C. During normal biological secondary treatment urea is wholly converted to ammonia (14) and the return of humus sludge and surplus activated sludge to primary settling tanks probably leads to an increase in the amount of urease which may hasten conversion.

Addition of lime to raw sewage is unlikely to affect the rate of hydrolysis provided the pH value is not greater than 10; at higher values, which are considered essential if ammonia is to be removed by air stripping, hydrolysis may be totally inhibited as indicated in Fig. 3.

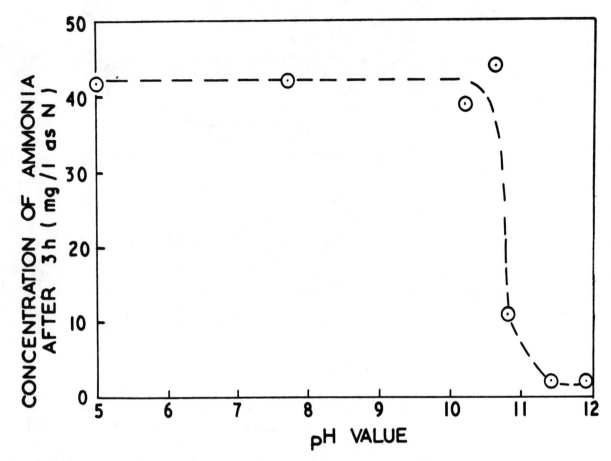

Fig. 3.   Relation between alkalinity and conversion of urea to
ammonia after 3 h in presence of urease.
          Initial concentration of urea, 5 mg/l as N; temperature
19.5-20°C

The persistence of urea in sewage poses an important problem in the
development of physico-chemical treatment systems if it is vital to remove
a large proportion of nitrogen, and at present it does not seem possible
to offer a satisfactory solution which does not depend on the use of some
form of biological treatment. Taking a long-term view it might be possible
to develop non-biological processes for removing urea from sewage but it
may prove more economical to develop systems for converting urea to ammonia,
perhaps with the aid of urease produced on site, so that one of the established
processes for removing ammonia may be employed.

Discussion

The likelihood of developing a technically feasible method for treating
sewage of the type normally encountered in the UK with the aid of lime and
activated carbon may be assessed in the light of the considerations and
results set out below.

Removal of Organic Material

During the first 10 days' operation of the laboratory-scale treatment plant the final effluent contained less than 25 mg organic carbon/l but since the tenth day of operation the level of organic carbon has normally been between 30 and 50 mg/l. At least two explanations may be offered to account for this early decline in performance; on the one hand only a limited number of suitable sites may be available for certain poorly adsorbed components and these may be saturated rapidly during the first few days' operation. On the other hand, and as Bishop, O'Farrell and Stamberg(7) maintain, colonization of the granular bed by micro-organisms after several days may result in a significant loss of available active surface; furthermore the **metabolism** of these organisms results in the production of non-adsorbable material which will be present in the final effluent. Observations made during the present work do not appear to help resolve this question; however they may throw some light on the general pattern of removal of soluble organic material by adsorption.

Examination of liquors discharged from each of the four carbon contactors used in the laboratory tests suggests that part of the organic carbon present in sand-filtered liquor was apparently very difficult to remove by adsorption and similar conclusions can be drawn from studying the results obtained during operation of pilot plants at Cuyahoga County(5) and Lebanon, Ohio (6). Material which was adsorbed was apparently removed at a rate directly proportional to the concentration of adsorbable material remaining in solution. The relation between the average concentration of organic carbon remaining in liquor discharged from each column or contactor and the corresponding detention time may be represented by a single differential equation:

$$dC/dt = - K(C-A)$$

which, after integration, leads to

$$C = A + (C_o - A)e^{-Kt}$$

where C is the average concentration of organic carbon (mg/l) after detention time t(h), K is an adsorption coefficient (h^{-1}), A is the concentration of organic carbon not readily adsorbed (mg/l), and C_o is the concentration of organic carbon present in liquor flowing into the adsorption system. This simple relationship takes no account of the progressive exhaustion of adsorptive capacity which presumably must occur; however reference to Fig. 2 indicates no readily discernible trend in the concentration of organic carbon present in final effluent and it has therefore been assumed that a steady-state condition exists after the initial decline in the removal of organic material already mentioned.

Numerical values for the constants in the integrated equation have been found by graphical methods and are shown in Fig. 4.

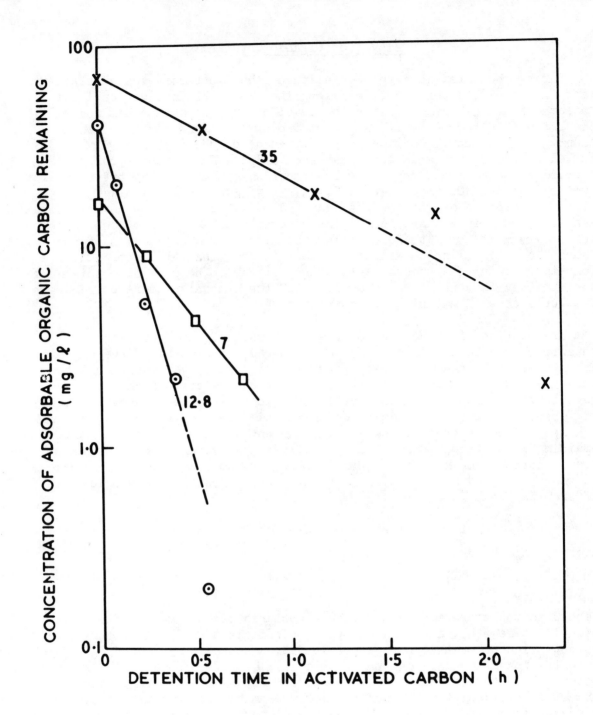

Fig. 4.   Relation between organic carbon remaining in chemically
treated sewage and time of contact with granular activated
carbon.
        Concentrations of organic carbon not readily removable
(in mg/l) is shown against each line.
        Circles, Rizzo and Schade (5)
        Squares, Villiers, Borg, Brunner, and Masse (6)
        Crosses, data obtained at WPRL

Experimental values agree fairly closely with the empirically derived curves which indicate wide differences in the rates of removal of organic material, the rate apparently being 7 to 8 times faster in the plant at Cuyahoga County than in equipment operated at Stevenage. The fractions of relatively non-adsorbable material, expressed as percentages of the total initial concentration of organic carbon, are within the range 24 to 32, the highest value coming from the English data.

At this point is it appropriate to compare the non-biological process used in the laboratory tests with a biological system capable of producing a final effluent of similar quality; experience gained by operating pilot plants treating Stevenage domestic sewage has shown that high-rate activated-sludge plants having a detention period of 1 to 2 h (based on the flow of sewage) would be able to produce an effluent in which the average concentration of BOD would be 50 mg/l. As capital costs are largely dependent on total tank capacities this comparison is especially significant as it implies that the conventional system would be less expensive to construct than the alternative non-biological plant.

At present the nature of the compounds which make up the relatively non-adsorbable material which is apparently present in Stevenage domestic sewage has not been identified but clearly this is a matter of importance. If an inexpensive system capable of removing or oxidizing this material can be developed, a suitably modified process using lime treatment followed by activated carbon might be suitable for treating the typically strong domestic sewage encountered in the British Isles, provided nitrogen removal was not required.

Removal of trade wastes

A great variety of trade waste waters may be discharged to sewers in a modern industrial town. Whereas it is obviously desirable to know how effectively such substances will be removed when considering a novel treatment process the scope of the present work is clearly limited and only nickel and cyanide will be considered.

Of the heavy metals used by industry in large quantities, and which may consequently be found in municipal wastes, nickel alone seems resistant to treatment with lime at a pH value of 9.5 (15); for this reason it was selected for study especially to see if treatment at higher pH values would be more effective. Results obtained in the experimental tests (Table 3) show that 98 per cent of the nickel added to raw sewage was removed, the major proportion being removed during sedimentation of the alkaline sewage at a pH value of 11.5. Passage of the recarbonated sewage through the sand-filter and carbon contactors probably accounted for further removal of nickel in suspension but more than 90 per cent of the dissolved nickel was also removed.

The ability of conventional sewage-treatment systems to remove nickel has been reported by Stones (16) and, though the results are not directly comparable with those given above, it is nevertheless worth noting that less than 60 per cent of the nickel present in an industrial sewage was removed in a full-scale treatment plant employing percolating filters (Table 5). According to Stones, activated-sludge plants at a nearby works removed

only 27 to 36 per cent of the nickel present in settled sewage - the
filters mentioned above removed 47 per cent of the nickel present in
settled sewage.

TABLE 5

Removal of Nickel During Treatment of Sewage Containing Industrial
Wastes at Salford, England (16)

	Raw sewage	Settled sewage	Percolating filter effluent (unsettled)	Percolating filter effluent (settled)
Concentration (mg/l)	0.19	0.15	0.09	0.8
Removal (per cent)	-	21	53	58

Cyanide was selected for study because of its widespread use and its
high toxicity; the high concentration added to raw sewage (80 to 85 mg/l
(as CN)) was intended to simulate conditions which might conceivably follow
an accidental massive discharge from a metal-finishing plant. The experimental results (Table 4) show that approximately 60 per cent of the cyanide
initially present passed through the system; only 11 per cent was removed
during passage through the sand filter and carbon contactors. It is
appropriate to point out that quite small amounts of cyanide, when
discharged intermittently, can have an adverse effect on biological stages
of conventional treatment systems but after being carefully acclimatized
such stages can oxidize cyanide at steady concentrations, at least up to 100
mg/l.

Removal of phosphorus

Phosphorus is normally present in crude sewage in a variety of forms,
the principal groups of compounds being ortho- and polyphosphate and organic
compounds such as nucleic acids; during conventional biological treatment
most of the organic phosphorus and polyphosphates are converted to orthophosphate which can be removed by precipitation as insoluble phosphates of
ferric iron, aluminium, or calcium. It is also well known that a high
proportion of the phosphorus in raw sewage may be removed by precipitation
with the aid of the chemical coagulants already mentioned but there seems
to be little information describing the fate of complexed phosphates when
relatively concentrated domestic sewage is treated with lime.

Results shown in Table 2 suggest that about 97.5 per cent of the polyphosphate initially present was removed by lime treatment followed by
recarbonation; the results also indicate a significant increase in the
concentration of soluble phosphorus during recarbonation of the alkaline
liquor as some suspended matter was redissolved.

Conclusions

Under laboratory conditions and in pilot-plant studies a variety of non-biological processes has been used for treating whole sewage. Systems employing membranes appear to be among the simplest methods yet derived for treating waste waters and their ability to produce high-quality final effluents has been well demonstrated, but operational problems associated with maintaining an acceptable rate of flow per unit area of membrane remain to be overcome and as yet no process for treating the reject liquor has been demonstrated.

Processes in which whole sewage is treated with a chemical coagulant, settled, and subsequently brought into contact with activated carbon have been demonstrated in several pilot plants in the USA and have produced final effluents containing 5-20 mg organic carbon/l when supplied with raw sewage containing 80-120 mg organic carbon/l. In tests carried out at the Water Pollution Research Laboratory, in which typically strong British sewage was treated with lime, settled, recarbonated in two stages, filtered, and passed through beds of activated carbon, several important problems were identified. Liquor supplied to the carbon contactors in these tests apparently contained 35 mg organic carbon/l which was virtually non-adsorbable; examination of results obtained at two pilot plants in the USA also indicated the presence of apparently non-adsorbable material but at substantially lower concentrations. The persistence of urea in sewage would also seem to present a major difficulty in the development of purely non-biological treatment processes as no physico-chemical system for removing urea or converting it to ammonia has been developed.

Little attention has been given to the question of trade wastes and, whereas it is almost certain that many heavy metals will be effectively removed by chemical treatment, substances such as cyanides may pass through a treatment plant employing lime and activated carbon. Consequently the adoption of some non-biological treatment processes would not necessarily relieve the appropriate management authority of the need to control the discharge of some industrial wastes and conceivably some substances might have to be controlled more rigorously than at present.

Acknowledgments

Mr. A. Lorimer operated the laboratory treatment plant and the many samples taken during the work were examined by the Laboratory's central analytical section.

British Crown copyright. Reproduced by permission of the Controller, H.M. Stationery Office.

References

1. F. D. Dryden, Industr. Wat. Engng, $\underline{8}$, 24 (1971).

2. D. H. Furukawa, Wat. Wastes Engng, $\underline{8}$, F 14 (1971).

3. K. A. Kraus, Application of **hyper** filtration to treatment of municipal sewage effluents. U.S. Federal Water Quality Administration Contract No. 14-12-423 (January 1970).

4. W. M. Conn, Raw sewage reverse osmosis; direct reclamation and phosphate removal. Paper presented at 69th annual meeting of the American Institution of Chemical Engineers, Cincinnati, Ohio, USA 17th May 1971.

5. J. L. Rizzo, and R. E. Schade, Wat. Sewage Wks, $\underline{116}$, 307 (1969).

6. R. V. Villiers, E. L. **Borg**, C. A. Brunner, and A. N. Masse, Wat. Sewage Wks, $\underline{118}$, R62 (1971).

7. D. F. Bishop, T. P. O'Farrell, and J. B. Stamberg, J. Wat. Pollut. Control Fed., $\underline{44}$, 361 (1972).

8. W. J. Weber, C. B. Hopkins, and R. J. Bloom, J. Wat. Pollut. Control Fed., $\underline{42}$, 83 (1970).

9. A. H. Molof, and M. M. Zuckerman, Proc. 5th Int. Conf. Wat. Pollut. Res., I-21 (1970).

10. A. L. Downing. Wat. Treat. Exam., $\underline{19}$, 223 (1970).

11. A. M. Hanson, and G. F. Lee, J. Wat. Pollut. Control Fed., $\underline{43}$, 2271 (1971).

12. S. H. Jenkins, J. Proc. Inst Sew. Purif., 149 (1950).

13. H. A. Painter, Wat. Waste Treat., $\underline{6}$, 496 (1958).

14. H. A. Painter, Unpublished work.

15. R. Nilsson, Wat. Res., $\underline{5}$, 51 (1971).

16. T. Stones, J. Proc. Inst Sew. Purif., 252 (1959).

17. B. A. Southgate, Wat. Sanit. Engr, $\underline{4}$, 213 (1953).

THE APPLICABILITY OF CARBON ADSORPTION IN THE
TREATMENT OF PETROCHEMICAL WASTEWATERS

Dr. Davis L. Ford, P.E.
Vice President, Engineering-Science, Inc.

There is considerable discussion at the present time on what constitutes
"best available treatment" for the refining and petrochemical industry. Although it is generally recognized that high rate biological systems represent
this treatment level based on present technology, there is a current interest
in the applicability of carbon systems. As 42 percent of the refineries recently surveyed by the American Petroleum Institute (API) have reported some
form of existing biological treatment (1), and activated sludge systems are
presently being built for most petrochemical complexes, it follows that carbon
systems should be investigated not only in terms of a total process but also
as a supplemental stage to biological treatment. As no refineries or petrochemical facilities in the United States are presently treating any of their
wastewaters other than storm runoff using the biological-carbon or carbon
systems, the purpose of this treatise is to underscore some of the problems
of utilizing carbon adsorption as a total process for treating refinery and
petrochemical wastewaters, demonstrate the compatability of biological-carbon
treatment, and discuss some of the practical realities and economics of
system adaptation. The concepts and conclusions presented herein are based
on extensive pilot work conducted by the author within the last two years at
refineries located in the industrialized Eastern and Southwestern regions of
the United States.

The efficacy of utilizing carbon adsorption for the treatment of refinery
and petrochemical wastewaters at any point in a process sequence can be determined only after a thorough investigation using continuous-flow pilot systems.
There is a tendency for investigators and equipment developers to oversimplify
the process adaptability for industrial wastewater applications. Specifically,
the translation of data from carbon systems receiving domestic wastes into
design criteria for industrial utilization has limited validity, and the use
of batch isotherm information under any testing condition as a basis for process selection is imprecise. The technical and economical justification for
including carbon adsorption as a treatment process in a refinery or petrochemical complex therefore must be predicated on pilot plant simulation, particularly in the absence of case histories and full scale operational experience. A proper interpretation of the results is then necessary to consummate

the process evaluation, determine the economics, and select the most appropriate treatment sequence.

The complexity of refinery and petrochemical wastewaters and the extremes in adsorbability of compound groups further mitigate the applicability of general rate equations. The influence of molecular structure and other factors on adsorbability, for example, are presented in Table 1 (2). This relative adsorbability, combined with unpredictable effects of process variables, forces an empirical approach for investigating carbon process applicability. Breakthrough curves defining contaminant removal rates and residuals (in terms of BOD, COD, TOC, color, etc.), carbon capacities, and influence of process variables, can therefore be developed using continuous flow columns. These pilot scale carbon columns are usually constructed of acrylic plastic (1.5 - 2.4 m) in height, properly piped and valved for series upflow or downflow operation. Facilities for backwashing the columns with external water or treated effluent are also provided.

The effectiveness of a carbon in removing selected contaminants can be predicted using equilibrium adsorption isotherms developed from batch tests. The isotherm as applied to refinery and petrochemical wastewaters is a plot of the contaminant (BOD, COD, TOC, color, etc.) versus the equilibrium residual remaining in solution. A linear representation using a log-log plot conforms to the Freundlich equation which relates the amount of contaminant in the adsorbed phase to that in solution by the expression:

$$x/m = K C^{1/n} \qquad\qquad (1)$$

where:

   x = amount of impurity adsorbed

   m = weight of carbon

   C = equilibrium concentration of impurity in solution

  K, n = constants

The Freundlich isotherm is limited in certain cases as shown in Figure 1. This occurs where concentrated and complex wastewaters are involved, with a significant portion of the organic impurities not amenable to adsorption, resulting in a constant residual regardless of the carbon dosage.

If Equation (1) does express the equilibrium limits of a wastewater, the constants are particularly sensitive to the nature of its constituents. Generally, "n" and "K" decrease with increasing wastewater complexity. High "K" and high "n" values indicate high adsorption throughout the concentration range studied. A low "K" and high "n" indicates a low adsorption throughout the concentration range studied. A low "n" value, or steep slope, indicates high adsorption at strong solute concentrations and low adsorption at dilute solute concentrations. It should be emphasized, however, that isotherm development for a particular wastewater has a limited application. This is underscored by the variations of isotherm constants developed by the authors and reported in the literature for refinery wastewaters following primary and secondary treatment, these values being listed in Table 2 (2, 3, 4, 5 & 6). Moreover, the carbon capacity, which generally increases with the COD gradient was higher

TABLE 1
Influence of Molecular Structure and
Other Factors on Adsorbability

1. Aromatic compounds are generally more adsorbable than aliphatic compounds of similar molecular size.

2. Branched chains are usually more adsorbable than straight chains.

3. Substituent groups affect adsorbability:

Substituent Group	Nature of Influence
Hydroxyl	Generally reduces adsorbability; extent of decrease depends on structure of host molecule.
Amino	Effect similar to that of hydroxyl but somewhat greater. Many amino acids are not adsorbed to any appreciable extent.
Carbonyl	Effect varies according to host molecule; glyoxylic and more adsorbable than acetic but similar increase does not occur when introduced into higher fatty acids.
Double bonds	Variable effect.
Halogens	Variable effect.
Sulfonic	Usually decreases adsorbability.
Nitro	Often increases adsorbability.

4. An increasing solubility of the solute in the liquid carrier decreases its adsorbability.

5. Generally, strongly ionized solutions are not as adsorbable as weakly ionized ones; i.e., undissociated molecules are in general preferentially adsorbed.

6. The amount of hydrolytic adsorption depends on the ability of the hydrolysis to form an adsorbable acid or base.

7. Unless the screening action of the carbon pores intervene, large molecules are more sorbable than small molecules of similar chemical nature. This is attributed to more solute carbon chemical bonds being formed, making desorption more difficult.

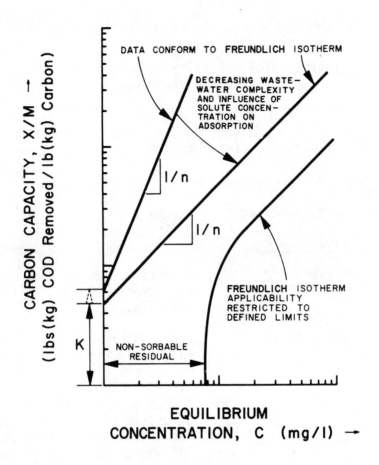

FIG. 1.   FREUNDLICH ISOTHERM APPLICATION

TABLE 2
Comparative Analysis of Batch Isotherm Data -
Refinery & Petrochemical Wastewaters

	K	n
OIL SEPARATOR (PRIMARY) EFFLUENT		
Refinery - Petrochemical Complex No. 1	.0290	0.77
Refinery - Petrochemical Complex No. 2	.0036	0.80
Refinery No. 3	.0140	0.36
SECONDARY (ACTIVATED SLUDGE) EFFLUENT		
Refinery - Petrochemical Complex No. 1	.0062	0.60
Refinery No. 3	.0043	1.00
Refinery Secondary Effluent	.0051	0.96
Refinery Secondary Effluent	.0038	1.08
Refinery Secondary Effluent	.0020	0.69
SINGLE ADSORBATES		
Phenol	.1110	5.80
Formic Acid	2.47	2.31
Acetic Acid	2.46	2.85
Succinic Acid	2.83	3.30
Adipic Acid	1.79	6.20
Citric Acid	0.73	4.93

when determined from continuous-flow carbon studies than that indicated from batch isotherms as shown in Figure 2 using the same conditions. This discrepancy has been reported previously although the higher capacity demonstrated in columnar operations has been attributed solely to biological removal in the columns (7). It is proposed, however, that a portion of this higher capacity is attributed to inherent differences in the continuous and batch tests. The capacity of the columnar operation is established by the continuously high concentration gradient at the interface of the adsorption zone with the virgin carbon as it passes through the column, while the concentration gradient decreases with time in the batch isotherm test. This difference in the mode of wastewater contact with the virgin carbon would cause a corresponding difference in carbon capacity.

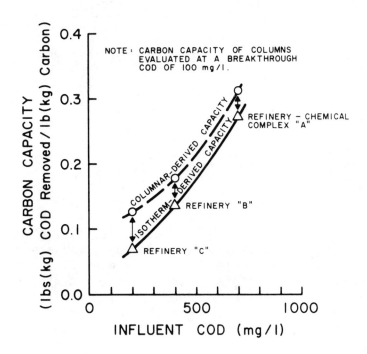

FIG. 2. CARBON CAPACITY AS A FUNCTION OF INFLUENT ORGANICS

The limitations of theoretical adsorption concepts relative to the practicalities of treatment requirements for refinery and petrochemical wastewaters necessitate that comprehensive process simulation studies precede the finalizing of design decisions. Because of these limitations and the idiosyncrasies of batch-isotherm columnar discrepancies, distorted breakthrough geometry, leakage of hydrocarbons through carbon beds, and economic contradictions, a more detailed discussion of these aspects is merited.

Application of The Carbon Adsorption Process Within A Treatment System

Consideration for placement of the fixed-bed carbon adsorption process includes biological-carbon series treatment, carbon-biological series treatment, and carbon adsorption as a total process. Each of these applications require primary treatment for the removal of oily substances and suspended matter using gravity separators and, in some instances, dissolved air flotation. A conceptual flow diagram for each of these candidate systems is shown in Figure 3.

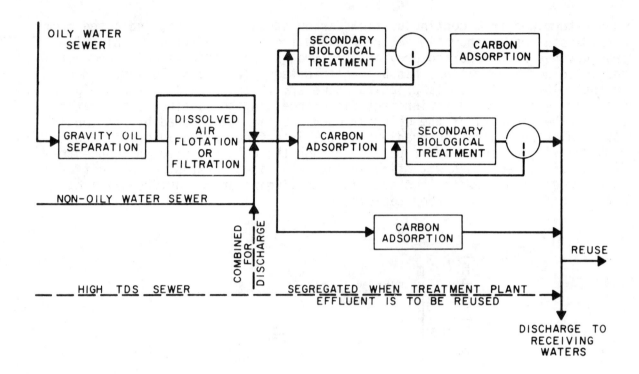

FIG. 3.   CANDIDATE SYSTEM APPLICATION FOR CARBON COLUMNS WITHIN A REFINERY
          TREATMENT COMPLEX

     Of the applications indicated above, the series biological-carbon treat-
ment scheme will probably be most prevalent in the immediate future. This is
true by necessity for many refineries as they have already made the capital investment in secondary biological plants and require a tertiary process to meet
new quality criteria. In the case of a new facility, this approach lends itself to phase construction by installing biological facilities to meet interim
effluent polishing when required.

     The series carbon-biological system is being considered by some refineries
and chemical plants, the apparent advantages being a more effective use of carbon, less chance of biological upset because of carbon removal of biotoxic substances with dampening of organic surges, and a reduction of excess biological
sludge inherent with the reduced organic loading. Although these stated advantages merit consideration, one must also recognize the disadvantages; namely,
a potential effluent suspended solids and color problem often associated with
biological systems, a less efficient biological removal of organics, and the
dependence upon a sometimes sensitive biological population to consistently
produce an effluent which will meet stringent quality requirements.

     The results of recently conducted pilot studies applying carbon as a total
process for refinery and petrochemical wastewater treatment were somewhat dis-
couraging. Although there are obvious advantages to eliminating biological
treatment altogether in favor of carbon adsorption, these studies consistently
indicated a "leakage" of organics (BOD and COD) regardless of the applied contact time, type of carbon, or linear flow velocity. This leakage and organic
residual as illustrated by the two-phase breakthrough curve shown in Figure 4
was actually higher than that observed in activated sludge reactors which were

operated in parallel. Because of these results, the author did not recommend carbon adsorption as a "total" system for the refineries where these results were observed.

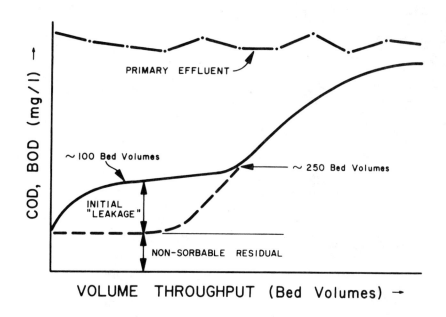

FIG. 4.   TWO-PHASE BREAKTHROUGH CURVE -
UNTREATED REFINERY WASTEWATER (PRIMARY EFFLUENT)

The Series Biological-Carbon System

The aforementioned columnar studies have demonstrated some interesting aspects when evaluating performance using primary effluent (from an API separator) for the column feed as compared to using biologically treated effluent as the feed. For example, the carbon appeared to be non-selective with respect to its affinity for adsorbing compounds responsible for BOD and COD respectively. This is evidenced by the fact that the BOD/COD ratio (the fraction of dichromate oxidizable compounds which are biodegradable) remained relatively constant throughout the carbon test series, regardless of the throughput volume. This was true when both the API separator effluent and the activated sludge effluent were applied to the columns. These results are plotted in Figure 5, and illustrate the magnitude of this ratio for both sequences of the biological-carbon series treatment. The reduction of the BOD/COD ratio through an activated sludge system is well documented and has been reported previously (8). Unfortunately, this ratio stability is responsible for unacceptable effluent BOD levels in many instances when applying the carbon system as a total process. The advantage of using carbon as a tertiary process following biological treatment is therefore apparent when considering stringent effluent BOD criteria.

Carbon Capacity

The carbon capacity as established by continuous column studies is particularly significant in that the design carbon capacity dictates the size of the columns and regeneration furnace as well as the carbon inventory requirements. These in turn affect the process economics in terms of capital and

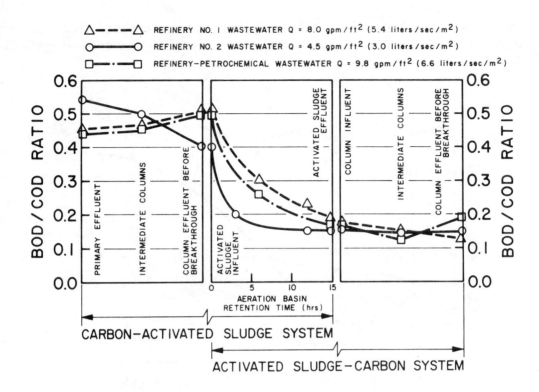

FIG. 5. ORGANIC SELECTIVITY THROUGH COMBINED SYSTEMS

annual costs. For example, the original carbon inventory alone accounted for over 15 percent of the estimated capital cost for a carbon system designed to treat a flow of 2 MGD (.09 CMS) (3).

It is recognized that capacity can be calculated several ways. This is shown in Figure 6 where the carbon capacity at breakthrough and exhaustion as a function of the depth of adsorption zone or wastewater complexity as illustrated. The influence of breakthrough curve geometry, particularly initial leakage, is shown in Figure 7. A true carbon capacity therefore can be assigned only when column studies using representative wastewater samples are operated for a sufficiently long period of time to fully develop the breakthrough geometry. Full utilization of the carbon in a column prior to regeneration can be realized by using a series of columns. For example, a series of three columns can be operated using one to complete exhaustion, one for polishing, and regenerating the third, with a sequential mode of operation.

It is recognized that carbon capacity based on COD or BOD breakthrough geometry may exclude consideration of selected critical contaminants such as phenols, D.D.T., surfactants, etc. If the breakthrough carbon capacity in the polishing column is less than that for COD or BOD, then effluent quality requirements for these constituents may necessitate selection of the lower carbon capacity value for design purposes

Water Reuse Considerations

Water reuse considerations within a refinery as a function of biological-carbon effluent quality is a broad and complex subject. A brief narrative of effluent quality as it relates to potential reuse is, however, considered pertinent to this discussion. A logical sequence of events in this regard would be:

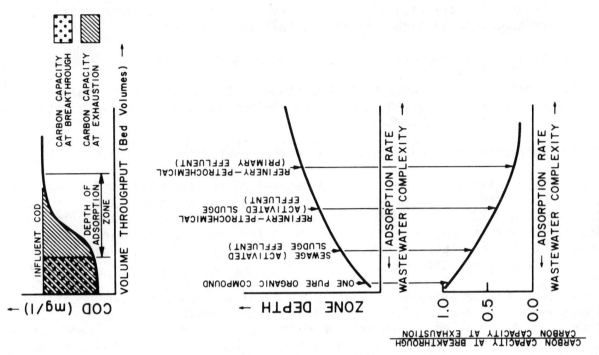

FIG. 6. CONCEPTUAL INTERRELATIONSHIPS OF WASTE-
WATER COMPLEXITY, ZONE DEPTH, & CARBON CAPACITY

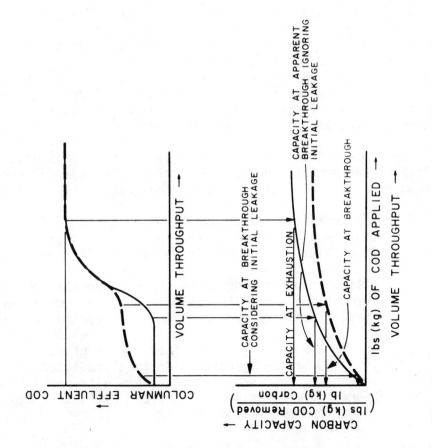

FIG. 7. INFLUENCE OF BREAKTHROUGH CURVE GEOMETRY
ON CARBON CAPACITY

1. Project the anticipated fate (or removal) of wastewater constituents through a total carbon or biological carbon system;

2. verify this projection from pilot plant quality data;

3. identify reuse-prohibitive constituents not removed by biological and/or carbon systems and investigate the possibility of in-plant pretreatment or segregation; and,

4. determine the economics for reuse as compared to direct disposal.

Based on pilot studies, the anticipated removal of refinery wastewater constituents by activated sludge, total carbon and combined treatment systems is tabulated in Table 3 (3, 4, 9 & 10). These levels of removal may not apply in all cases as most of the data presented herein is from integrated refineries (crude distillation, lubes, petrochemicals). However, they are indicative of general treatment effectiveness.

A review of the effluent quality which might be expected from the biological and/or carbon treatment of refinery wastewaters indicates total dissolved solids (TDS) to be a persisting problem with respect to reuse. As there is presently no efficient and economical method for lowering TDS in the combined effluent stream by treatment, in-plant segregation of low and high TDS wastes is required. If high TDS streams such as those resulting from water-treating ion exchangers, boiler and cooling tower blowdowns, desalter units, and brine dewatering operations can be separated and bypassed around the treatment facility to discharge, reuse of the treatment plant effluent for cooling tower makeup and other purposes is more attractive. If this is unrealistic, the segregation and in-plant use of high TDS water for process washing, firewater, and desalter makeup is also a possibility. In the event that no high TDS wastewater segregation is possible, a partial effluent reuse program might be explored. This would involve taking a pre-determined slipstream volume from the combined treatment plant effluent back to selected processes and estimating the corresponding increase in TDS at equilibrium. If this resultant concentration satisfies both effluent discharge and process reuse quality criteria, and no other conservative pollutant will concentrate to critical levels, then the economic incentive for reuse can be evaluated and the appropriate wastewater management decisions finalized.

The summary statements and conclusions from this discussion are presented as follows and pertain to the application of fixed-bed carbon and biological systems for the treatment of refinery/petrochemical wastewaters (10):

1. The technical and economic justification for applying carbon adsorption in a treatment system for an industrial wastewater must be predicated on a comprehensive pilot plant evaluation of the process.

2. Adsorption theory described by rate equations and the distinct breakthrough curve geometry seldom apply to complex wastewaters such as those discharged from refineries and petrochemical plants.

3. The carbon capacity determined from continuous-flow column studies is 10 to 80 percent greater than that predicted by the batch isotherm tests for the cases studied. This difference is attributed to the higher concentration gradient and biological degradation which prevail in columnar studies.

Table 3

Estimated Effluent Quality for the Activated Sludge, Carbon, and Combined Treatment of Refinery Wastewaters*

Constitent	Mean Value Range Primary Effluent	Activated Sludge Effluent	Total Carbon Effluent	Combined Activated Sludge-Carbon Effluent	Remarks
COD	500-700 mg/l	100-200 mg/l	100-200 mg/l	30-100 mg/l	Exact COD residuals vary with complexity of refinery & design contact times in the Act.S. and Carbon Treatment Plants
BOD$_5$	250-350 mg/l	20-50 mg/l	40-100 mg/l	5-30 mg/l	BOD residual depends on BOD/COD ratio which characterizes relative biodegradability of wastewater.
Phenols	10-100 mg/l	<1 mg/l	<1 mg/l	<1 mg/l	Phenols(ics) are generally amenable to biological and sorption removal.
pH	8.5-9.5	7-8.5	7-8.5	7-8.5	pH drop in Act.S. systems attributed to biological production of CO_2 and intermediate acids. pH change in carbon columns depends on preferential adsorption of acidic and basic organics.
SS	50-200 mg/l	20-50 mg/l	<20 mg/l	<20 mg/l	Primary effluent solids depend on design and operation of oil removal units. Act.S. effluent solids depend on effectiveness of secondary clarifier. Low effluent solids characterize carbon column effluent.
TDS	1500-3000 mg/l	1500-3000 mg/l	1500-3000 mg/l	1500-3000 mg/l	TDS is essentially unchanged through all three treatment systems.
NH$_3$-N	15-150 mg/l	5-100 mg/l	10-140 mg/l	2-100 mg/l	Exact concentration depends on pre-stripping facilities, nitrogen content of crude charge, corrosion additive practice and biological nitrification.
P	1-10 mg/l	<1-7 mg/l	1-10 mg/l	<1-7 mg/l	Only removal attributed to biological synthesis.

*Based on wastewater characterization data and treatability studies conducted by the author at eight refineries and petrochemical installations.

4. The series biological-carbon systems will probably be the most preva-
    lent application of carbon for refinery/petrochemical wastewater
    treatment in the immediate future. Pilot studies recently conducted
    at several installations indicated a persistent "leakage" of BOD or-
    ganics when fixed-bed carbon columns were applied as a single treat-
    ment system.

5. The BOD/COD ratio of the wastewaters tested remained essentially un-
    changed as the liquid passed through the carbon bed, regardless of
    the volume applied.

6. A more pronounced breakthrough curve and smaller adsorption zone
    depth was observed for the carbon adsorption removal of TOC and COD
    when biologically treated effluent was used as the column feed as
    compared to primary (oil separator) effluent feed.

6. A more pronounced breakthrough curve and smaller adsorption zone
    depth was observed for the carbon adsorption removal of TOC and COD
    when biologically treated effluent was used as the column feed as
    compared to primary (oil separator) effluent feed.

7. The carbon capacity in a fixed-bed column increases with influent or-
    ganic concentration. However, the breakthrough/exhaustion capacity
    ratio decreases with wastewater complexity, resulting in a sizeable
    portion of unused carbon in the column when breakthrough occurs.

8. If economics favor reuse over direct discharge, biological-carbon and
    total carbon systems are capable of producing effluents of acceptable
    quality for selective reuse providing conservative constituents such
    as TDS, chlorides, and certain heavy metals do not concentrate to pro-
    hibitive levels within the system.

References

1, "1967 Domestic Refinery Effluent Profile," Report by the Committee
for Air and Water Conservation, American Petroleum Institute, September (1968).

2. Hassler, J. W., Activated Carbon, Chemical Publishing Company, New
York (1963).

3. Confidential Report submitted to the Suntide Refining Company, Corpus
Christi, Texas, by Engineering-Science of Texas, February (1971).

4. Interim Report, "Deepwater Pilot Plant Treatability Study," submitted
to the Delaware River Basin Commission by Engineering-Science, Inc.,
January (1971).

5. "Adsorption as a Treatment of Refinery Effluents," Report for the CDRW
Subcommittee on Chemical Wastes, American Petroleum Institute, (1969).

6. Snoeyink, V. L., Weber, W. J., and Mark, H. B., "Sorption of Phenol
and Nitrophenol by Active Carbon," Environmental Science & Technology,
October (1969).

7. "Appraisal of Granular Contacting. Phase I, Evaluation of the Literature on the use of Granular Carbon for Tertiary Wastewater Treatment. Phase II, Economic Effect of Design Variables," Report No. TWRC-11, Robert A. Taft Water Research Center, Cincinnati, Ohio, U. S. Department of the Interior.

8. Eckenfelder, w. W., and Ford, D. L., Water Pollution Control - Experimental Procedures for Process Design, Pemberton Press, Austin (1970).

9. Confidential Report submitted to the Humble Oil and Refining Company, Baytown, Texas, by Engineering-Science of Texas, October (1968).

10. Ford, D. L. and Buercklin, M. A., "The Interrelationship of Biological-Carbon Adsorption Systems for the Treatment of Refinery and Petrochemical Wastewaters" presented to the 6th International Association of Water Pollution Research Conference, Jerusalem, June, 1972.

General Symbols and Abbreviations

BOD = biochemical oxygen demand

C_s = concentration of impurity in solution

$\dot{C}$ = equilibrium adsorbate concentration

cm = centimeters

CMS = cubic meters per second

COD = chemical oxygen demand

D = adsorbent bed depth

K = constant

$k_a r$ = overall mass transfer coefficient

kg = kilograms

m = weight of carbon

MGD = million gallons per day

n = constant

q = flow rate

ρ = packed density of carbon in the column

TOC = total organic carbon

TDS = total dissolved solids

W = weight of carbon in the column

x = amount of impurity adsorbed

Applications of New Concepts of Physical-Chemical
Wastewater Treatment
Sept.18-22, 1972

TREATMENT OF WASTES FROM METAL FINISHING
AND ENGINEERING INDUSTRIES.

By: R.K.Chalmers, B.Sc., M.Chem.A.,
F.R.I.C., M.Inst.W.P.C.
Managing Director, Bostock Hill and Rigby Ltd.,
Consultants, Birmingham, England.

It is a sobering thought that large scale plant is often installed to remove from large bodies of water very small amounts of contaminants which are themselves materials of value to the industry throwing them away. The metal finishing industry seems peculiar in this respect: that the majority of its problems are caused not by losses of by-products which must be washed away, but by losses of high quality process materials, and losses which generally need not arise.

1).   Minimising the Problem.

It is usually worthwhile to re-examine some of the older concepts in waste treatment from metal-based industries, since these are frequently ignored, before applying new ones.

a). Minimising Water Flows and Effluent Discharges.

Today the need for water economy is plain. The question of re-use generally resolves itself into one of cost-benefit. The case for simple care in the use of water is, however, one of plain common sense, to be applied before any other measures are considered.

Slightly less obvious than the direct cost-benefit of reducing water flows is the fact that strong solutions of effluent are more effectively and more economically treated than weak ones, and the volumes of sludge precipitated are much less from strong solutions than from weak ones, as the following table of results on aircraft engine manufacturing wastes shows.

147

TABLE I

Neutralisation of strong Chrome^{3+} solution with 10% Milk of Lime.

Volume Treated	Dilution	2 hour Settled Sludge Volume. (volume treated = 1.0)
1	0	1.75
1	1+1	2.44
1	1+2	2.80
1	1+3	3.40
1	1+4	3.60
1	1+9	6.20

Methods for reducing water consumption in metal-finishing processes are well known and have been widely reported.(1),(2), (3). Most of them are elementary; many of them are still frequently ignored. It is still unusual to find all the steps that can be taken being applied together in a logical interrelated pattern. The obvious steps which should be taken are summaried in Table II.

TABLE II

Examination of the need for water

1). Construct a 'water balance' sheet for the factory. Relate input flows to output and to need. Record water intake with all taps closed: in one factory this amounted to 1 m. gal./week.

2). Exclude cooling and uncontaminated waters from direct discharge to drain, initiate or extend re-use and recycle and examine bleed off rates.

3). Examine timing on automatic flushing systems, and spring loaded taps or pedal controls for domestic supplies. This resulted in a reduction at one factory of 2 m.gal/week.

4). Compare day and night flows. Higher pressure in mains gave an unnecessary increase from 40,000 to 50,000 gal/h. at one Birmingham factory.

5). Stop rinse flows when no work is present, through on/off pedals at tanks with hand operated processes; mechanical linkages controlling water supply against plant cycles on automatic plants with a time delay if required; and conductivity controllers.

6). Examine efficiency of water use, through: avoiding short circuiting in rinse tanks; using air agitation whenever possible; and counter-current rinsing.

7).   Determine precisely the water flow needed to maintain
       the required quality in each rinse tank. Control the
       maximum flow by pipeline restrictors.

8).   On all new process plant incorporate water and waste
       economy measures at the planning stage.

       At the planning stage it often pays to examine the water
requirement quoted by the manufacturers of process plants. In
competitive tendering price may militate against inclusion
of water economy measures.   In some cases it has been found
that £100 additional expenditure on process plant results in
a capital saving of about £1,000 on effluent treatment plant
and additionally lowers water costs and effluent acceptance
charges.

       The water requirements of a process plant before and
after applying simple economy measures,   showing a saving of
67% have been reported.(4).

       Having minimised the need for water flow, it may also be
possible to apply sequential use of rinse waters from the
critical through progressively less critical processes. The
example shown in FIG.I is taken from a copper-nickel-chrome
plating line in the lock manufacturing industry.

FIG. 1.

SEQUENTIAL USE OF RINSES.

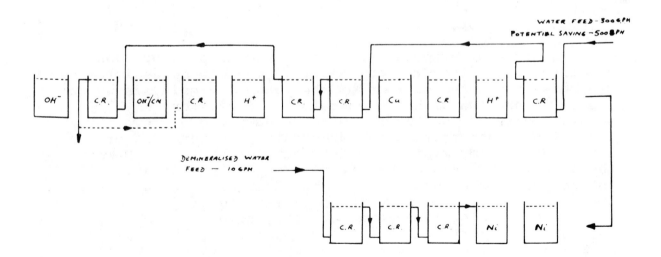

OH⁻	–	Hot Soak Cleaner.

OH⁻   -   Hot Soak Cleaner.
OH⁻/CN   -   Cold Anodic Cathodic Cleaner.
H+   -   Sulphuric Acid Dip.
Cu.   -   Copper Plate.
Ni.   -   Nickel Plate.
C.R.   -   Cold Rinse.

b). <u>Minimising amounts of contaminants discharged to drains.</u>

The use of static recovery rinses following process baths is generally applicable. The example quoted by Harris (5) is still one of the clearest justifications:

<u>Drag Out Recovery</u>

Plating solution strength	50 oz/gal.
First Rinse concentration	0.3 oz./gal.
Drag out	1.5 gal/h
Maximum strength of drag-out	25 per cent of plating tank
Ratios	Process tank 100
	Drag out tank 25
	First rinse 0.6
Discharge without drag out recovery	75oz. salts + 250 gal. water/h.
Discharge with drag out recovery	18.75 oz. salts + 62.5 gal.water/h. at maximum drag out contamination.

The contents of the drag out recovery tank may be disposed of either by discharge as a low volume dump solution for destructive treatment along with the irreducible minimum of other discharges, or preferably by return to the process tank.

The possibility of the return to the process tank depends upon the rate of evaporation loss from that tank exceeding the drag out return, the absence of contaminating salts in the water used, and prevention of the accumulation of degradation products in the main tank.

The rate of evaporation loss in nickel plating tanks usually exceeds the drag out return, and can be assisted by draughts across the surface of the tank, by maintaining the process tank at a higher temperature, consistent with solution safety, during non processing periods, e.g. at night, by concentrating the drag out by evaporation or ion exchange, and by minimising drag out to the tank. Drag in to the tank may be reduced by air blasts, by paying attention to the position of work on suitably designed jigs and making provision for removal from the process line of jigs not in use, by smooth and slow withdrawal of work (time on withdrawal may be more important than time on drainage), by draining pauses and drainage returns, and by low surface tensions. These are affected by the composition of the bath, the temperature, and by wetting agents. The same principles apply to drag out. Spray or intermittent rinses also help. One of the largest aluminium finishing shops in the U.K. operates with a total effluent disposal of 2800 gals/h., based on spray and fog rinses recycled many times before disposal.(6) This is a reduction of 90% on the water requirement without such measures.

The absence of contaminating salts in the water used can be ensured by supplying a small amount of demineralised water.

The concentration build up of a substance dissolved in water entering a process tank is determined by evaporation loss, drag out loss and initial concentration. If all drag out is returned the effect may be substantial.

The accumulation of process break down products in the main tank is automatically prevented on some plating plants by the routine bath maintenance, and with nickel plating by low current density purification and carbon filtration normally emmployed. It may be achieved on chrome plants by limited local ion exchange purification.

Application of these principles is possible at new plating factories, and is of potential benefit at all factories.

A novel arrangement in use at one factory manufacturing automobile headlamps is the continuous feed back to nickel plating of a low flow of 10 gal./h of demineralised water through three counter-current swills directly into the plating tank. This eliminates both the discharge of trade effluent and the need for a static recovery rinse.

Surge effects due to discharge of large volumes of strong spent solutions may be minimised by using controlled bleed-off at regular intervals. Large scale dumping from one 18,000 gal anodising tank has been eliminated, together with the need for large-scale dump holding tanks, by bleeding off 200 gals. of solution once per day. The process tank is allowed to build up to a concentration of 18 to 20 g/l Al, and maintained at this level by the bleed off.

Amounts of sludge produced may be further minimised, in some cases, by polyelectrolyte treatment.

TABLE III

Sludge from Neutralisation of Acids in Turbine Blade Manufacture

Polyelectrolyte added; mg/l	2 hour settled sludge volume as % v/v of the liquor treated.
0	82
1	78
5	65

c). Modifications to Production Processes:

Chemical modifications to processes should be examined as well as possible physical changes. Among the possibilities, all of which are in use, are:

i). substitution of low concentration zinc cyanide solutions
     for traditional baths

ii).Elimination of silicates from alkaline cleaner formulations,
     to obviate incompatibility with cyanides and aid free
     rinsing. One factory blends its own cleaners to ensure
     that they are silicate-free.

iii).Maintenance of a process bath concentration by feed forward
     from the preceding swill via the 'drag-in'   This is
     applied, for example, to compensate for the loss of nitric
     acid in chemical polishing.

iv).Total elimination of some rinses where process baths have
     been made compatible with following baths. In one large-
     scale copper plating installation no rinses are employed
     after the alkaline cleaner, acid dip, or cyanide dip
     stages. Copper flash is from a cyanide solution after which
     there is the only running rinse to the drain in the large
     plating shop - as a spray rinse. Subsequent copper
     plating is by copper pyrophosphate.

     After nickel plating the work is washed in a 500 gallon
     static rinse tank which is discharged once per day. The
     carry over of this into the chrome plating tank is
     significant only in upsetting the critical sulphate balance.
     In order to make this acceptable the chrome bath is
     treated with a few pounds per day of barium hydroxide
     added directly to the bath and the barium sulphate allowed
     to settle. It is not necessary to clean out this bath
     more than once or twice per year.

2).   Treatment of the Irreducible Minimum of Effluents

        Rinse waters of any desired quality can generally be
produced by controlling contaminant carry-over and water flows.
Treatment of these waters may be by batch treatment or flow-
through methods, in each case with or without segregation of
flows needing particular pre-treatments. Batch treatment is
frequently restricted to relatively low flows. In flow-through
treatment for rinse waters it is sometimes possible to avoid
the construction of large settlement tanks and the cost of sludge
handling and disposal by adjusting the flow to reach the
required acceptance standard after simple neutralisation and
flow recording.

        Disposal of strong spent solutions is then the dominating
problem. At the Rolls-Royce, Glasgow, aircraft engine factory
the treatment plant was constructed essentially for the
treatment of the strong spent solutions, and the flow diagram
has been published. (7).   The solutions for treatment have
been kept as strong as possible, consistent with the treated
liquors remaining pumpable. In developing this treatment it
was found that:

a). strong spent solutions, collected weekly to their points of origin, required controlled dilution, and are never discharged without automatic addition of diluting water - at 3:1 for electropolishing and hard chrome solutions.

b). chrome solutions could be treated at a strength which produced a sludge needing no consolidation, and that could by-pass settlement.

c). Wet sludge production could amount to 100% to 400% of the original waste solution treated, requiring sludge filtration for economic disposal.

d). Lime was preferable to sodium hydroxide as a general neutralising agent, although precautions were necessary to prevent a build-up of calcium phosphate or sulphate scales.

e). Polypropylene was the pipeline material of choice near points or origin of strong solutions, and limits were necessary on the concentrations of oxidising agents.

f). Hydrofluoric acid solutions should be segregated for slow rate transfer to the spent acid flow in order to minimise effects upon pumps and pipelines.

g). Full 'fail-safe' instrumentation was advisable, with monitoring systems to recycle flows requiring further treatment.

h). In very strong solution treatment control electrodes need careful siting and cleaning

3).   Approaching 100% Water Re-Cycle.

Some Companies now aim for total re-use of all water discharges. While this is possible in a few cases, the rapidly escalating costs of treatment for re-use between 90% and 100% recycle militate against its complete adoption.

TABLE IV

Costs of Water Re-Use

Manufacture	Water Usage m^3/m	Water Saving %	Capital Cost	Running Cost per 1,000gals (4.55 m^3)
Locks	32	28.5	Nil	Nil
		66	£35,000	£0.12
		90	£50,000	£0.35
Data Processing Equipment	23	78	£50,000,	£0.075
		95+	£200,000	-

It is necessary to ensure that there is a real need for
returned water.   It is very much cheaper to reduce water
consumption wherever possible than to treat unnecessarily
inflated volumes for re-use.   It pays also to consider using
treated effluent to replace other water uses.   In television
tube manufacture, with an effluent derived from high quality
demineralised water it was found possible to route 66% of
the effluent as make up to evaporative cooling towers, saving
both water purchase costs and effluent acceptance charges.

Where very low metal contents are required in effluent
products, these can be attained usually at very considerable
expense, by the application of final polishing treatment.
Such treatments have been applied by sand filtration, by
paper filtration in filter presses and by selective ion
exchange in the Na or Ca form.   Results are generally less
than 0.1 mg/l from the two other processes.

Reverse osmosis treatment does not appear to have
practical applications for this purposes.   It does have
applications to rinse waters such as those from nickel plating,
leading to 99% recovery of nickel losses for re-use and a
waste effluent containing 32 mg/l Ni. (8).

Concentration of drag out solutions by evaporation
processes is a tried and tested technique.   Experiments carried
out at a Joseph Lucas factory have shown that by using simple
evaporation on a constant feed basis the process can be carried
out in the process shop.   From a large automatic zinc
plating plant the drag out solution after zinc plating was
transferred to an open evaporator at rates between 10 and 21
gals/h.   Both rates were found adequate to maintain satisfactoy
conditions in the running swill (90 gals/h.) following the drag
out, and permit the use of a very small evaporator.

4)   Centrifuging

Centrifuges have well-known applications for dewatering
sewage sludges, recovery of oil from waste liquors, recovery
of blood from slaughterhouse wastes, and many other
separations.

The have the advantages of higher capacities per unit area
than other solids removal techniques, the ability to take
feeds of widely varying concentration, and possibilities of
continuous flow treatment in place of batch treatment for
processes that are markedly slow by static separation measures
such as the treatment of soluble oil emulsions.

Continuous flow operation for soluble oils separation on
small scale plant has been under investigation during 1972 at
one of the C.A.V. factories near London.

The Outline Flow diagram in Fig. II and the Alfa Laval pilot
plant is shown in Fig. III.

FIG. 2.

EMULSIFIED OIL TREATMENT.

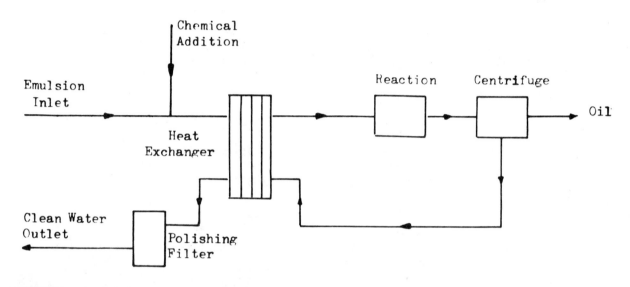

In this process the oil emulsion is sucked from the storage tanks, via a filter, by a pump with a variable capacity to the plate heat exchanger where it is pre-heated. A metal salt - normally magnesium chloride - is added and mixed with the process liquid when passing through the preheater. The liquid passes to a reaction tank and then to the centrifuge. The centrifuge pumps the separated water through the heat exchanger and if necessary through the polishing filter to the drain.

With a throughput of 180 gals/h. the plant is small enough to mount on a trolley, and yet capable of handling about 10,000 gals/week of waste oil emulsions.

Operating conditions:

Feed rate	180 gals/h
$MgCl_2$ (38.4%w/v) addition	1.0 to 4.0% by volume
Temperature	95°C
Hold times	3 to 15 minutes
pH value of effluent	6.0
Oil content of feed	Variable 2.0 to 4% v/v
Oil content of effluent	22 to 89 mg/l

The process is rapid and has the advantage of producing a nearly neutral effluent, and requires no toxic additives. Further tests are being undertaken.

With the current emphasis upon saving water and contaminants discharged there is some attraction in applying waste treatment techniques that are similar or related to the production processes. Centrifugal separators in the engineering industry is one example; electrolysis following concentration by reverse osmosis or evaporation is clearly indicated for the electroplating industry. In the food industry treatment plants, operating substantially as recovery plants are in operation using purely physical processes - screening, centrifuging, filtration and evaporation.

References

(1). Kushner, J.B. <u>Plating</u> , 1949, <u>36</u> 798-915

(2). Tallmadge, J.A. and Buffham, B.A. J.Wat.Pollution Control Fed., 1961, <u>33</u>, 817

(3). Chalmers, R.K. Water Pollution Control, 1967 <u>1</u> 49-55

(4). Chalmers, R.K. Paper presented at the 6th International Conference of the International Association for Water Pollution Research, Jerusalem, 1972.

(5). Harris E.P. 'A survey of nickel and chromium recovery in the electroplating industry' 1960, <u>London</u> D.S.I.R.

(6). Clarke M Chemistry and Industry, London 1970, Oct. 31

(7). Chalmers R.K. 'The use of water and the treatment of effluent in the metal-finishing industry. Chemistry and Industry, London 1970, Oct. 31

(8). Golomb, A. Paper presented at the 6th International Conference of the International Association for Water Pollution Research, Jerusalem, 1972

PHYSICAL AND BIOLOGICAL INTERRELATIONSHIPS IN
CARBON ADSORPTION

W. Wesley Eckenfelder, Jr., Thomas Williams, &
George Schlossnagle

Vanderbilt University

Introduction

There is a considerable question as to the mechanism of carbon adsorption treating organic wastewaters, particularly with respect to the biological interrelationships. Batch isotherm studies have usually yielded less adsorptive capacity than that attained in granular carbon columns (2)(8). It has been commonly accepted that biological activity within the carbon column enhances the adsorption of the carbon by reopening new adsorption sites (7). The purpose of this study was to investigate biological interrelationships in adsorption by laboratory studies involving aerobic and anaerobic biological flora seeded on the activated carbon. Sterile columns were used as a basis of comparison.

Two divergent views of the biological effects on carbon adsorption have been reported. Bishop (1) indicated that an anaerobic growth in the carbon bed decreases adsorption. He theorized that biological growths covered the pores thereby reducing physical adsorption. Contradictory results were presented by Parkhurst (4) when he concluded that an active biological flora contributed extensively in reducing the organic content of the wastewater as it passes through carbon columns.

Weber (8) and Hopkins (3) described a 7,200 gpd pilot plant to test the feasibility of direct physical-chemical treatment of domestic wastewater, without secondary biological processes. They found that over a four month period the carbon in the first column of both expanded and packed bed contractors removed approximately 60 percent by weight of organics, a much higher value than was expected from adsorption studies. This was attributed to biological action. Anaerobic conditions were encountered during one period. Aerating the column influent resulted in heavy growth and clogged the packed bed adsorbers. Hopkins (2) concluded that the expanded bed mode of contacting offers several distinct advantages; one of these being the potential for exploiting biological activity without clogging.

It is commonly accepted today that biological activity acts as a

regenerant of the adsorption sites (7). Rodman and Shunney (5,6) have outlined
a process which would biologically regenerate exhausted carbon. Weber,
Friedman, and Bloom (8) have recently described a study with improved removal
of organics from a waste comprised of approximately 75 percent domestic waste
and 25 percent industrial waste. Aerobic and anaerobic conditions in
expanded bed carbon adsorbers were compared. Two conclusions were drawn from
this study. Trouble free and effective operation of the biological adsorption
systems were achieved. The effective operating period of the adsorbers was
increased in the presence of biological activity. It should be noted that the
concentration of organics applied to the columns was low (TOC < 30 mg/l);
therefore, the rate of biological buildup and activity could be expected to
be low. Even though the aerobic adsorbers produced a better effluent, the
rate of exhaustion was no greater than for the anaerobic adsorbers. Weber
postulates that the controlling microbiologic growth responsible for surface
regeneration appears to be anaerobic.

Equipment and Procedures

A synthetic wastewater was employed in the studies to determine the
biological effects in columnar adsorption. The waste mixture of one study
contained 100 mg/l glucose, 50 mg/l ABS, and 100 mg/l phenol. A second mixture
contained 100 mg/l aniline and 100 mg/l phenol. Additional studies used
single components of 200 mg/l aniline and 200 mg/l phenol. These concentrations are in the order of what might be expected from industrial wastewaters
receiving no biological treatment.

After a culture of microorganisms was fully acclimated to the wastewater,
the culture was seeded in the carbon columns. The TOC retained by the columns
during this seeding process was small as compared to the total capacity of the
carbon. After seeding, upflow column operations were conducted using an
aerobic, anaerobic, and sterile column. To ensure that one column would
remain sterile, a small amount of mercuric chloride was added to the feed on
alternate days. Nutrients in the form of ammonium phosphate were supplied in
excess to all feeds. Breakthrough data was recorded for TOC and each of the
waste constituents.

Glass columns were designed for minimum clogging and maximum convenience
during sampling and maintenance. Each column contained 150 grams of WV-W
20 x 50 granular activated carbon. Air was supplied to the aerobic and
sterile feed, and the anaerobic feed was purged with nitrogen 10 minutes per
day and kept under a nitrogen blanket to ensure low dissolved oxygen concentrations (Figure 1). Feed was supplied to the columns via variable speed
masterflex pumps with viton tubing.

Experimental Results

A biological growth was developed under anaerobic and aerobic conditions
on the granular carbon. The carbon under aerobic conditions is shown in
Figure 2 and under anaerobic conditions in Figure 3. The biological activity
is considerably greater under aerobic conditions than under anaerobic conditions as might be expected.

The first column study employed a mixture of 100 mg/l phenol, 50 mg/l
ABS, and 100 mg/l glucose. The columns were operated at a flow rate of 23
ml/minute, a hydraulic loading of 2.0 gpm/ft^2, and a detention time of 16
minutes. The TOC breakthrough profiles and the component concentrations of

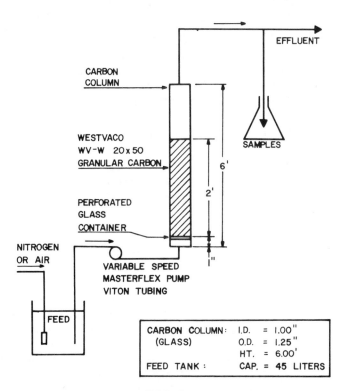

FIG.1
Schematic of Laboratory Equipment For One Carbon Column

FIG. 2
Activated Carbon Granule in Contact With Seed Acclimated
To An Aerobic Mixture of 100 ppm Phenol and 100 ppm Aniline
(300 X, 1/125 sec exposure, TRI-X, ASA 400, Red Corning Filter 2112).

sterile, aerobic, and anaerobic columns are shown in Figure 4.

The anaerobic column broke through earlier than the other columns and reached a steady state condition with a low removal rate due to biological reduction. Phenol and ABS reached a total breakthrough, the TOC reduction due to anaerobic activity being related to the glucose. The breakthrough in the aerobic column occurred after the sterile and anaerobic columns. Contrary to the previous results of Weber (9) the rate of .exhaustion of the aerobic column was much lower than for the anaerobic column. In the aerobic column glucose is essentially completely removed by biological oxidation. Phenol is partially removed (5%) after breakthrough and ABS completely breaks through. The sterile column breakthrough patterns indicate less carbon adsorption capacity than for the aerobic column, but more capacity than for the anaerobic column for all organic constituents.

A second study employed a mixture of 100 mg/1 phenol and 100 mg/1 aniline. The columns were operated at a flow rate of 12 mg/1 minute, a hydraulic loading of 0.59 gpm/ft^2, and a detention time of 26 minutes. The TOC breakthrough profiles for the sterile, aerobic, and anaerobic columns are shown in Figure 5. As in the first study the anaerobic column begins breakthrough early and reaches a steady state condition before breakthrough in the other columns. Contrary to the previous study, the aerobic column begins breakthrough before the sterile column, but the rate of exhaustion of the aerobic column was much lower than for the anaerobic or sterile columns. The aerobic column reaches a steady state condition where the amount of TOC removal is 30 percent. The sterile column reached complete breakthrough and during the 220 hour operation of this study the sterile column had 13 percent more cumulative adsorption than the aerobic column.

In another study 100 mg/1 aniline was employed. The columns were operated at a flow rate of 31.3 ml/minute, a hydraulic loading of 1.52 gpm/ft^2, and a detention time of 9.9 minutes. The results of this study were not surprising since the aniline is the least biodegradable substance used in these column seeding studies. Aerobic and sterile columns had the same breakthrough profile and the anaerobic column broke through slightly before the other columns. Biological seeding had little effect in this study.

The significant differences between the results in this study relates to the composition of the synthetic mixtures. In the aerobic columns phenol was observed to break through before and after the sterile column. When phenol was accompanied by aniline, a more biologically resistant compound, breakthrough began in the aerobic column before the sterile column. The opposite was true when phenol was accompanied by glucose, a highly degradable compound. The decreased activity of the biomass in the phenol-aniline mixture reduced the column capacity. In contrast, the more degradable waste had a more active biomass which enhanced removal.

Inspection of the anaerobic profiles indicates that biological activity probably covered the pores of the carbon and inhibited adsorption. In the phenol-aniline mixture the anaerobic column did not breakthrough as quickly as the phenol-ABS-glucose micture. This is probably due to the fact that the biomass was not as active and the film development over the carbon was slower to form.

In all the biological columns, a steady state condition was reached. In this last phase the controlling mechanism of removal is by biological means.

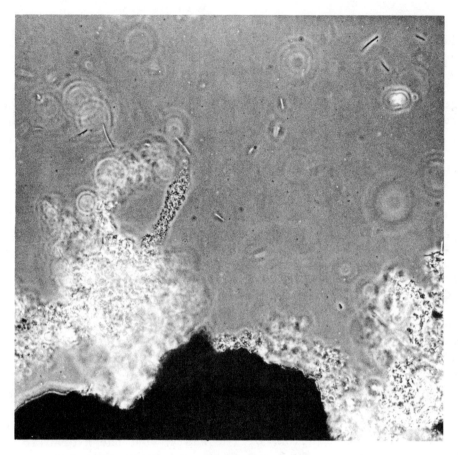

FIG. 3
Activated Carbon Granule in Contact with Seed Acclimated
To An Aerobic Mixture of 100 ppm Phenol and 100 ppm Aniline
(300 X, 1/125 sec exposure, TRI-X ASA 400, Red Corning Filter 2112).

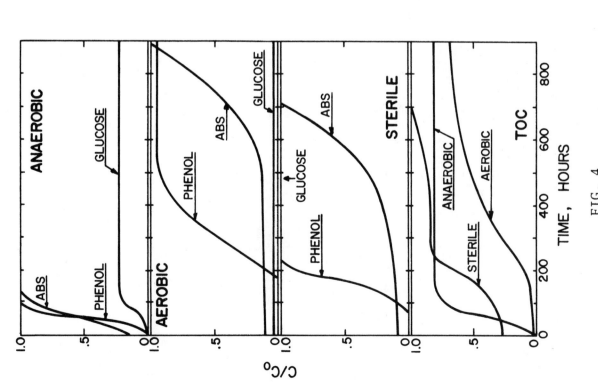

FIG. 4
TOC and Component Breakthrough Profiles
For Glucose-ABS-Phenol Synthetic Waste

The more degradable the waste, the more organic carbon will be removed, thus acting like a trickling filter.

Discussion

A general trend of the removal phenomena observed in this study is shown in Figure 6. Depending on the waste composition, the sterile column may have a high initial leakage due to poorly adsorbed organics such as glucose (A). The anaerobic column may reach earlier breakthrough for easily biodegradable organics such as glucose (C) and later breakthrough for biologically resistant compounds such as aniline (D). The aerobic column may begin breakthrough before the sterile column due to biologically resistant compounds as aniline (E) or exhibit higher removal than the sterile column for degradable compounds such as glucose (F). The anaerobic breakthrough reached completion before the sterile column. The less biodegradable the waste the closer it shadows the sterile column breakthrough profile. Contrary to Weber (8), the rate of exhaustion of the anaerobic column was always higher than in the sterile or aerobic columns. A significant difference between this study and Weber's is that the concentrations employed are considerably higher, leading to a greater rate of biological activity. In some cases it was observed that the aerobic breakthrough occurred before that of the sterile column depending on the nature of the organics.

Two conclusions can be drawn from these studies. One is that anaerobic cultures acclimated to the wastes inhibit carbon adsorption at high concentrations. This conflicts with Weber (8), but the difference may be explained by the nature and concentrations of the wastes and the hydraulic parameters. It appears from this study that the anaerobic microorganisms cover the carbon granules and inhibit adsorption. It can also be concluded that aerobic cultures may or may not enhance carbon adsorption depending on the character and concentration of the waste to be treated.

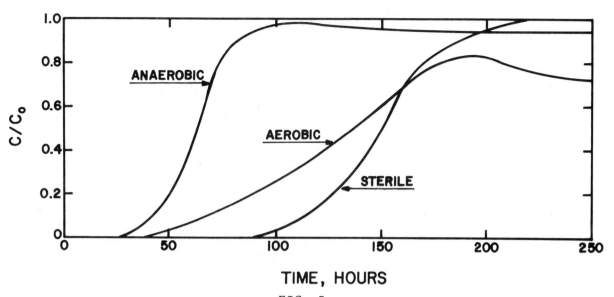

FIG. 5
TOC Breakthrough Profiles For Aniline-Phenol Synthetic Waste

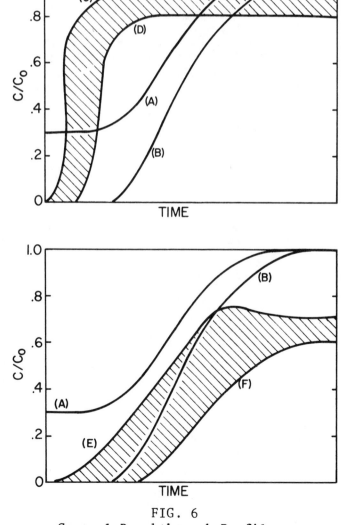

FIG. 6
General Breakthrough Profiles

References

1. Bishop, D.F., et al; Journal WPCF, 39:188 (February, 1967).

2. Ford, D.L., Buercklin, M.A., "The Interrelationship of Biological-Carbon Adsorption Systems for the Treatment of Refinery and Petrochemical Wastewaters," Presented at the 6th International Conference on Water Pollution Research, June, 1972.

3. Hopkins, C.B., et al; Granular Carbon Treatment of Raw Sewage Washington, D.C.: Federal Water Quality Administration, May, 1970.

4. Parkhurst, J.D., et al, Journal WPCF, 39:R70 (October 1967).

5. Rodman, C.A., and Shunney, E.L., Bio-Regenerated Activated Carbon Treatment of Textile Dye Wastewater, Washington, D.C.: Environmental Protection Agency, January, 1971.

6. _____, Water and Wastes Engineering, 8:E-18 (September 1971).

7. The Swindell-Dressler Company, Process Design Manual for Carbon Adsorption, Washington, D.C.: Environmental Protection Agency, 1971.

8. Weber, W.J., Jr., et al, Journal WPCF, 42:83 (January 1970).

9. Weber, W.J., Jr., Friedman, L.D., and Bloom, R., Jr., "Biologically Extended Physiochemical Treatment", Presented at the 6th International Conference on Water Pollution Research, June 1972.

REGENERATION OF ACTIVATED CARBON

G. L. Shell[1], L. Lombana[2], D. E. Burns[3], and H. D. Stensel[4]

Introduction

The need to produce high quality wastewater discharges in recent years has stimulated developments in the application of activated carbon for wastewater treatment. Granular activated carbon has been used on plant scale in the treatment of wastewater and more recently powdered activated carbon has been evaluated on a pilot scale. The cost of activated carbon generally requires regeneration and reuse. Only when carbon usage or liquid volumes treated are relatively small can a throw-away carbon system be economically justified.

The type of regeneration system employed depends on the carbon used; granular or powdered. Carbon handling systems are necessary to remove the carbon from the adsorption system, store and feed it to the regeneration system in an acceptable state and convey it back to the adsorption system.

1 Director of Sanitary Engineering Research and Development, Eimco Processing Machinery Division, Envirotech Corporation.

2 Manager, Process Engineering, Envirotech Systems Inc., Brisbane, California.

3 Senior Research Engineer, Eimco Processing Machinery Division, Envirotech Corporation.

4 Senior Research Engineer, Eimco Processing Machinery Division, Envirotech Corporation.

The purpose of this paper is to discuss general aspects of carbon regeneration, granular and powdered carbon regeneration systems, and carbon handling systems. Carbon regeneration performance and costs are also discussed.

Process of Thermal Activated Carbon Regeneration

Activated carbon has a highly developed porous structure with a wide range of pore sizes (from less than 10 to over 1,000 angstroms). The porous structure results in an extremely high surface area (e.g., about 1,000 square meters per gram of carbon) for adsorption. In wastewater treatment, carbon may absorb as much organic material as its original weight. Since the vast majority of the carbon's surface area is contained in the pore structure, much of the adsorbed material is contained within the carbon. Thermal regeneration is used to destroy the adsorbed organics and renew the carbon's adsorption properties.

The thermal regeneration process involves three steps:
1. Drying
2. Baking
3. Activation

In a multiple hearth furnace used for granular carbon regeneration, the regeneration process steps occur as the carbon moves from hearth to hearth. For powdered carbon regeneration devices, it is not practical to distinguish between these above steps since they all occur within seconds.

During the drying step, the water entrained in the carbon pores is evaporated and driven off. This occurs at granular carbon temperatures in the range of 60°F to 212°F. During the baking step, the organic adsorbate is pyrolyzed, resulting in the evolution of gases and formation of a free carbon residue in the micropores of the activated carbon. This occurs at granular carbon temperatures in the range of 212°F to 1500°F. In the activation step, the objective is to oxidize the residuals with minimal damage to the basic pore structure. The granular carbon temperature for activation is between 1500°F to 1650°F. Flue gas containing a small percentage of oxygen is the activating gas with varying amounts of steam used.

The combustion of natural gases with air supply the heat required, while carbon dioxide, oxygen and steam are the activating agents. This is shown in the following equations:

Oxygen	$C + O_2 \rightarrow CO_2$	(1)
Carbon dioxide	$C + CO_2 \rightarrow 2CO$	(2)
Steam	$C + H_2O \rightarrow CO + H_2$	(3)

The quantity of oxygen present is normally less than 1-2% of the gas volume to prevent uncontrolled combustion of the carbon. The important factors affecting the rate of activation are temperature, and concentration of carbon dioxide and steam in the activating gas mixture.

The success of the granular carbon regeneration process may be
monitored by analyzing for apparent density. Apparent densities
of regenerated carbon larger than virgin carbon indicate incomplete activation, whereas smaller values indicate that some
burning of the carbon is occurring. Iodine Number (1) determinations are often made for the regeneration product and compared to that for virgin carbon. Low Iodine Numbers may indicate that the regeneration temperature is too low and adsorbate residues are being left in the pore structure. Adsorption
isotherms may be conducted to compare the adsorption capacity
of virgin and regenerated carbon.

Regeneration of Granular Carbon

Multiple Hearth Furnace

Figure 1 shows a cross sectional view of a multiple hearth
furnace used to regenerate granular activated carbon. The
furnace consists of a steel shell lined with refractory material.
The interior spaces are divided by horizontal brick arches into
separate compartments called hearths. Alternate hearths have
holes at the periphery or at the center for the carbon to drop
through. A variable speed drive located at the bottom of the
furnace rotates the center shaft. It is sealed at the top and
bottom to prevent air or gas leakage. The center shaft is
hollow to allow for cooling air and has sockets where rabble
arms are connected. The arms are fitted with rabble teeth
placed at an angle so that as the shaft rotates, carbon is moved
in or out on the hearths. Dewatered spent carbon enters the
top of the furnace on Hearth Number 1 and is rabbled in a
spiral pattern through the furnace. Hot steam and flue gases
travel upward through the hearths counter-current to the movement of carbon.

Burners are located at several hearths in the furnace to maintain the desired temperature. These burners can be of the
nozzle mixing or pre-mixed type and are normally positioned on
hearths 4, 5, and 6. Steam inlet tubes are also provided at
these hearths. The temperature on these hearths are independ-
ently controlled by an automatic controller which can maintain
the temperature within 10°F of the desired temperature. A proportional flow meter is used to supply the desired mixture of
air and gas to the burners.

About 1 lb. of steam per lb. of carbon regenerated is often
recommended for use. Normal gas flow rates including steam
are about 50 ft^3 per lb. of carbon. Table 1 shows the flue gas
temperatures above various hearths for a typical operation (2).

Drying occurs on the first three hearths for a period of about
15 minutes. Baking occurs on the 4th hearth for about 4 minutes
and activation occurs on the fifth and sixth hearths for about
11 minutes.

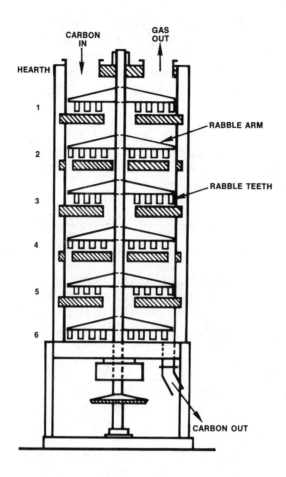

FIG. 1.   CROSS-SECTIONAL VIEW OF MULTIPLE HEARTH FURNACE

TABLE 1

Furnace Gas Temperatures

Hearth Numbers	Temperature F°
1	800
2	1000
3	1300
4	1680
5	1680
6	1680

A number of safety devices are included in the furnace design.
An ultraviolet scanner is used on each burner to detect a flame-
out in which case the furnace is automatically shut down.   The
furnace is also automatically shut down in the event of high or
low gas pressures, high combustion air pressures, low scrubber
water pressure, high stack gas temperature and inoperation of
the shaft cooling air fan.   An air fan supplies cooling air
through the hollow shaft to the rabble arms to prevent overheat-
ing.

The gases from carbon regeneration exit from the top of the
furnace to an afterburner to burn off volatile and noxious components. A wet scrubber is provided to collect dust and odorous
substances. It also cools the gases so that the induced draft
fan can handle them. The induced draft fan is necessary to draw
the gases through the system. The wet scrubber is fabricated of
stainless steel due to the corrosion and abrasion properties of
the wet carbon dust.

Thermal Regeneration Results

A certain quantity of carbon is lost during the thermal regeneration cycle. This can be attributed to attrition in transporting the carbon, gasification and burning in the regeneration furnace or thermal shock in the quench tank. Regeneration may also
result in a decrease in the Iodine Number, and carbon adsorption
capacity. Carbon losses of 7 to 10 percent and 5 to 8 percent
have been reported (2,4).

The average Iodine Number has been observed to decrease from
about 1,000 to a value of about 800 at Pomona for the first regeneration. The carbon adsorption capacity was observed to decrease 20-25% after 5 regeneration cycles (3). Results at Lake
Tahoe showed taht little change in operating adsorption capacity
occurred after 4 regeneration cycles. However, the Iodine
Numbers did decrease by about 100 (4).

Biological Granular Carbon Regeneration

Biological regeneration has been attempted on granular activated
carbon. Biological regeneration of granular carbon applied to a
textile waste has been reported (5). Spent carbon is subjected
to a recycled aerated biological suspension. At best, this
approach is only applicable to biodegradable organics which will
readily desorb. The presence of adsorbed refractory organics
would allow their buildup on the carbon resulting in decreased
effectiveness. In such a case, provision for another regeneration method capable of removing refractory material would be
necessary.

Powdered Carbon Regeneration

Due to its small particle size, powdered activated carbon (PAC)
can be dryed, baked and activated very rapidly. Several approaches of thermal regeneration have been evaluated. These include: indirect heating (5), partial wet air oxidation (7),
transport reactor (8, 9, 10) and fluidized bed furnace (9, 11).

The fluidized bed furnace approach has been extensively evaluated
on a pilot plant scale.

A sketch of the furnace is shown in Figure 2. A fluidized bed of
sand serves as a controlled constant temperature zone for regeneration. Fluidization occurs by the action of hot combustion
gases passing upward through a dsitribution plate. The operating
temperature of the bed is from 1300-1600°F. Dewatered carbon is
pumped directly into the fluidized bed.

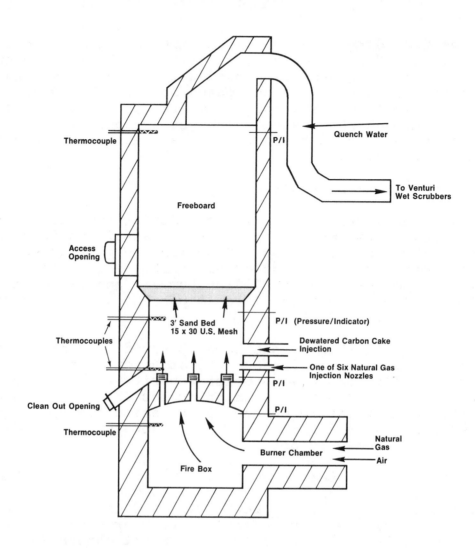

Thermocouple

Quench Water

P/I

Freeboard

**To Venturi
Wet Scrubbers**

**Access
Opening**

**3' Sand Bed
15 x 30 U.S. Mesh**

P/I (Pressure/Indicator)

Thermocouples

**Dewatered Carbon Cake
Injection**

**One of Six Natural Gas
Injection Nozzles**

P/I

Clean Out Opening

P/I

Thermocouple

**Natural
Gas**

Burner Chamber

Air

Fire Box

FIG. 2. FLUIDIZED BED REGENERATION FURNACE

To prevent structural failure, the temperature in the fire box must be maintained at less than 2100°F. The fire box temperature is controlled by using 150 percent of the stoichiometric amount of air required for combustion. The regenerated carbon and gases are cooled to about 200°F as they exit the furnace. This cooling is accomplished by spraying water directly into the exit gas duct. After cooling, the gases and regenerated carbon are passed through venturi scrubbers. The scrubber water is collected in a carbon recovery and scrubber water recycle tank.

The fluidized bed furnace operation can be highly automated. Burner air, and injection and burner gas flows are manually set to provide the desired fluidization velocity and exit oxygen concentration. Carbon is automatically fed to the furnace at a rate necessary to maintain a preset bed temperature.

Initial regeneration studies in Salt Lake City on a pilot plant fluidized bed furnace with a capacity of 60 lbs/hr of 23% dry solids carbon cake showed minimum carbon losses of 17%.

Additional tests are presently in progress. Carbon adsorption
effectiveness was not changed after regeneration (11). A
similar but smaller fluidized bed pilot plant operation at
Lebanon, Ohio resulted in a 15% carbon losses (9).

Other Powdered Carbon Regeneration Methods

An indirect heating device has been developed and to the authors
knowledge no public disclosures of the system are available (6).
In private communications, representatives of the firm indicate
PAC losses across the entire full scale regeneration system of
less than 5%. The regenerated product is used in a chemical
processing system.

The wet air oxidation system involves subjecting a thickened
PAC slurry (about 6-8% solids) to detention in a pressurized,
steam heated reactor (7). The reactor product (PAC slurry)
is passed through a decant tank prior to reuse. Results of
a laboratory evaluation indicated effective recovery of PAC
adsorption properties.

The transport reactor approach involves direct injection of
thickened spent PAC (about 10% solids) into a direct flame
reactor held at about 1400-1600°F. The PAC detention time in
the reactor is approximately one second. The PAC is captured
in water and air cooled jet condensers. Evaluation of a 4 lb/
hr (dry PAC feed) unit indicated PAC recoveries of 90% and near
complete recovery of adsorption characteristics (8). Another
evaluation of the same transport reactor indicated average
PAC recoveries of 84% (9). In this latter study, considerable
difficulty was encountered in maintaining a "choke" ring in the
gas burning chamber. A recent study on a 10 ton/day unit resulted in an 80 to 90 percent carbon recovery (10),

Methods Of Carbon Handling

A carbon regeneration system includes more than the regenerating
device. The carbon must be removed from the adsorption reactor,
stored and transported to and from the regeneration furnace.

Granular Carbon

Figure 3 shows a carbon regeneration flow scheme for a granular
carbon contacting system. The carbon must be transported from
the carbon contactor, to the spent carbon storage tank, to the
dewatering screw and from the furnace back to the contactor. It
is conveyed in a slurry form which can be handled by a number of
devices. These include eductors, centrifugal pumps, diaphragm
pumps and torque flow pumps (12). An eductor mixes carbon and
water and accelerates the fluid. An auxilary source of motive
power is required, such as 60 to 80 psig water supply. Centrifugal pumps are rubber lined to protect the pump and minimize
degradation of the carbon particles. A torque flow pump is a
special type of centrifugal pump with a recessed impeller that
effectively minimzes carbon particle degradation. A diaphgram
pump needs an air or water pressure source to actuate the

diaphragm. These units also require a 3 foot suction head which may present installation problems.

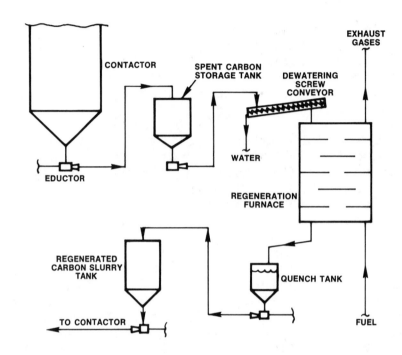

FIG. 3.   GRANULAR CARBON REGENERATION SYSTEM

Carbon slurry velocities should not be less than 3 ft/sec to prevent settling out, and not more than 6 ft/sec to minimize carbon abrasion, pipe wear and mechanical attrition of carbon particles. About 0.5 gallons of water per pound of carbon is recommended for transport design (13). Piping used for carbon transport should include long radius bends. A bend radius equal to twice the pipe diameter should be used. The regeneration equipment should be placed as close as possible to the carbon adsorption reactors to minimize pipe lengths.

The carbon transport rate from the spent carbon storage tank can be controlled by either a rotary valve feeder or diaphragm pump. Since it is transported in slurry form, a dewatering screw is used to convey the carbon to the furnace. The carbon is dewatered to about 50 percent moisture before it is fed to the furnace. For small installations the spent carbon storage tank may be located directly above the dewatering screw. Figure 4 shows an example of this type of design. A variable speed motor on the dewatering screw is used to control the feed rate.

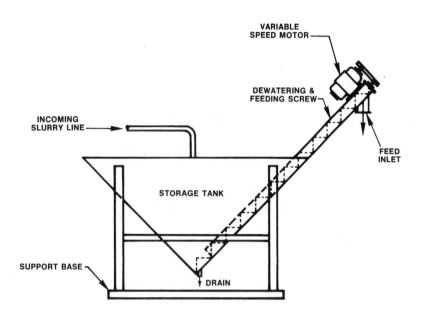

FIG. 4.   DEWATERING SCREW CONVEYOR WITH STORAGE TANK

The screw conveyor seals the furnace by using a covered trough and flanged connection to the feed tank. The carbon itself serves as the sealing medium. Therefore, a rotary air lock feeder is not necessary. The screw conveyor can be used as either a carbon feed device or a dewatering - feeding device.

As the carbon discharges from the furnace it drops into the quench tank. This is merely a small tank filled with water up to a level set by a level controller or float valve. The discharge from the furnace is normally through a stainless steel chute. The same conveying equipment already discussed is used to transport the carbon from the quench tank.

Powdered Carbon

Figure 5 shows the carbon regeneration flow scheme for the powdered activated carbon pilot plant at Salt Lake City (5). The powdered carbon slurry was thickened and dewatered prior to feeding it to the fluidized bed furnace. Since the powdered carbon is in a slurry form, it is easily pumped to and from the carbon contactor. The carbon gravity thickens to about 10% solids at a loading of 10 lbs/day-ft^2. Vacuum filtration with a slight polyelectrolyte addition for improved solids capture resulted in a cake with a 75-80 percent moisture content. A vibrating screw feeder or moyno pump can be used to feed the cake to the fluidized bed furnace. The regenerated carbon is thickened and pumped back to the carbon feed system.

Cost Of Regeneration

The total cost of regenerating activated carbon depends on the capital cost of the furnace and auxilary equipment and the cost

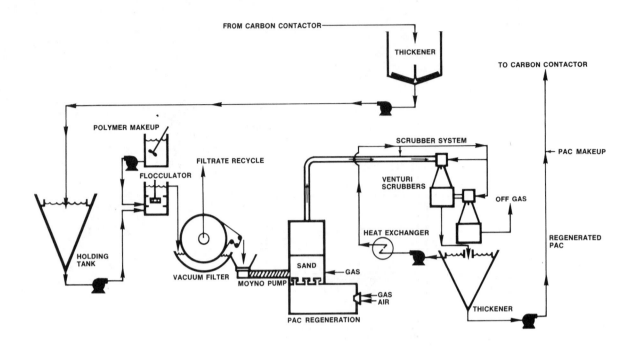

FIG. 5.   POWDERED CARBON REGENERATION SYSTEM

for carbon make-up, operation and maintenance. Since granular
carbon costs about 25-30 cents per pound and powdered carbon
costs about 8-12 cents per pound, total regeneration cost must
be less than the above virgin carbon costs for economic justification of the regeneration system.

For the Lake Tahoe system, a total regeneration cost of about
10 cents per pound has been reported. This includes about 3.2
cents per pound for operating costs and 6.8 cents per pound for
amortization and carbon make-up (8 to 9% losses). As plant
size increases, amortization costs will decrease, thus reducing
the total regeneration costs.

The cost of regenerating powdered carbon has been reported by
several investigators (9)(10)(11). A range of 2 to 6 cents per
pound was indicated for the total regeneration and make-up costs.
These figures are based on relatively short term pilot plant
studies. Long term plant scale experience is needed for sub-
stantiation.

Conclusions

Regeneration of activated carbon is required for most carbon
applications to wastewater treatment. Regeneration experience
of granular carbon using the multiple hearth furnace approach
has been based on long term plant scale operation. Carbon
losses, regenerated carbon characteristics and costs are all
well documented.

Numerous pilot plant powdered activated carbon regeneration
methods have been evaluated. Several of these methods show a

great deal of promise for plant scale operation. The overall
economic of powdered carbon concerning first and regeneration
costs would seem to indicate a bright future for its application
to wastewater treatment. Further study and application experience is needed to verify the regeneration costs reported.

References

1. R. L. Culp and G. L. Culp, Advanced Wastewater Treatment
Van Nostrand Rienhold Company, New York (1971).

2. A. J. Juhola and F. Tepper, Regeneration of Spent
Granular Activated Carbon, Report No. TWRC-7, Federal Water
Pollution Control Administration, Cincinnati, Ohio February
(1969).

3. Appraisal of Granular Carbon Contacting, Report No.
TWRC-11, Federal Water Pollution Control Adminstration, Cincinnati, Ohio, May (1969).

4. Advanced Wastewater Treatment as Practiced at South
Tahoe, Environmental Protection Agency, Project 17010ELQ, Washington, D. C., August (1971).

5. Fram Corporation, "Bio-Regenerated Activated Carbon
Treatment of Textile Dye Wastewater", Final Report EPA, WQO
Grant, Project No. 12090, (1971).

6. Corn Products Company, Private Communication, Argo,
Illinois, (September 17, 1970).

7. P. V. Knoop and W. B. Gitchel, "Wastewater Treatment
With Powdered Activated Carbon Regenerated By Wet Air Oxidation",
Presented at the 25th Purdue Industrial Waste Conference,
(May, 1970).

8. R. Bloom, Jr., R. T. Joseph, L. D. Friedman and C. B.
Hopkins, New Techniques Cuts Carbon Regeneration Costs, Environmental Science and Technology, Vol. 3 No. 3 p. 214, March (1969).

9. E. L. Berg, R. V. Villiers, A. N. Masse and L. A.
Winslow, Thermal Regeneration of Spent Powdered Carbon Using
Fluidized-Bed and Transport Reactors, Chemical Engineering
Progress, Vol. 67, No. 107, p. 154 (1970).

10. C. F. Koches, S. B. Smith, "Reactivate Powdered Carbon"
Chemical Engineering p. 46, May 1, 1972.

11. G. L. Shell and D. E. Burns, Powdered Activated Carbon
Application, Regeneration and Reuse in Wastewater Treatment
Systems, Presented at the 6th International Water Pollution
Control Association Conference, Tel Aviv, Israel, June 1972.

12. L. A. Lombana, "Carbon Regeneration Systems" Envirotech
Systems Inc. Brisbane, California (1971).

13.   Swindell-Dressler Company, "Process Design Manual For Carbon Adsorption", Environmental Protection Agency Technology Transfer, (October, 1971).

PRINCIPLES AND PRACTICE OF GRANULAR CARBON REACTIVATION

C. E. Smith
Calgon Corporation

Granular activated carbon has been used in many applications to
adsorb organic materials. Among these are chemical and gas purification,(1,2,
3) solvent recovery, (4) decolorizing of solutions, (5) as well as water
and wastewater purification. (6,7,8) Of its many uses, that of wastewater
purification is the most recent and may become the most important. In 1965
the first full scale domestic wastewater treatment plant using granular
activated carbon and thermal reactivation was placed in operation at South
Lake Tahoe, California. (9) Since then several industrial wastewater
facilities using the same process have been installed.

Granular activated carbon is effective in removing a wide variety
of organic pollutants from water. As with any material used in the treatment
of water or wastewater, its applicability is highly dependent upon economic
factors. In many wastewater treatment systems, operating cost is the major
factor to be considered. To reduce operating cost of wastewater treatment
with granular activated carbon, it is necessary to reactivate the granular
carbon for reuse. The purpose of this paper is to discuss the methods which
are presently used to reactivate granular carbon and the factors affecting
the feasibility of those methods.

Activated carbon is a black porous material produced by charring
carbonaceous materials such as coal or nut shells. The carbon may be
powdered or granular, similar in size to fine or medium sand. Whether
powdered or granular, each particle of carbon contains an exceptionally
large number of pores and fissures. Consequently activated carbon has a
very large surface area per unit volume. For instance, the surface area of
one pound of carbon is greater than 125 acres. Since adsorption is a
surface phenomenon, carbon with its tremendous surface area is an excellent
adsorptive medium. Most of the pores are in the range of 5 to 1000 angstroms
depending on the method of activation, base material, and desired purpose.
Most carbons used for water and wastewater treatment contain a large
percentage of pores in the 5 to 40 angstrom range and have surface areas in
the range of 900 to 1100 square meters per gram. Carbons with these
properties are usually made from coal.

Adsorption occurs as organic molecules pass from solution to the surface of the carbon. The molecules are held to the carbon surface by physical forces (Van der Waals force). Although a certain amount of adsorption occurs at the exterior surface of the particle, this is relatively insignificant, since greater than 99% of the surface area available is contained within the internal porous structure. (10)

As more and more molecules are adsorbed, the available surface area decreases and the carbon loses its ability to function as an adsorbent. It becomes "spent" or "exhausted" and must be discarded or reclaimed for reuse. Unlike powdered carbon, granular activated carbon may be reactivated and reused in practically all instances. During reactivation the previously adsorbed organics are removed and the available surface area of the carbon is usually completely restored.

Several processes have been used or attempted for reactivation of carbon. These are basically biological, chemical and thermal.

To date little success has been reported with attempts at biological reactivation of granular carbon. Biological reactivation has been attempted by encouraging the growth of bacteria within the bed of carbon. (11,12) As organic molecules are adsorbed they are assimilated by the bacteria so that theoretically additional adsorption sites would be provided for organic materials. Although the useful life of activated carbon has been found to be extended by this operation, the carbon eventually requires reactivation by another method.

A similar system would permit bioactivity in a separate vessel where spent carbon is aerated for a required period of time. Here aerobic bacteria utilize the adsorbed organic molecules as a substrate. The cleaned carbon is then returned to service.

Biological reactivation of granular activated carbon suffers many limitations. One of these is the fact that it would only be feasible for use in a waste which is biodegradable. Organics which are not readily biodegradable or are toxic would not be amenable to such a process. Additionally in the case of biodegradable organics, end products of biological activity are not amenable to further biodegradation and are generally readily adsorbable. At this time, no biological reactivation process is used on a commercial scale.

Reactivation of granular carbon with chemicals has been accomplished in certain specific applications. It is particularly desirable in industrial systems where the adsorbed organics can be reclaimed for further processing. If the adsorbed organics can be effectively stripped (chemically) from the carbon then we may conclude that the carbon may be directly reusable without additional reactivation. In order to accomplish this the organics must be physically drawn from the carbon surface into solution (desorbed) or react with the chemical solution so that the original molecular form no longer exists.

An example of chemical reactivation of granular carbon is the removal of phenol which has been adsorbed on carbon. (13) Adsorption occurs at low pH (below 7.0). To reactivate chemically, a solution of caustic soda (NaOH) is passed through the bed. This converts the phenol to sodium phenolate which is released into solution.

After stripping the phenol from the carbon, a washing step is provided prior to placing the carbon back on stream. The caustic reactivation may be accomplished in the adsorption vessel or in a separate treatment tank.

Desirable features of chemical reactivation are threefold:

1.   Reactivation can be accomplished without removing the carbon from the adsorber vessel. Expensive equipment is not required, thus reducing capital costs for the treatment system.

2.   If successful, the reactivated carbon is directly reusable.

3.   Product can be recovered for processing rather than be destroyed or lost.

It should be emphasized that at the present time, chemical reactivation of activated carbon is feasible in only a very few specific applications.

Thermal reactivation of granular carbon is the most widely used method at the present time because it can effectively reactivate carbon regardless of the nature of the organic materials adsorbed. Thermal reactivation involves heating the carbon in a furnace in which the atmosphere is controlled so that the organics are volatilized and driven off and adsorption sites are restored.

Two types of furnace have been used for reactivating granular carbon. The one most often used is the multiple hearth furnace which may contain from four to eight hearths. Rotary kilns also have been used where large furnace capacity is not required.

The multiple hearth furnace consists of a series of circular hearths placed one above the other. The furnace interior is lined with refractory material similar to that used for the hearths. A central rotating vertical shaft extends from bottom to the top of the furnace. Horizontal rabble arms are attached to the shaft which move the carbon across each hearth as the shaft turns.

Burners are provided at two or more hearths depending on the furnace size. On a six hearth furnace for example, hearths 4 and 6 (numbering is from top to bottom) are usually fired. Each of these hearths contains two or three automatically controlled burners. Fuel may be gas or oil. In order to provide a means of controlling the furnace atmosphere to reactivate the carbon without destroying it, air and steam inlets are provided on the two fired hearths.

An additional fired hearth may be installed at the top of the furnace to function as an afterburner to meet air pollution control standards. Alternatively an external afterburner may be provided. A stack gas scrubber may also be provided for removal of particulates.

For those installations where smaller quantities require reactivation (up to about 4000 pounds of granular activated carbon per day) a rotary kiln is often preferred. This consists of a refractory lined rotating cylinder placed with its axis nearly horizontal. It is sloped slightly downward from the feed end to the discharge end. Firing is accomplished by a single burner installed at the lower (discharge) end. Carbon enters at the upper end and moves slowly in a tumbling fashion downward toward the discharge point. Residence time in the kiln is controlled by the speed of rotation. As in the multiple hearth furnace, air and steam are added to control the atmosphere.

An afterburner and/or scrubber may be placed at the exhaust end of the kiln to meet air pollution control requirements.

Rotary kilns supplied for granular carbon reactivation are usually skid mounted and assembled before shipment. This simplifies installation and results in a lower capital cost. In addition to the lower initial cost, kilns are simpler to operate and maintain than the multiple hearth furnaces. However, where a large kiln is required such that skid mounting would be impractical, the multiple hearth furnace would probably provide the lower cost installation.

When granular carbon is to be thermally reactivated, it must first be transported from the adsorber vessel to a storage tank. This may be accomplished by an eductor, blowcase, pump or gravity flow. The furnace and adsorber are usually located relative to each other such that some force other than gravity must be used when transporting spent carbon. If carbon flow is by gravity from the adsorber to the spent carbon storage vessel, it must be lifted by an eductor or pump to a dewatering device before entering the furnace. If, on the other hand, the spent carbon is transported to an elevated spent carbon storage vessel, it can be fed by gravity to the dewatering device or can be dewatered in the storage vessel and fed by gravity directly to the furnace.

More recent installations include a dewatering screw conveyor which serves a dual purpose of dewatering the spent carbon and charging the furnace. Control of the spent carbon feed rate is maintained at the outlet of the spent carbon tank. Free water (water not contained in the porous structure of the carbon) overflows at the lower end of the dewatering conveyor and the carbon particles are carried upward by the conveyor and deposited in the furnace or kiln.

Gas temperatures are usually maintained at about 600° F at the first hearth of the furnace or the charging area of the kiln. This serves to dry the carbon and tends to drive off some of the more volatile materials.

In the multiple hearth furnace the carbon is moved alternately outward and inward across successive hearths by the rabble teeth which constantly turn the carbon so that it is uniformly exposed to the heat and gases. As it moves downward through the furnace, temperature gradually increases until at the bottom hearth it reaches approximately 1700°F. During the movement of the carbon granules some abrasion occurs and fines are generated. Most of these are light and are carried by the hot gases out of the furnace to air pollution control devices where they are destroyed or removed.

The combination of heat and gases in the furnace causes adsorbed organics to be volatilized and driven off. In this process some new pores are created which may actually result in higher activity than virgin carbon. This is not usually desirable however since it may tend to reduce the structural strength of the granule making it more susceptible to attrition.

The atmosphere is closely controlled within the furnace to prevent loss of structural strength, destruction, or insufficient reactivation. Steam is generally added at a rate of one pound of steam per pound of carbon and air is added to assist in controlling the atmosphere.

Reactivated carbon is then discharged from the furnace directly into a water filled quench tank which provides a water seal on the furnace, cools the carbon and stores it for transport as a slurry to a storage vessel or back to the adsorber vessel. Again, as with spent carbon, the reactivated carbon can be transported via an eductor or blowcase.

Reactivation in the smaller rotary kiln proceeds in a fashion similar to that in the multiple hearth furnace. Carbon charging equipment, air pollution control equipment, and water quench are identical for both systems.

The design of carbon reactivation systems depends on many factors but is carried out in a straightforward manner using common engineering principles. Factors affecting the design include but are not necessarily limited to:

1.   Amount of spent carbon requiring reactivation.

2.   Type of organic adsorbed.

3.   Expected rate of reactivation and required residence time.

The amount of carbon to be reactivated depends on the amount of organics adsorbed (weight per day) and efficiency of adsorption (weight of organics per unit weight of activated carbon). The type or types of organics adsorbed will determine what special materials of construction are required if any and may provide some idea of other equipment which may be required.

Some organics are so readily adsorbed as to make the carbon loading greater than normal requiring a longer reactivation period. Additionally those organic materials which are more tightly held to the carbon will require a longer residence time in the furnace. The design retention time can be established experimentally in the laboratory. In order to do this a sample of spent carbon is reactivated in a laboratory furnace under controlled conditions. The time required to reach virgin carbon characteristics is established as the design reactivation time.

The required reactivation time determines the design furnace hearth loading (pounds of carbon per square foot of effective hearth area). Normal hearth loadings for carbon used in treatment of domestic sewage approximate 100 pounds per day per square foot of hearth area. Carbon used for treating industrial wastes often requires more hearth area and longer reactivation time.

When the engineer has established these conditions he can then size the reactivation unit. He must of course keep in mind other conditions which may affect the design (i.e., future expansion, required excess capacity, expected variation in carbon exhaustion, desired frequency of reactivation, etc.).

The reasons for using thermal methods for reactivating granular activated carbon include the following:

1. It has been proven through many years of use at many installations. Most of the difficulties have been eliminated and many improvements have been developed making it reliable and relatively simple.

2. Thermal reactivation is universally applicable in that it is effective regardless of the type or nature of organic materials adsorbed on the carbon.

3. It not only removes the materials and destroys them, it also restores carbon to near virgin condition.

4. Carbon, when thermally reactivated on site, usually costs 10-20% of that for virgin carbon. The actual cost depends upon the size of the system.

In summary, the technical and economic feasibility of reactivating granular carbon has been developed to the extent that several full scale installations are presently in operation. Most of the existing installations use multiple hearth furnaces for reactivation with rotary kilns being used in a few of the smaller installations.

References

1.   "Activated Carbon for Effective Control of Evaporative Losses" by
      R. S. Joyce, P. D. Langston, G. R. Stoneburner, C. B. Stunkard,
      and G. S. Tobias.
      Presented at the International Automative Engineering Congress,
      Detroit, Michigan, Jan. 13-17, 1969.

2.   "Purification of Ethanolamines"
      Pittsburgh Activated Carbon Division of Calgon Corporation,
      Vol. 1, No. 2, 1966

3.   "Effective Removal of CO_2 and H_2S at Low Concentrations" by
      G. R. Stoneburner, R. S. Joyce, R. F. Sutt and G. S. Tobias.
      Presented at the Industrial & Engineering Chemistry Division of
      the American Chemical Society Meeting, Chicago, Illinois,
      Sept. 13-18, 1970.

4.   "Activated Carbon" by J. W. Hassler.
      Chemical Publishing Company, New York, 1963.

5.   "Activated Carbon Reclaims Textile Industry's Waste Waters"
      Environmental Science And Technology,
      Vol. 3, April 1969, p. 314-315.

6.   "Granular Carbon System Purifies Water at Del City, Oklahoma" by
      Kenneth Klaffke.
      Water and Wastes Engineering, March 1968.

7.   "Recovery of Coagulant, Nitrogen Removal, and Carbon Regeneration
      in Waste Water Reclamation" by C. E. Smith and R. L. Chapman.
      Final Report Federal Water Pollution Control Administration,
      Grant WPD-B5, June 1967.

8.   "Adsorption/Filtration -- A Technical Breakthrough for Industrial
      Wastewater Treatment" by T. B. Henshaw and B. T. Bawden.
      Presented at the 160th National American Chemical Society Meeting,
      Chicago, Illinois, Sept. 13-18, 1970.

9.   "The Lake Tahoe Water Reclamation Plant" by R. E. Roderick and
      R. L. Culp.
      Journal Water Pollution Control Federation,
      Vol. 38, p. 147, February 1966.

10.   "Basic Concepts of Adsorption on Activated Carbon"
      Pittsburgh Activated Carbon Division, Calgon Corporation,
      Publication PAC-5M-867-9.

11.   "Bio-regenerated Activated Carbon Treatment of Textile Dye Wastewater"
        by Fram Corporation.
        Environmental Protection Agency Water Pollution Control Research
        Series,
        Grant Project No. 12090 DWM, Jan. 1971.

12.   "Physiochemical Treatment of Wastewater" by W. J. Weber, Jr.,
        C. B. Hopkins and R. Bloom, Jr.
        Presented at the 42nd Annual Conference of the Water Pollution Control
        Federation, Dallas, Texas, October 5-10, 1969.

13.   "Temporary Water Clarification System" by M. Gould and J. Taylor.
        Chemical Engineering Progress, Vol. 65, No. 2, Dec. 1969, p. 47-49.

CHEMICAL REGENERATION OF ACTIVATED CARBON

J.M. ROVEL
Direction Technique Degrémont France

The Table (Fig. 1) below is derived mainly from a study made by Mr. F.P. Sebastian[1] and shows, as a function of the quantity of pollutants to be eliminated (g of COD/l) under the current price conditions applicable in the U.S.A., the various factors affecting the cost for the treatment of water on activated carbon, whether or not such treatment would include a thermal regeneration unit.

TABLE
FIG. 1

Installations 1 MGD
All Costs in Cents/1,000 g

Elimi- nated COD (g/m3)	Carbon + adsor- bers (capi- tal)*	Furnace (capi- tal)*	Make up carbon (8%)	M.O. power fuel	Total cost with out regen	Total cost with regen	Incidence of Furn- ace cap. (%)	Make up (%)
180	1.7	2.2	3.95	1.5	51	9.34	23.5	42
80	1.82	2.35	1.75	1.73	24.43	7.66	31	23
20	1.85	2.35	0.45	1.51	7.93	6.16	38	7.1

* 6% - 25 years

From this table we see that :

- The use of lost carbon is out of economics as soon as we leave the field of potable water

- With thermal regeneration the major elements of the cost are:-

  . Carbon addition (calculated at 8% per regeneration)

  . Amortization of the regeneration furnace.

The interest of searching for a process improving both factors is thus evident. This is why Degrémont has been studying this problem for 2 years and developed a process covered in France by the patents:-

- No 2 094 336 registered : 17th June 1970
- No 2 042 212 registered : 20th June 1970

and in the U.S.A., by the patent:-

- No 47 599

I - Conducting the Research Work

It has long been recognized that:-

(a) The lower the pH value, the better most of the organic molecules adsorb together

(b) Regenerations (at least partial) were made by means of hot caustic soda (desorption of phenols in particular).

Besides, the study of these mechanisms showed us that, in addition to the adsorption process modifications due to an inversion in the polarity of carbon surface oxides, soda also provokes an hydrolysis of some compounds (amino acids...).

Since this action was not complete, we tried to combine it with the action of a solvent which, to be satisfactory, had to possess the following qualities:-

- Important solvation power with large spectrum

- Adsorption power in alkaline medium in order to displace the substances worse adsorbed than it in this pH range

- Easily eluted, in particular under the action of vapour (thus involving a low boiling point and a very low adsorption capacity in neutral medium)

- Low cost and easy recovery from the eluate

A study of numerous systems of solvents (alcohol, chlorinated solvents, DMF, etc...) was made using in particular a microcalorimeter which enabled us to estimate, from the measurements of adsorption temperatures, the affinity of the various solvents for carbon, as a function of ambiental conditions (temperature, pH ..).

Thanks to this study, we could select the isopropyl alcohol as meeting the best of all the above criteria.

Besides, we checked over a wide range of examples that the use of soda and alcohol in combination, as described below, enables to eluate most of the products adsorbed on a carbon more completely than with any of both components used separately or combined.

II - Description of the Degrémont Process

It includes 5 main stages, 2 of them optional (1 and 5).

(1) Initial treatment by acid, intended to eliminate any mineral gang which would have deposited on carbon during fixation.

  - 0.5 to 1 volume of solution of HCl at 1 to 2% (which can be re-used several times) - Duration : 1 hour.

(2) Hot treatment (heating by vapour at 100 - 110° C) by a soda solution at 10%, allowing carbon to macerate during 1 hour.

  - 1.25 volume of soda 10% - Total time $2\frac{1}{2}$ hours.

(3) Injection of a solvent at the top of the column and very slow percolation through the bed at 80° C.

  - 0.5 volume of solution at 50% alcohol - Duration : 3 - 4 hours.

(4) Evaporation (120 - 140° C) - Duration : 1 hour, i.e. 0.75 to 1 kg of vapour per kg of carbon.

(5) If required, re-acidification of carbon before use

  - Solution with 1 to 2% HCl - 10 mn

The sodic and alcoholic effluents are stored then:-

  - 90 to 95% of the alcohol is recovered through distillation.

- The sodic effluent and the solids on the bottom of the tank of the previous distillation are either rejected after neutralization or, more currently, concentrated and brought to 500° in order to obtain an alkaline fusion; during this operation, the organic matter is pyrolised, producing an easily sedimented coke and gases which are re-introduced into furnace burner.

The recovered soda (5% losses appr.) is put again into a solution for the next regeneration. It contains only 50 to 100 mg/l of COD.

III - Performing the Regeneration

Two procedures can be applied:-

(1) Semi-mobile beds

The sketch at Fig. 2 below shows the major components of such an installation:-

The fixation column (1), the regeneration column (2), a water/alcohol separator (3), a condenser (5), a reagent storage unit (6), a reagent heating exchanger (4) and the alkali fusion tank (7).

Part of the carbon is transferred hydraulically to the regeneration tank. After passing through the said tank, the carbon is re-injected on top of the contact bed; the tank itself is used later as distillator for the alcoholic phase and as concentrator for the sodic phase.

(2) Fixed beds

In this example : fixation columns manufactured in order to allow in situ regeneration of the carbon (stainless steel made, rated at 3 bars with the necessary valves and fittings for reagent inlet and outlet, thermal insulation).

Usually three columns will be used which can be regenerated alternatively and put back into operation in the third position.

Though heavier from an investment point of view, this process presents the great advantages of:-

- Involving no carbon transfer, hence no losses by attrition

- Requiring labour only at the regeneration stage, i.e. one work post per column to be regenerated

Figure 2

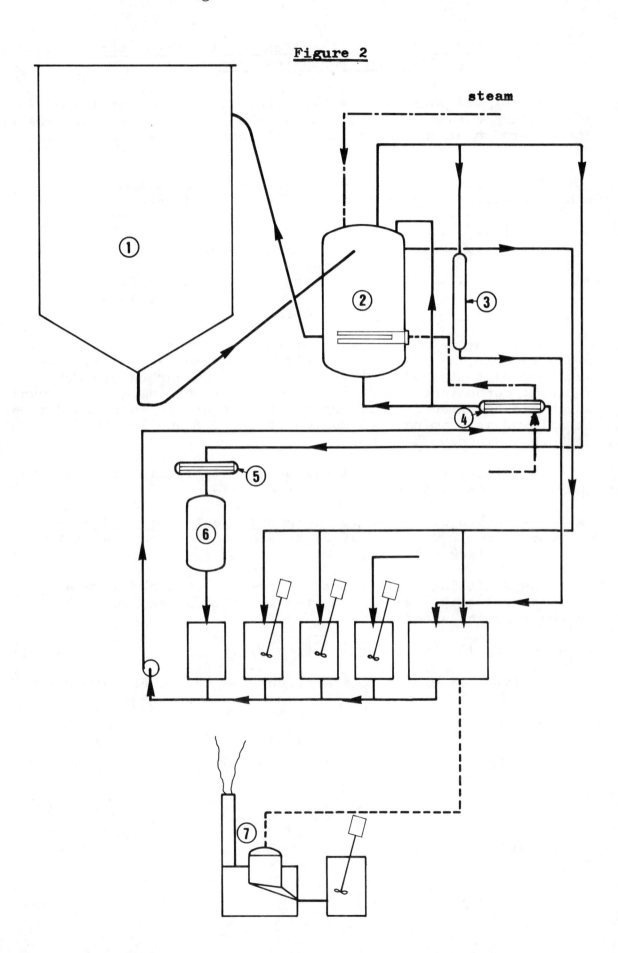

IV - Example of Application : Pilot Study for a PTBB Acid Production Shop

One of the shops at the Cuise Lamotte Factory of the NOBEL BOZEL Company synthetizes the PTBB (para-tertio-butyl-benzic acid) through oxidation of the tertio-butyl-benzene by nitric acid.

The effluent rejected by this shop contains numerous aromatic nitrated by-products and is featured by:-

- Output 12 m3/h
- COD .... 1700 to 2000 mg/l
  average value : 1800
- Colour .... strong yellow still noticeable after dilution in 100 volumes of clear water.

Since the main components of this effluent were not biodegradable, an adsorption treatment became necessary. The preliminary tests showed that a treatment on carbon did not allow to economically reduce the COD until obtaining an effluent which can be directly discharged but that under proper conditions it allowed to eliminate 60 to 70% of the COD and all the coloured components.

Therefore, a pilot unit has been installed in this factory and two series of tests were carried out:-

(a) Utilization of a fixation column on semi-fluidised beds, containing 230 litres of Pittsburgh carbon and a 42 litre regenerating column.

The following Table shows the results obtained with a fixation velocity : 250 1/h (retention time 55 mn) and two carbon renewing speeds:-

- Tests 1 to 10 with 1 regeneration i.e. 42 1/d

- Tests 11 to 18 with 1 regeneration i.e. 42 1 every other day

TABLE
FIG. 3

Test No.	Treated Volume	COD mg/l of O_2
1	5,250 l	310
2	5,250 l	380
3	5,250 l	385
4	5,250 l	685
5	5,250 l	640
6	5,250 l	570
7	5,250 l	525
8	5,250 l	470
9	5,250 l	550
10	5,250 l	465
11	10,500 l	670
12	5,950 l	
13	10,500 l	730
14	10,500 l	635
15	10,500 l	560
16	10,500 l	550
17	10,500 l	580
18	10,500 l	660

One should note that after 18 cycles the carbon has been regenerated 3.3 times and that if a visible COD increase appears for cycles 1 to 5 which correspond to the time to reach the column normal operation conditions (regeneration of 210 l on 230 l), then the performances remain very stable and the obtained capacity is excellent:-

- Average for tests 11 to 18:-

    . COD at the inlet .... 1,800 mg/l
    . COD at the outlet .... 625 mg/l

i.e. 1.18 kg of COD fixed by treated cubic meter or 615 g COD by kilo of saturated carbon.

(b) <u>Utilization of a column with 6.6 1 fixed beds</u> (∅ 80 mm)
    to be able to check the constancy of the result on a
    greater number of regenerations.

In this case, the retention time was shorter (25 mm, i.e.
15 1/h) which explains that on a average the results in the
following Table correspond to a smaller capacity than those
obtained in the previous test:-

TABLE
FIG. 4

No.	Vol. cycle (1)	Velocity (1/h)	COD after 300 1	End of cycle	Colouration
1	291	15		250	Nil
2	600	15	150	500	"
3	437	15		400	"
4	500	15	200	420	"
5	463	15		400	"
6	787	30		780	"
7	1070	30		1300	pale yellow
8	903	30	400	1300	yellow
9	685	15	50	300	trace
10	531	15	310	1000	yellow
11	500	15	270	400	Nil
12	350	15	270	400	"
13	500	15	250	500	"

It has been established that after a stabilization
obtained on the first 5 cycles, an attempt to increase the
velocity to 30 1/h (12 mn of retention time) showed a rapid
colouration leakage which entailed perturbations until the
10th cycle.

On the contrary the last three cycles have performances
close to those shown for the first five.

However, it shows a slight capacity decrease which
corresponds to 10 to 15% approx. of the initial capacity. Yet,
one must note that no carbon was added during these 13 cycles
and that the total carbon lost by dragging small carbon is
lower than 10%.

The regeneration eluate analysis shows that the maximum
pollution is extracted from coal when the solvent injection
begins (see curves Fig. 5).

The above results and other obtained from similar experiments run on other type of industrial waste waters, such as :

- pharmaceutical wastes (at Roussel Uclaf Vertolaye plant)

- beet sugar juice for discolouration treatment (at Sucreries du Soissonnais)

- textile wastes such as bleaching and printing sections (at Teintureries Steinheil Dieterlen in Rothau)

have disclosed the efficiency of the proposed regeneration process they lead us to conclude that the best regeneration proceedure is always the application of caustic soda followed by alcohol for it restore at least 80 % of the initial capacity of the carbon.

V - ECONOMICS

Table (fig. 6) summarizes the economics of plants designed for the regeneration of 10 m3/day, with application of either chemical or thermal proceedures.

Costs of chemicals and other entries are expressed in french francs per one cm3 of treated carbon.

Figure 6

Chemical regeneration			Thermal regeneration	
Investment costs				
- All needed equip^t	350.000 F		-Furnace	430.000 F
- Boiler (if no vapor			-Accessories (storage	
available locally)	50.000 F		drainage, transfert	
- Soda melter	50.000 F		line, regulations,	
			mounting)	550.000 F
	450.000 F			980.000 F
Running costs				
reagents				
-Caustic(1) 6,2kg 0,32F/kg	2	F		
-Fuel (1) 450kg 0,12F/kg	54	F		
-Alcohol(2) 16kg 0,80F/kg	12,8	F		
-Steam (heating +				
displacement)	40	F	∿ 100 F	
-Softened water	2,7	F		
Labor	20	F		
Carbon Make up				
2 % = 20 1 × 2,25	45	F	- 8 % = 80 1	180 F
	176,5 F			280 F

(1) If caustic effluent can be wasted - no fuel needed but we used
    125 kg of caustic soda = 40 F (cost reduced by 14 F)

(2) Assumption : 90% recovery by distillation.

It appears that not only prime costs are 50% cleaper but also
running cost are some what 35% lower.

In addition to these essential conclusions it can be said that for
small plants I up to 5 m3 day of regenerated carbon, labor costs
are even more favorable as the plant can be designed such as to be
operated only every few days (in situ regeneration) meanwhile
thermal regeneration requires always a continuous operation and
care.

We see other advantages such as an easier temperature control and
also the fact that final wastes of the plant do not contain any
toxic chemical. The latter remark is to be opposed to the large
amount of CO gas which is released to the atmosphere by furnaces.

It is also avoided that inert chemicals (such as iron, alcaline
salt compounds...) which eventualy during thermal process would
melt and cause the activated carbon grains to aggregate with
subsequent carbon loss and troubles.

More over following factors would be mandatory for the use of
chemical regeneration; such as :

  - The possibility of using low grade (mechanical characteristics)
    activated carbon of low price such as NORIT PKST   which
    nevertheless has been effectively regenerated up to 3 times
    during our experiments.

  - Special industry wastes containing nitrate compounds (see
    above NOBEL BOZEL wastes) which can be found also in synthetic
    dies or polyurethane industries, and would create explosion
    hazards if brought to elevated temperatures.

However for the time being all applications of this chemical rege-
ration process are not yet all investigated and we must admit that
there would be some instances where refractory chemicals present
in the water would offer strong affinities to carbon and hamper
the efficiency of this chemical regeneration process.

This is the reason why we feel that preliminary survey must be
performed over any specific effluent to be treated in order to
determine whether this proceedure is applicable or not.

Conclusions

The economic survey as expanded here above, and the favourable
results obtained for some specific effluents show that this
chemical regeneration process is promised to a brilliant future
and that it will help to minimize the actual treatment costs
which, at least presently in Europe, confine the use of activated
carbon to very specific cases.

Ref. (1) F.P. Sebastian (Envirotech) - Water & Waste Treatment
                                                    (May, 72)

Fig. 5 : Typical regeneration of 42^e saturated carbon -

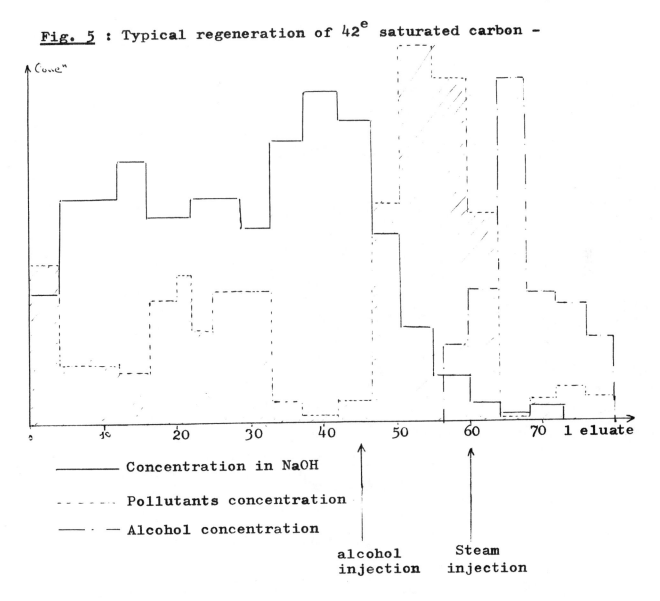

_____ Concentration in NaOH

- - - - - Pollutants concentration

— · — Alcohol concentration

alcohol
injection

Steam
injection

Applications of New Concepts of Physical-Chemical
Wastewater Treatment
Sept.18-22, 1972

AUTOMATION AND CONTROL OF PHYSICAL-CHEMICAL TREATMENT FOR MUNICIPAL WASTEWATER

John J. Convery, Joseph F. Roesler, Robert H. Wise
Environmental Protection Agency
National Environmental Research Center
Advanced Waste Treatment Research Laboratory
Cincinnati, Ohio

Introduction

As the national commitment to achieve adequate environmental protection increases, the need for improved performance and more cost-effective operation of existing and planned wastewater-treatment facilities will intensify. In anticipation of this need, the National Environmental Research Center of EPA in Cincinnati, Ohio, is accelerating its research activities relating to instrumentation and automation of waste-treatment processes.

A specific research program to develop and encourage the utilization of instrumentation and automation of wastewater-treatment processes was initiated last year at the Advanced Waste Treatment Research Laboratory of NERC-Cincinnati. The current investment in wastewater-treatment instrumentation and automation has been estimated by Smith [1] to be 1 to 1.5% of capital expenditures. Comparable expenditures in the water-treatment industry and in the chemical and petroleum industry average 5% [1] and 12.5% [2], respectively. As the capability of instrumentation to achieve greater process reliability and reduced operating and capital costs is demonstrated, the proportion of capital invested in instrumentation will increase. The primary objective of the instrumentation program is to demonstrate this capability. A grant to the City of Palo Alto, California, has been awarded to demonstrate the cost-effectiveness of seven separate control strategies to achieve improved activated sludge treatment performance. A contract to develop a "State of the Art on Instrumentation and Automation of Wastewater Treatment Facilities" has also been awarded. The development and evaluation of alternative control strategies for physical-chemical treatment is being conducted primarily at the laboratory's Blue Plains Pilot Plant in Washington, D. C. The purpose of this paper is to present the results of these pilot plant investigations, as well as the cost-effectiveness analyses that will be used to guide further research.

Physical-Chemical Treatment Systems

The basic physical-chemical process sequence being evaluated at Blue Plains is shown in Figure 1; it consists of two-stage (high pH) lime precipitation with intermediate recarbonation, dual-media filtration, pH control with chlorine, CO_2 stripping, breakpoint chlorination and granular carbon adsorption. In the first stage of the process, powdered lime (CaO 350 mg/l), raw wastewater, and recycled solids (10%) are rapid-mixed, flocculated and settled at 1000 gpd/ft^2 to remove bicarbonate, phosphate and magnesium.

The pH of the clarified effluent is reduced from 11.5 to 9.5 in a recarbonation tank (17 min) using a CO_2 dose of 145 mg/l, then recycled second stage solids (10%) and a 5 mg/l dose of ferric ion are added to aid flocculation of the calcium carbonate precipitate.

The effluent from the second stage clarifier flows by gravity through dual-media filters consisting of 24 inches of 0.9-mm coal and 6 inches of 0.45-mm sand. The filters are air scrubbed (5 min) and backwashed (10 min) automatically when the headloss exceeds 9 feet of water.

After filtration, chlorine is added in a static mixer (1.6 seconds) to reduce the pH to 6-7. Carbon dioxide is air stripped from the filter effluent to reduce the alkali required for the breakpoint reaction. Breakpoint is achieved by adding chlorine (Cl:NH_3-N ratio is 9:1 by weight) and alkali (NaOH) to the wastewater ahead of a second static mixer (1.6 seconds) and contact tank (1 minute). The base is added to control the pH and prevent formation of undesirable end products such as nitrate and nitrogen trichloride.

Soluble residual organics are removed from the wastewater with two-stage downflow carbon columns using 8 by 30-mesh granular carbon and a contact time of 40 minutes.

It has been demonstrated at Blue Plains that sludges from the lime precipitation units can be thickened and classified in a centrifuge to separate the calcium carbonate for recalcination. This, however, is not the normal procedure.

A high quality effluent is produced by this treatment sequence. Typical water quality includes 5 mg/l of BOD, 15 mg/l of COD, 0.15 mg/l of P and 2.6 mg/l of N (3).

Control Strategies for Physical-Chemical Processes

The fundamental control logic reported herein was developed by the Blue Plains staff and has been presented in various forms elsewhere (3,4).

Lime Clarification

A dry lime feeder maintains a constant-concentration lime slurry which is circulated through a head tank above the primary reaction zone of the clarifier. An electro-pneumatic valve meters the slurry into the reaction zone on demand signals from the lime-feed controller. The four alternative strategies studied for lime-feed control were conductivity-ratio, flow-proportional, pH plus flow-proportional, and alkalinity plus flow-proportional.

Figure 2 schematically represents the basic components used in the four alternative control schemes. The conductivity-ratio control scheme involves the measurement of conductivity in the primary reaction zone and in the influent wastewater. The ratio of these conductivity measurements generates a control signal(C) for the lime-feed valve. For flow-proportional control, the influent flow rate is measured, and this signal (properly modulated at MT and PS) is transmitted directly to the control valve. For pH plus flow-proportional control, the pH is measured in the primary reaction zone, and this signal is used to adjust the signal generated from the flow-proportional loop. For alkalinity plus flow-proportional control, a sample is pumped from the clarified zone of the clarifier through a porous rock filter to an automatic titrator. The resulting alkalinity signal is transmitted to the multiplying transmitter in a flow-proportioning control system for final adjustment of lime addition.

Performance of the four control systems was monitored by manually titrating grab samples from the primary reaction zone every two hours. The conductivity and flow-proportional systems were considered to be semi-automatic controls and were adjusted by the plant operators when a deviation from grab-sample values exceeded 10%. The pH and alkalinity systems were considered to be fully automatic and were not adjusted after initial startup.

The results of a seven-day test run are shown in Table 1.

TABLE 1

Percentage Deviation From Target
During Seven-Day Test Run

Control Scheme	Ranges of Deviation from Target Alkalinity, %	Comments
Conductivity-ratio	+ 16 to -20	
Flow-proportional	+ 15 to -15	
pH plus flow-proportional	+ 10 to -10	1st 2 days
	+ 10 to -15	Entire 7-day test period
Alkalinity plus flow-proportional	+ 7.5 to -7.5	

Conductivity-ratio control was found to be the least accurate control system, but it would be a good backup control system since it is dependable and requires little maintenance. The flow-proportional control system was very sensitive to any change in lime-slurry concentration and was very much dependent on the accuracy of the flow measurement device. After seven days of operation, the pH electrodes were coated with a calcium carbonate scale which was approximately 1/16th-inch thick. This coating was removed in 2% hydrochloric acid, and the electrode regained its initial response characteristics. By scheduling electrode cleaning every 2 days, pH control will work satisfactorily. Placement of the pH probe in a separate rapid-mix tank reduces the maintenance requirements associated with placement in the primary reaction zone of a single-unit clarification system.

While alkalinity plus flow-proportional control produced the closest alkalinity control of all the systems studied, the equipment malfunctioned repeatedly because of filter clogging. The inability to filter high solids concentrations efficiently required relocation of the sample point from the reaction zone to the clarified zone. This resulted in a two-hour lag in the response time which caused large swings in process effluent quality when the lime-slurry concentration changed. Until the solids handling problem for alkalinity plus flow-proportional control is solved, the recommended control system is pH plus flow-proportional, with conductivity-ratio control as a backup.

The solids wasting control loops for the first and second stage clarifiers are simple feedforward systems where periodic pulses, proportional to flow, produce a discharge (1.5% of influent flow). The discharge from this tank is controlled by a level switch sensing the fixed volume.

The control loop for the $FeCl_3$ feed in the recarbonation tank is a feedforward flow-proportional system which changes the duty cycle of the dosing pump.

The carbon dioxide dosing control system is the same pH plus flow-proportional control system used for the lime addition control. The chemical dose is controlled by a feedforward signal proportional to flow and a feedback signal generated by pH error.

Filtration

Operation of the dual-media gravity filters was controlled with four, alternative, backwash initiation and control schemes. Alarm schemes used to initiate backwash had time-delay circuits to prevent accidental or momentary events from triggering the backwash cycle prematurely. The four modes used were headloss, high-level (influent level), programmed time-interval, and manual. The headloss sensor initiates the backwash cycle when the available head decreases to a preset minimum value (H.L. = 9 ft H_2O). When the level tends to change, the high-level indicator opens an effluent control valve so that a constant level is maintained. When the control valve is 100% open, backwash is initiated. The programmed time-interval controller will initiate backwash at the expiration of a preselected number of operating hours. The operator may override any of the above controls at any time with the manual mode. A turbidity breakthrough alarm was considered; however, filter effluent turbidity changed with clarifier efficiency, thus eliminating turbidity from consideration for filter backwash initiation.

The effluent from clarification was distributed equally to the operating filters by a mechanical splitter box. As a filter was isolated for backwash, the flow to that filter was redistributed to the remaining operating filters. If the headloss alarm was used and if the filter backwash occurred at peak flow rates, the redistribution caused the already stressed operating filters to be overstressed. The final result was a chain reaction resulting in the need to backwash all available filters in a relatively short period of time which increased the requirements for backwash-water pumps and storage capacity.

The programmed time-interval controller was used to schedule filter backwashing at different hours during periods of low flow; this reduced backwash-water pumping and storage requirements, and it eliminated overstressing of the system. The headloss indicator was then used as a backup alarm to prevent flooding when system upsets caused increased solids loading and shorter filter runs than the programmed time interval. The high-level alarm was connected to an audio-visual alarm and was used to indicate equipment failure. This system has provided peak operating efficiency at the lowest possible operating cost.

Breakpoint Chlorination

The control scheme developed to control the breakpoint-chlorination process is shown schematically in Figure 3. The chlorine dosage-control loop employs a feedforward signal proportional to the mass of influent ammonia and a feedback signal based on the free residual chlorine concentration error. The feedforward signal is derived from the concentration of ammonia in the influent, the influent flow rate, and a preselected weight ratio of chlorine to ammonia. If digital control of the system were practiced, this feedforward signal would be adjusted by the amount of chlorine used for pH control during prechlorination. The control loop for alkali addition (NaOH) is derived from a feedforward signal based on the chlorine dosage used and a feedback signal based on the pH error. The on-stream analysis of ammonia by a colorimetric analyzer, both before and after breakpoint chlorination, has been accurate and dependable. Free residual chlorine is also continuously measured by a colorimetric analyzer. Preliminary operating experience has been favorable. Breakpoint chlorination reduces operating problems with the carbon adsorption system by reducing biological slime growths. The carbon adsorption process is a good backup system for the breakpoint chlorination process because of the dechlorination potential.

Carbon Adsorption

The carbon adsorption process presented control problems very similar to the filtration system; however, the carbon adsorption process was only semi-automated. A level-controller regulated the flow through the system, while an automatic pressure-controller on the discharge side of the carbon-column feed pump maintained a constant pressure at the inlet to the first carbon column. There were five carbon columns interconnected in series by headers and automatic valves to allow any number of columns to be operated in sequence and any column to be the lead column. Column sequencing and backwash initiation were determined by an operator. The lead column is backwashed daily in the raw wastewater-treatment applications.

Direct Digital Control

An IBM Systems/7 computer will be installed at the EPA Blue Plains Pilot Plant this November to provide total systems control for the independent physical-chemical treatment sequence.

Preliminary work (3) has been initiated to define the response characteristics of the plant sensors, actuators and unit processes to permit development of the needed computer-control algorithms. This computer installation will assist plant operators and at the same time serve as a valuable research tool utilizing process simulation techniques.

The process-control computer will also perform a data-acquisition function. Plant data from approximately 100 wastewater-process sensors will be collected, converted into engineering units and stored for later analysis and evaluation of the systems performance.

Alarm modes will also be included to alert the operator of plant conditions that require his attention and correction. Examples are out-of-range alarms for such parameters as pH and chemical doses to the wastewater, as well as equipment-failure alarms for such items as sludge-blowdown equipment and pneumatic pumps.

Direct digital computer control will no doubt be used in the larger wastewater treatment facilities where the economies realized in data acquisition, reporting, preventive maintenance scheduling, load programming and improved performance can offset the additional costs of specialized manpower and the capital costs of the initial hardware and backup system. The capability to control individual unit processes and indeed complete physical-chemical treatment systems is rapidly becoming a reality. The successful development of this capability is due in large measure to our ability to measure key operating variables accurately (pH for example) and the relatively fast response time of these unit process to permit feedback trimming.

Development of control strategies for alternative treatment processes such as combined biological-chemical-physical processes (like mineral addition to the aerator for phosphorus removal and biological denitrification) are also underway. Complete systems-control capability for these processes is not as well developed. Current capability does exist, however, for controlling the addition of supplemental chemicals such as methanol. The control scheme utilized is similar to the feedforward mass-proportional control technique described for the breakpoint chlorination process and is shown schematically in Figure 4. The capability to implement these control strategies depends upon the analytical capability to measure key load variables such as the nitrate concentration to a denitrification tank. Kamphake and Williams (5) have recently reported on the design of AutoAnalyzer manifolds for the continuous monitoring of nitrate-nitrite nitrogen and orthophosphate. Successful utilization of this technique for control of the denitrification process has been reported by Smith et al (6). Mineral addition to the final settler of a trickling filter plant has also been successfully achieved at Richardson, Texas, utilizing a feedback loop based on the residual phosphorus concentration in the final effluent. Work is continuing to modify the phosphorus determination procedure so that the phosphorus concentration in raw wastewater and primary effluent can be determined. The feedforward control strategy shown in Figure 4 can then be utilized to control mineral addition to the aerator.

Cost-Effectiveness of Instrumentation and Automation

A preliminary attempt to quantify the economic justification for installing the needed instrumentation and control accessories to provide feedforward mass-proportional chemical dosing of alum for phosphorus removal and methanol addition for denitrification is presented in the following analysis. The economic consequences of adopting four different control strategies for a 10-mgd facility were determined. The four control schemes evaluated include dosing to meet the peak daily chemical requirements, periodic operator adjustment, flow-proportional dosing, and mass-proportional dosing.

Alum Addition

Instrumentation requirements will vary depending on the control strategy adopted. Dosing to meet the peak daily chemical requirement (no control) or for periodic operator adjustment requires only a flow meter. Dosing levels are based on historic records of the phosphorus concentration. The flow-proportional control strategy only incurs the incremental cost of a transmitter and controller. The mass-proportional control strategy imposes the additional requirements of an analytical device, signal conditioner and multiplying transmitter.

The phosphorus loading for a typical 10-mgd plant is shown in Figure 5. The loading varies from about 175 to 1180 lbs phosphorus/day, based on hourly readings. The flow for the 10-mgd plant varies from 5 to 10.6 mgd. If operator control is used, it is assumed that the plant operates from midnight until noon under a low phosphate loading of 700 lbs P/day, and from noon until midnight under a high phosphate loading of 1200 lbs P/day.

The alum dosing requirement was established at 1.5 moles of aluminum per mole of phosphorus. Liquid alum, consisting of 8.3% Al_2O_3 and costing \$46.90/ton on an equivalent of 17.15% dry Al_2O_3, was used for the cost estimates.

The capital costs for chemical storage, housing and pumping were amortized at 6% for 25 years and expressed in August, 1971 dollars. Capital costs for instrumentation were not included. The capital and operating costs for the various methods of controlling alum addition are shown in Table 2.

TABLE 2

Costs for Alum Addition At A 10-mgd Plant
for Various Control Schemes

Control Scheme	lbs alum/day	Capital Cost, $	Cost of Alum, ¢/K gal.	Total Cost of Plant, ¢/K gal.	Annual Cost, $1000/yr
Mass-proportional	16290	112,000	1.66	2.22	81
Flow-proportional	26850	128,000	2.74	3.33	121
Operator Control	28260	134,000	2.88	3.48	127
No Control	35700	145,000	3.64	4.27	156

Periodic operator control provided a saving of $24,000/yr. This saving could be increased by more frequent operator attention. Similarly, if flow-proportional control were employed, an additional saving of $5,500/yr would result. Feedforward mass-proportional control provided an annual saving of $40,500 compared to flow-proportional control alone. This cost saving does not include the significant savings in sludge handling costs that would also accrue and is more than adequate to justify the purchase of a phosphate analyzer.

Methanol Addition

Similar calculations were also made for the denitrification process. Figure 6 compares the diurnal variations of flow for a typical 10-mgd plant with the corresponding variations in nitrate concentration.

For an impartial analysis, the quality of the effluent achieved under each control scheme should be kept reasonably constant. This is somewhat difficult since each mg of methanol is equivalent to 1.5 mg of COD or about 0.88 mg BOD. Thus, in the "no control" scheme, some additional aeration is required to remove the excess COD caused by the unused methanol. In the other control schemes, the unused methanol is minimal and the quality of effluent is acceptable without the additional aeration. Both columnar and dispersed growth denitrification were considered. It was also assumed that 2.5 mg of methanol were required to remove each mg of NO_3-N.

It was estimated that if the no-control scheme were used, the BOD caused by the excess methanol would vary from 0 to 20 mg/1 and would require an additional one-hour aeration of 733 SCF/min for methanol degradation. It was also assumed that the additional solids generated from columnar denitrification would be about 30 mg/1, which would be acceptable since the dispersed growth process is typically followed by a settler. The capital cost of aeration for a 10-mgd plant is about $129,000; if this is amortized by a factor of 0.08, the yearly cost is $13,500. Operating and maintenance costs are $6360/year. These costs are based on Black and Veatch's (7) estimates and are adjusted to January, 1971. The cost of the additional methanol is $4,380/year. Thus, the total excess cost to achieve comparable effluent quality is $24,250/year for the no-control scheme.

A sampling and periodic operator-adjustment frequency of 1 hour was used in preparing Figure 6 and in estimating the analytical and operating costs. The excess yearly cost of an operator-control scheme is $21,800 based on an annual operator's salary of $12,000, three-shift operation, 52% utilization of operator time and the annual cost of analytical equipment and space ($800). To provide methanol addition based on influent flow rate, the previous history of the plant must be studied to obtain some relationship between methanol and flow. The following equation was derived from the data in Figure 6.

$$\text{mg/1 } CH_3OH = 4.09 \text{ (mgd)} + 12.04$$

The cost for flow-proportional control includes the cost of the control devices and the cost of obtaining and updating the plant history. The cost of control equipment is estimated to be $1500, or $355 per year amortized at 6% for 5 years. The cost of updating the plant history on a yearly basis would include a two-day hourly sampling and analysis program costing $1200 ($25 per sample), plus $300 for data evaluation. This means that the total excess annual cost is $1855. The cost of a flow meter was neglected since it was assumed all plants, regardless of control scheme, will buy a flow meter for this application.

For mass-proportional control, a Technicon AutoAnalyzer (cost $8000) would be used, together with the control system described above (cost $1500). The estimated annual cost (6% for 5 years) is $2254, plus $300 (chemicals), for a total of $2554.

Table 3 summarizes the excess costs for denitrification, using each of the control schemes described above.

TABLE 3

Excess-Cost Comparisons for Denitrification
Using Various Control Schemes

Control Scheme	Excess Costs, $/Year
No Control	$24,250
Operator Control	21,800
Flow-Proportional	1,850
Mass-Proportional	2,550

These excess costs include the costs of instrumentation and control devices. Note that when these latter costs are considered, mass-proportional control is slightly more expensive than flow-proportional control. The primary reason for this is the relatively small variation in nitrate concentration entering the denitrification tank. Based on the accuracy of these estimates, mass-proportional control would appear to be worthwhile since it is almost a breakeven proposition compared to flow-proportional control.

Summary and Conclusions

Process-control strategies and instrumentation are available for reliable operation of an independent physical-chemical treatment system consisting of two-stage lime clarification, dual-media filtration, breakpoint chlorination and carbon adsorption. Also available is the needed instrumentation to control methanol addition for denitrification.

Additional analytical development work is required to permit monitoring of the total phosphorus concentration in both raw wastewater and primary effluents to achieve feedforward control of phosphorus removal processes using alum and iron.

Additional research should be directed toward development of the needed instrumentation and control theory for sludge-handling unit processes since these can account for 40 to 50% of the total operating costs of a treatment facility.

REFERENCES

1. R. Smith, "Wastewater Treatment Plant Control," presented at the Joint Automatic Control Conference, Washington University, St. Louis, Missouri (August 11-13, 1971).

2. J. Andrews, Water & Sewage Works, 2, 26 (February 1971).

3. D. F. Bishop, W. W. Schuk, R. Bernstein, and E. Fein, "Computer Control of a Chemical Clarification Waste Treatment Pilot Plant," presented at the First International Meeting - Pollution: Engineering & Scientific Solutions, Tel Aviv, Israel (June 12-17, 1972).

4. W. W. Schuk, "Control Systems in Advanced Waste Treatment," presented at the 68th National AIChE Meeting, Houston, Texas (March 1, 1971).

5. L. J. Kamphake and R. T. Williams, "Automated Analyses for Environmental Pollution Control," presented at the Technicon International Congress, New York City (June 12-14, 1972).

6. J. M. Smith, A. N. Masse, W. A. Feige, and L. J. Kamphake, Jour. Env. Sci. & Tech., 6, 260 (March 1972).

7. W. L. Patterson and R. F. Banker, "Estimating Costs and Manpower Requirements for Conventional Wastewater Treatment Facilities," EPA Water Pollution Control Research Series, Report No. 17090 DAN 09/71 (In preparation).

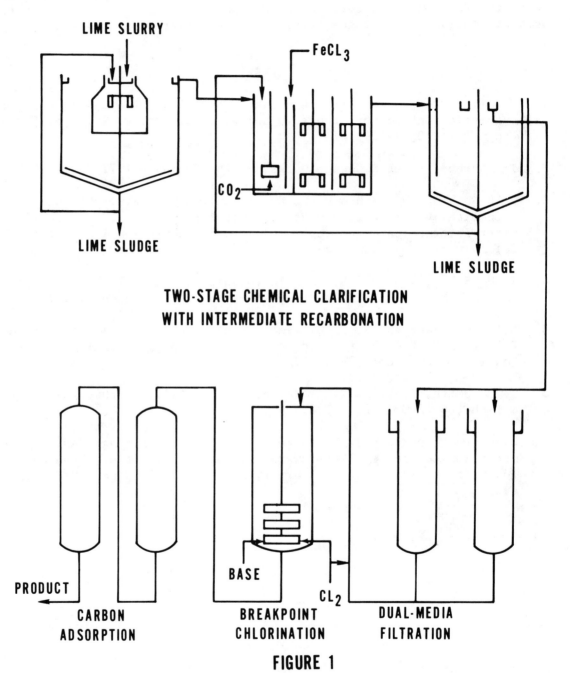

TWO-STAGE CHEMICAL CLARIFICATION
WITH INTERMEDIATE RECARBONATION

FIGURE 1

PHYSICAL CHEMICAL TREATMENT

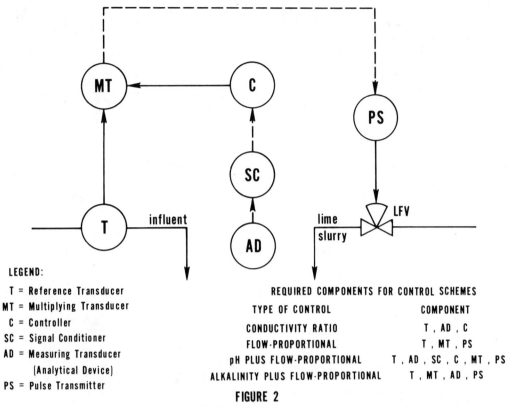

LEGEND:

 T = Reference Transducer
MT = Multiplying Transducer
 C = Controller
SC = Signal Conditioner
AD = Measuring Transducer
     (Analytical Device)
PS = Pulse Transmitter

REQUIRED COMPONENTS FOR CONTROL SCHEMES

TYPE OF CONTROL	COMPONENT
CONDUCTIVITY RATIO	T , AD , C
FLOW-PROPORTIONAL	T , MT , PS
pH PLUS FLOW-PROPORTIONAL	T , AD , SC , C , MT , PS
ALKALINITY PLUS FLOW-PROPORTIONAL	T , MT , AD , PS

FIGURE 2

LIME FEED CONTROL SCHEMES

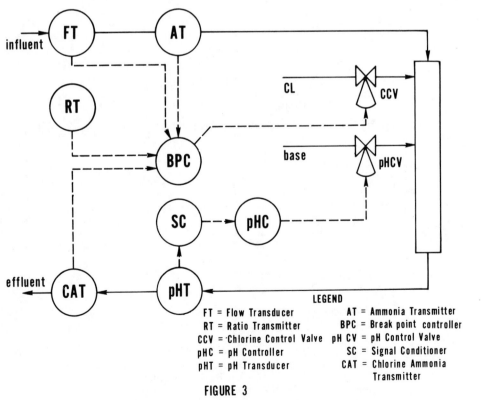

LEGEND

FT = Flow Transducer
RT = Ratio Transmitter
CCV = Chlorine Control Valve
pHC = pH Controller
pHT = pH Transducer

AT = Ammonia Transmitter
BPC = Break point controller
pH CV = pH Control Valve
SC = Signal Conditioner
CAT = Chlorine Ammonia
      Transmitter

FIGURE 3

BREAK POINT CHLORINATION CONTROL

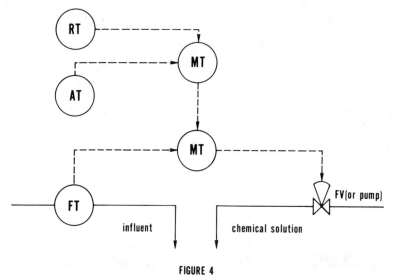

FIGURE 4

CHEMICAL ADDITION RATIO CONTROL

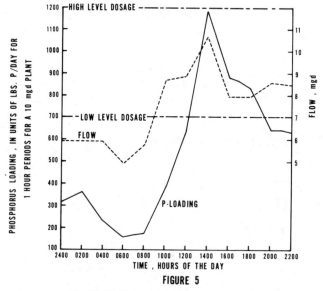

FIGURE 5

PHOSPHORUS LOADING AND FLOW FOR A 10 MGD PLANT

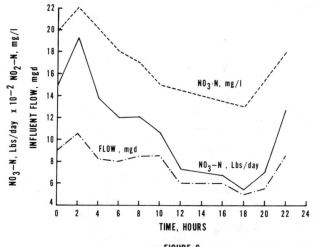

FIGURE 6

NITRATE LOADING AND FLOW FOR A 10 MGD PLANT

RECENT STUDIES OF CALCIUM PHOSPHATE PRECIPITATION IN WASTEWATERS

David Jenkins
Associate Professor of Sanitary Engineering
University of California, Berkeley
Arnold B. Menar
Doctoral Candidate in Sanitary Engineering
University of California, Berkeley
John F. Ferguson
Assistant Professor of Geography and Environmental Engineering
The Johns Hopkins University

Under the conditions and concentrations encountered in wastewaters the stable (thermodynamically predictable) calcium phosphate solids are hydroxyapatite ($Ca_5(PO_4)_3OH$) or fluorapatite ($Ca_5(PO_4)_3F$). If these ionic crystalline solids were actually to form then dissolved phosphate concentrations would be exceedingly small. Thus for the typical wastewater conditions of pH 8, $[Ca^{2+}]$ = 2 mM (200 mg $CaCO_3$/ℓ) hydroxyapatite with pK_{sp} = 57.5 is in equilibrium with a PO_4^{3-} concentration of $10^{-10.7}$ M or at this pH a total orthophosphate concentration of $10^{-6.9}$ M or 4 μg P/ℓ. Such phosphate levels are at least an order of magnitude lower than those commonly encountered in wastewater calcium phosphate precipitation processes conducted at pH values several units higher than pH 8. It must therefore be assumed that, in general, the thermodynamically stable crystalline phase does not control the dissolved phosphate residual in wastewater calcium phosphate precipitation processes. The levels of dissolved phosphate are indeed controlled by the kinetics of calcium phosphate precipitation rather than the thermodynamically predictable state of the system. This paper will present a discussion of calcium phosphate precipitation that will illustrate the importance of precipitation kinetics in the precipitation of calcium phosphates from wastewaters.

A general picture of the kinetics of calcium phosphate precipitation has been presented by Leckie and Stumm [1]. A supersaturated solution may exhibit a period of induction before any measurable decrease in dissolved phosphate takes place. Precipitation then may proceed to form a metastable amorphous phase and a removal of phosphate from solution occurs. This metastable phase undergoes a phase transformation to form the thermodynamically stable crystalline phase (no net dissolved phosphate removal occurs during this stage). Then, finally, the stable crystal grows at the expense of both the metastable phase and solution components; more phosphate disappears from solution during this step. The length in time and position in concentration at which each of these four general stages (Figure 1) (induction, precipitation, phase transformation, and crystal growth) in the precipitation of a crystalline calcium phosphate occurs is a function of precipitation conditions such as pH, reactant concentration, temperature, presence of other anions and cations, presence of preformed crystal nuclei, etc. Some stages may be eliminated entirely. For

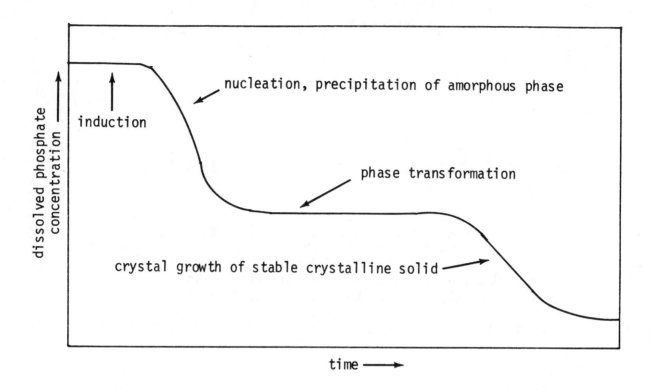

FIG. 1
Idealized Scheme for Calcium Phosphate Precipitation Kinetics

example the addition of preformed crystalline solids to a supersaturated solution
may eliminate the first three stages and crystalline growth may take place at
once (Figure 2). This is illustrated by experiments in which calcium phosphate
was precipitated from defined solutions at pH 7.6. After a steady state level
of dissolved phosphate had been reached the solids were separated by filtration
and transferred to a medium of the same initial composition as that from which
they were formed. The induction period was completely eliminated. Alternatively the induction period may be eliminated and an amorphous phase formed by
the addition of nuclei or the presence of highly supersatured solutions. This
is illustrated in Figure 3 which are the steady state results of continuous flow
calcium phosphate precipitation experiments [2] conducted with the indicated
initial constituent levels and at various reactor mean detention times at pH
8.0.

The initial solutions in the results shown in Figure 3 are substantially
more supersaturated than those in Figure 2. Rapid formation of an amorphous
"steady-state" phase takes place which in nonrecycle reactors of up to 3 hr
mean residence time controls the dissolved phosphate level. The solid calcium
phosphate phase formed in some of these experiments was shown by X-ray diffrac-
tion analysis to transform very slowly from its amorphous form to a tricalcium
phosphate.

Using this general model for the kinetics of calcium phosphate precipitation
the results of two studies are discussed. The work of Menar [2] has shown that
the steady-state phase that forms in short-term calcium phosphate precipitation
processes (e.g. lime addition to sewage with no precipitate recycle) has a
Ca/P stoichiometry of 1.44 (Figure 4). The constancy of the concentration
products for Menar's residual dissolved calcium and phosphate data on chemically

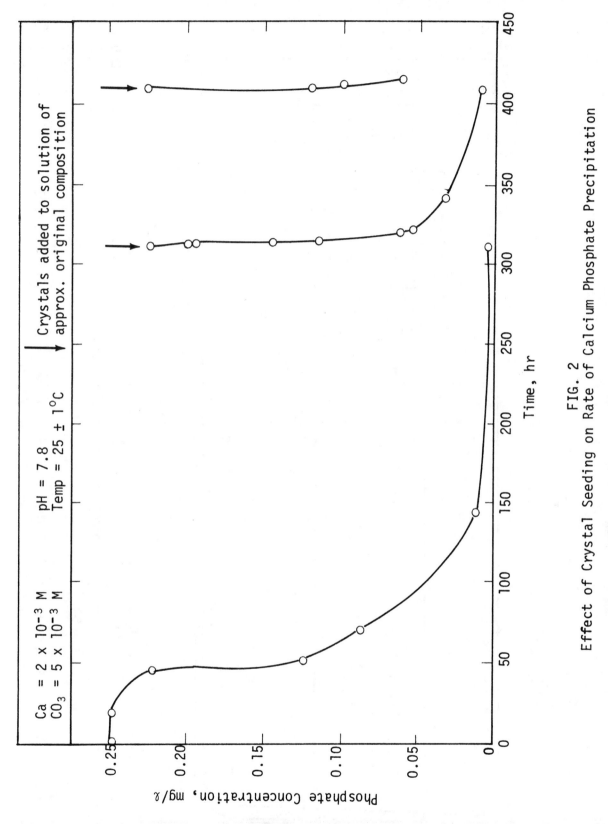

FIG. 2

Effect of Crystal Seeding on Rate of Calcium Phosphate Precipitation

defined systems has been tested for the solids octacalcium phosphate ($Ca_4H(PO_4)_3$), hydroxyapatite ($Ca_5(PO_4)_3OH$), dicalcium phosphate ($CaHPO_4$), and tricalcium phosphate ($Ca_3(PO_4)_2$) (Figure 5). The constancy with pH of the activity product of tricalcium phosphate (TCP) with pH is an indication that this solid, or a solid with its stoichiometry, controls dissolved phosphate residuals. The value of 23.56 for -log concentration product obtained

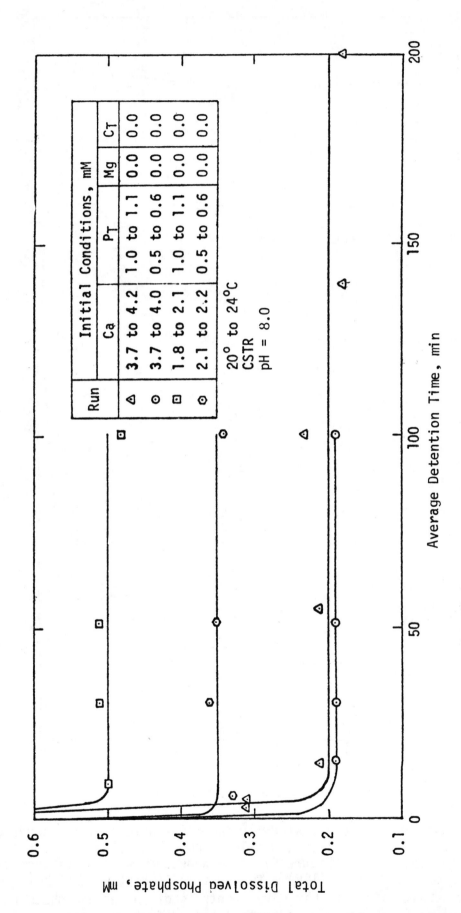

FIG. 3

Dependence of Steady State Total Dissolved Phosphate Concentration on Average Detention Time in a CSTR Reactor

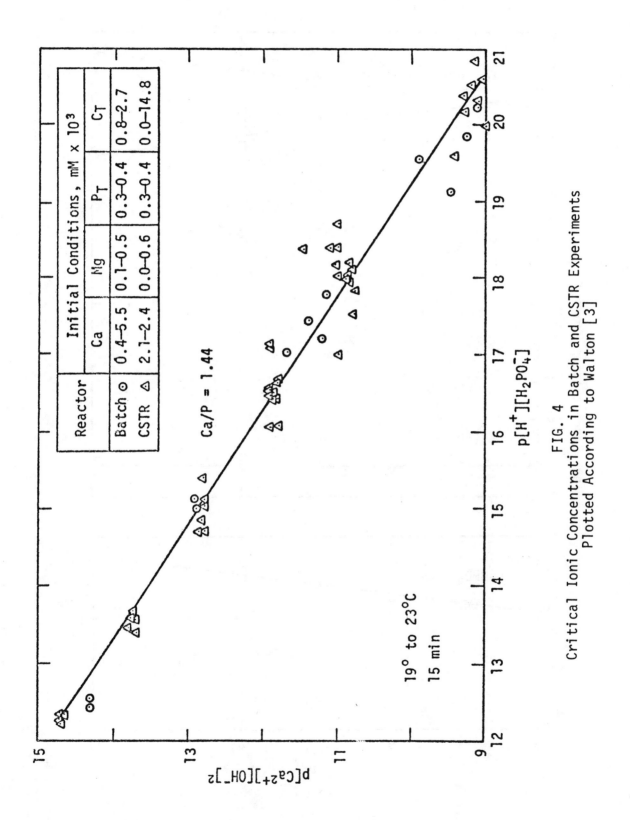

FIG. 4

Critical Ionic Concentrations in Batch and CSTR Experiments
Plotted According to Walton [3]

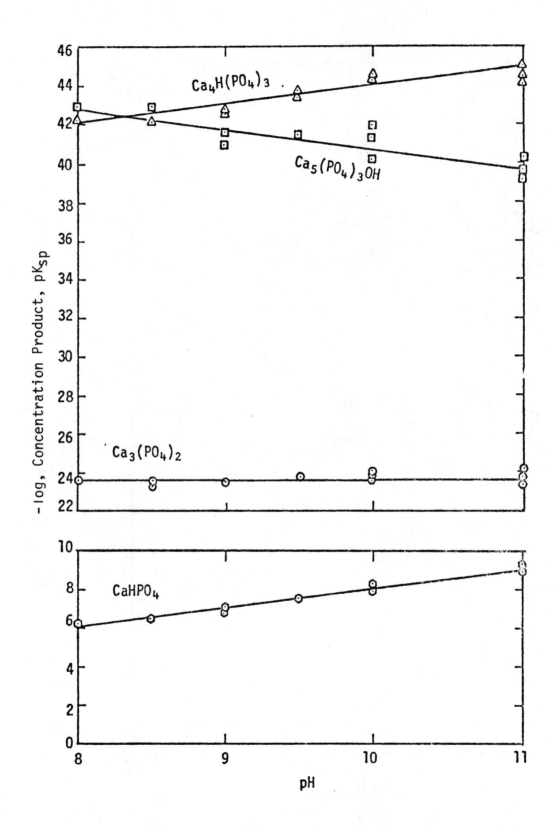

FIG. 5
Effect of pH on pK$_{sp}$ for Various Calcium Phosphate Solids

for TCP (Figure 6) is lower than the values of 25, 26, and 28.5 that have been reported in the literature [4,5,6]. The concentration product of TCP obtained in chemically defined solutions predicts data obtained from nonprecipitate recycle wastewater calcium phosphate precipitation experiments on primary and secondary wastewaters with good precision up to pH 10.5 (Figure 7).

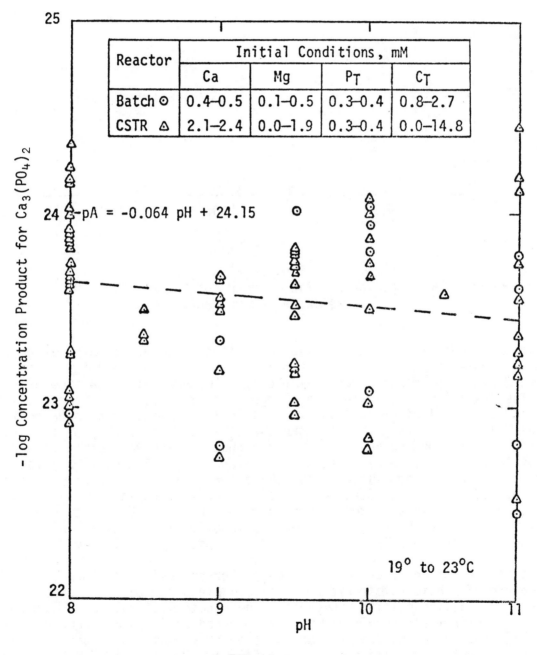

Reactor	Initial Conditions, mM			
	Ca	Mg	P_T	C_T
Batch ⊙	0.4–0.5	0.1–0.5	0.3–0.4	0.8–2.7
CSTR △	2.1–2.4	0.0–1.9	0.3–0.4	0.0–14.8

$-pA = -0.064\ pH + 24.15$

19° to 23°C

FIG. 6
Effect of pH on Concentration Product of Tricalcium Phosphate

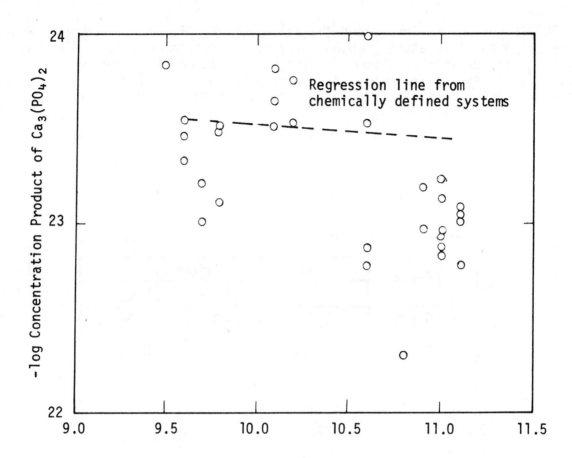

FIG. 7
Dependence of Concentration Product of Ca₃(PO₄)₂ on pH
for Wastewater Experiments

In these wastewater precipitation experiments either primary sewage effluent or activated sludge effluent was dosed with lime and precipitated in a
stirred 3 or 4 compartment reactor-flocculator with a detention time of 42-47
min. The reactor was followed by a sludge blanket clarifier, a recarbonation
unit with 5 min detention time and then clinoptilolite columns for ammonia
removal. It was observed that the phosphate removal efficiency of the process
train (precipitation, sedimentation, recarbonation, and filtration through the
clinoptilolite) was determined at the overflow weir of the sedimentation basin.
The data in Table 1 show that any phosphate-containing particulates that one
might hope to remove by filtration are dissolved very rapidly in the recarbonation basin — the values of total phosphate leaving the sedimentation basin are
almost identical to the values following clinoptilolite columns. After filtration, however, the total phosphate is made up entirely of dissolved material.

In precipitation processes that employ longer residence times and include
the feature of precipitate recycle it is possible under certain conditions to
obtain a crystalline calcium phosphate. The kinetics of calcium phosphate
precipitation in these conditions has been studied by Ferguson, Jenkins, and
Stumm [7], and Ferguson, Jenkins, and Eastman [8]. Batch precipitation experiments from chemically defined solutions of CaCl₂, NaH₂PO₄ and NaHCO₃ at close
to pH 8 show an induction period followed by a crystal growth where phosphate
disappeared from solution (Figure 8). The introduction of preformed solids

TABLE 1

Effect of Unit Processes on Total and Dissolved Phosphate Residual in Wastewater

Sample Location	Experiment 1			Experiment 2			Experiment 3			Experiment 4		
	Phosphate		pH	Phosphate		pH	Phosphate		pH	Phosphate		pH
	Total mM	Diss. mM		Total mM	Diss. mM		Total mM	Diss. mM		Total mM	Diss. mM	
Primary Effluent	0.37	0.31	7.4	0.39	-	7.4	0.37	0.28	7.4	0.35	0.28	7.4
Activated Sludge Effluent	0.36	0.30	7.2	0.35	0.34	7.3	-	-	-	-	-	-
Lime Precipitation Effluent	-	0.0039	10.8	0.053	0.018	9.8	0.0089	0.0032	11.0	0.018	0.0074	10.2
Recarbonation Effluent	-	0.016	8.5	0.056	0.053	8.2	0.0076	0.007	7.7	0.027	0.023	-
Clinoptilolite Effluent	0.017	0.016	8.5	0.056	0.054	8.4	0.0076	0.007	7.8	0.021	0.019	-

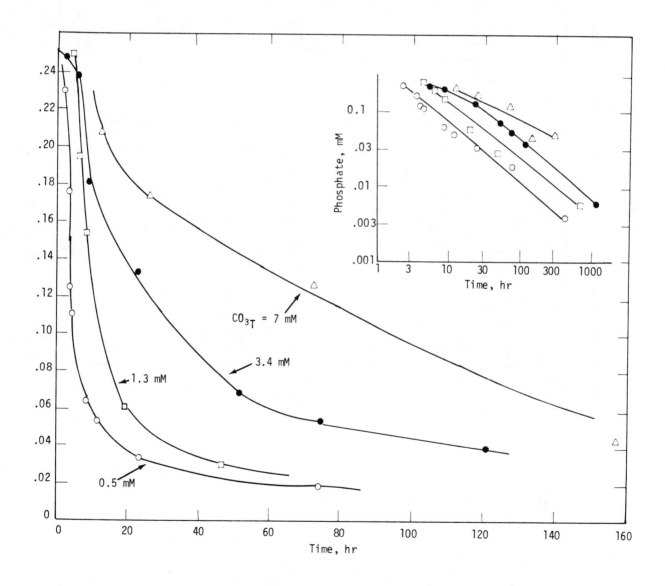

FIG. 8
Effect of Bicarbonate Concentration on Calcium Phosphate Precipitation at pH 8
Initial Condition: Ca = 2.0 mM, PO_4 = 0.25 mM, T = 26 $\pm 1^{\circ}$C

into a precipitating solution of the same composition as that from which the solids were formed caused the elimination of the induction period (Figure 2) and suggested that solids recycle processes would exhibit more rapid precipitation than processes not employing this feature.

The effect of increasing bicarbonate (alkalinity) from 0.5 mM to 7 mM was to lengthen the induction period and decrease the rate of phosphate removal following the induction period (Figure 8). The rate of phosphate removal following the induction period can be expressed as,

$$\frac{dc}{dt} = - ks \, (c - c_0)^n \tag{1}$$

where

n is the order of reaction, s is the available crystal surface area, c is concentration of dissolved orthophosphate (mM), c_0 is its equilibrium concentration (mM) and t is the time from the end of the induction period (hr).

Further assumptions are justified for calcium phosphate precipitation from wastewaters: The available crystal surface was assumed to be constant during precipitation so that s was constant and ks was lumped into a new constant, K. The term c_0 was assumed to be negligibly small. Thus for hydroxyapatite precipitation at pH 8 with $[Ca^{2+}]$ = 2 mM, and pK_{sp} = 57.5, the PO_4^{-3} concentration at equilibrium would be $10^{-10.7}$ M giving at this pH a total orthophosphate concentration of $10^{-6.9}$ M or 0.004 mg/ℓ as P — a value that is negligible when compared with observed values. Using these approximations, the rate equation can be written:

$$\frac{dc}{dt} = -Kc^n \tag{2}$$

A least squares analysis of a double logarthmic plot of phosphate removal rate and phosphate concentration for the lowest bicarbonate concentration (Figure 9) yielded values of the kinetic parameters of n = 2.7 and K = 8.2.

Equation (2) gave excellent prediction of phosphate removal rate in solutions containing 1.3 and 3.4 mM bicarbonate but not when the solution contained 7 mM bicarbonate (Figure 10).

The effect of bicarbonate on phosphate removal rate (following the induction period) could be empirically incorporated into the rate equation as follows:

$$\frac{dc}{dt} = \frac{-K'}{CO_{3_T}} \, c^{2.7} \tag{3}$$

where

$$K' = KCO_{3_T} = 4.1$$

The precipitates from this series of experiments showed greatly decreasing calcium phosphate crystallinity with increasing bicarbonate concentration. The calcium phosphate precipitate was a poorly crystalline apatite and at the highest bicarbonate concentration calcite was formed.

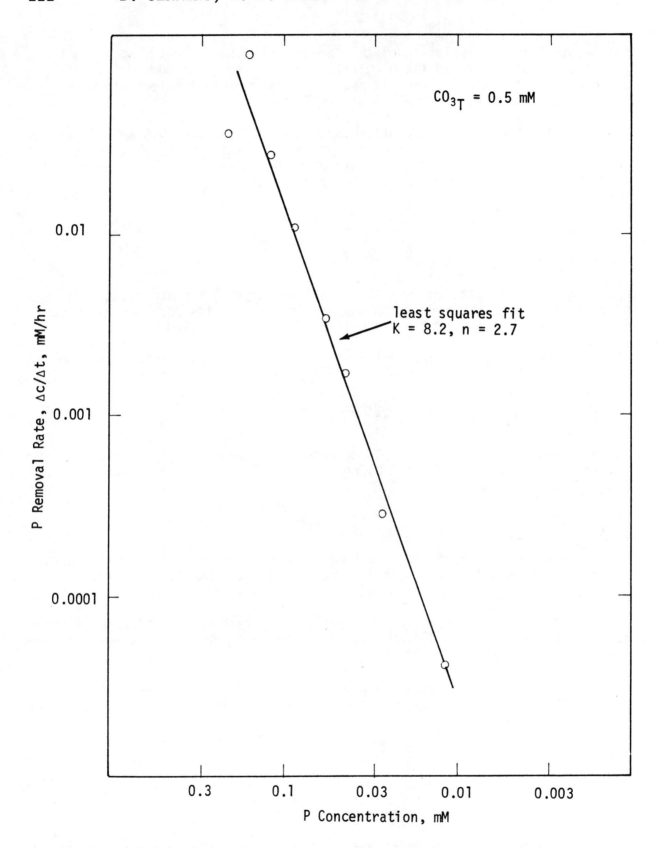

FIG. 9
Phosphate Removal Rate Versus Phosphate Concentration for
0.5 mM Bicarbonate Concentration

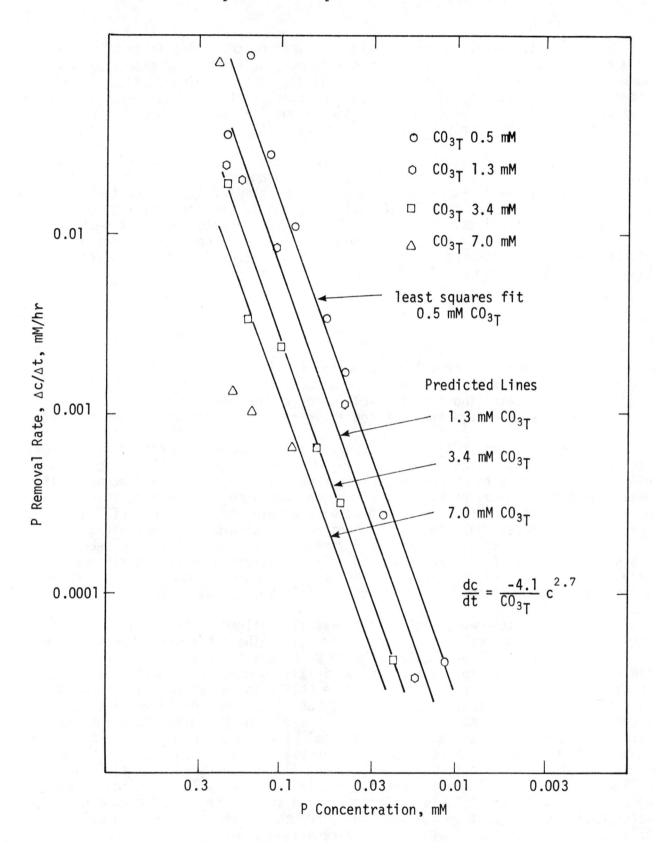

FIG. 10
Phosphate Removal Rate Versus Phosphate Concentration — Comparison
of Predicted Curves to Experimental Data for Bicarbonate Concentration
of 1.3 mM, 3.4 mM, and 7.0 mM

If phosphate precipitation at slightly alkaline pH values is to be applied in a continuous process the induction period must be eliminated either by backmixing or by recycle of precipitated solids. Recycle, in addition, may increase the surface area of precipitate in the reactor. The phosphate concentration exponent of 2.7 also indicates that a tubular reactor should be more efficient that a completely stirred tank reactor (CSTR).

The empirical rate equation derived from batch data was combined with mass balances and hydraulic characteristics for four reactor types that form the basis of most wastewater treatment reactors — the completely stirred tank reactor with and without recycle of solids and the tubular reactor with and without recycle of solids. In reactors with solids recycle the relationship of the rate constants for phosphate removal to the rate constant in reactors without recycle was assumed to be in direct proportion to the solids detention times of the two reactors. Thus:

$$K'' = K' \, \theta_s / \theta$$

where

K'' = rate constant for solids recycling reactor
K' = rate constant for nonrecycle reactor
θ_s = solids detention time for solids recycling reactor
θ = solids detention time for nonrecyle reactor = hydraulic detention time.

For the conditions: θ_s / θ = 30, recycle ratio q/Q = 0.2, K' = 4.1, n = 2.7, Ca^{2+} = 2.0 mM, PO_{4T} = 0.25 mM, CO_{3T} = 1.3 mM, and pH 8, a tubular reactor with recycle is predicted to produce phosphate removals in excess of 80% at hydraulic detention times typical of those at which the activated sludge process is operated in practice (Figure 11) even at bicarbonate concentrations of 7 mM (Figure 12). Indeed, 90% removal is predicted for bicarbonate concentrations of 1.3 mM in a 4 hr hydraulic residence time. For a single CSTR with recycle, phosphate removals of 80% are predicted only for bicarbonate concentrations below about 1.3 mM (Figure 13). For this type of reactor, 90% phosphate removal is predicted to be impossible for practical (<10 hr) hydraulic residence times.

These predictions were tested on chemically defined solutions in a laboratory continuous flow reactor with solids recycle. The experimental results are in close agreement with predictions (Table 2, Figures 12 and 13). Increasing the pH from 7.2 to 7.8 to 8.1 at constant initial bicarbonate ($\sim$1.2) and calcium (2.0 mM) concentrations in a single CSTR with solids recycle produced successively lower phosphate residuals — 0.086, 0.061, and 0.053 mM, respectively (Table 2, Experiments 1, 2, and 3). At a pH of 8 and the same initial calcium concentration, an increase in initial bicarbonate from 1.3 to 7.0 mM in a single CSTR with recycle produced a phosphate residual increase of from 0.053 to 0.099 mM (Table 2, Experiments 3 and 4). By changing from a single CSTR with solids recycle to a series of 4 CSTR's with recycle but with the same nominal residence time, (Table 2, Experiments 4 and 5) the 7.0 mM bicarbonate water could be reduced from 0.099 mM to 0.060 mM at a pH of 7.7 — some 0.2 pH units lower than in the single CSTR with recycle.

The effect of precipitating conditions and other wastewater constituents on the rate of dissolved phosphate removal has been assessed both by review of previous work and by experiment. Only the effects of magnesium concentration and pH were judged to be of major significance. In chemically defined systems the phosphate residual was found to increase as the pH was increased in the

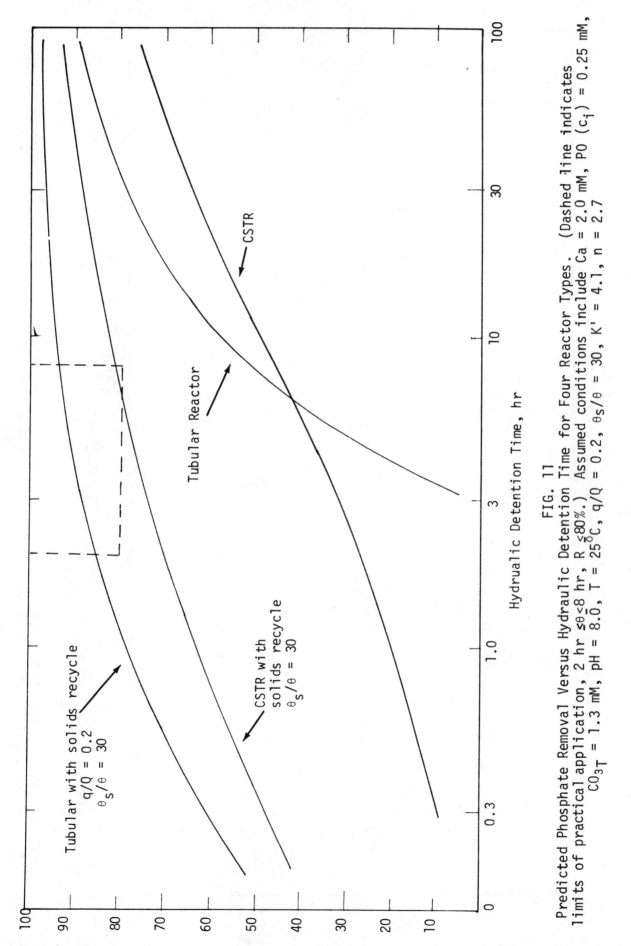

FIG. 11

Predicted Phosphate Removal Versus Hydraulic Detention Time for Four Reactor Types.   (Dashed line indicates limits of practical application, 2 hr $\leq \theta < 8$ hr, R $\leq 80\%$.)   Assumed conditions include Ca = 2.0 mM, P0 (c_i) = 0.25 mM, CO_{3T} = 1.3 mM, pH = 8.0, T = 25°C, q/Q = 0.2, θ_S/θ = 30, K' = 4.1, n = 2.7

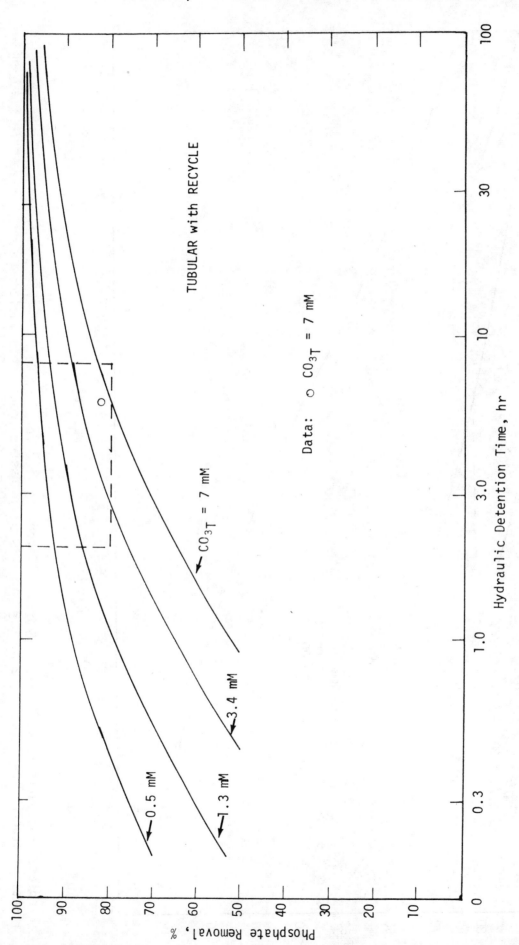
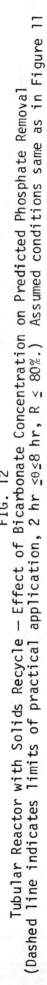

FIG. 12

Tubular Reactor with Solids Recycle — Effect of Bicarbonate Concentration on Predicted Phosphate Removal
(Dashed line indicates limits of practical application, 2 hr $\leq \theta \leq 8$ hr, R $\leq$ 80%.) Assumed conditions same as in Figure 11

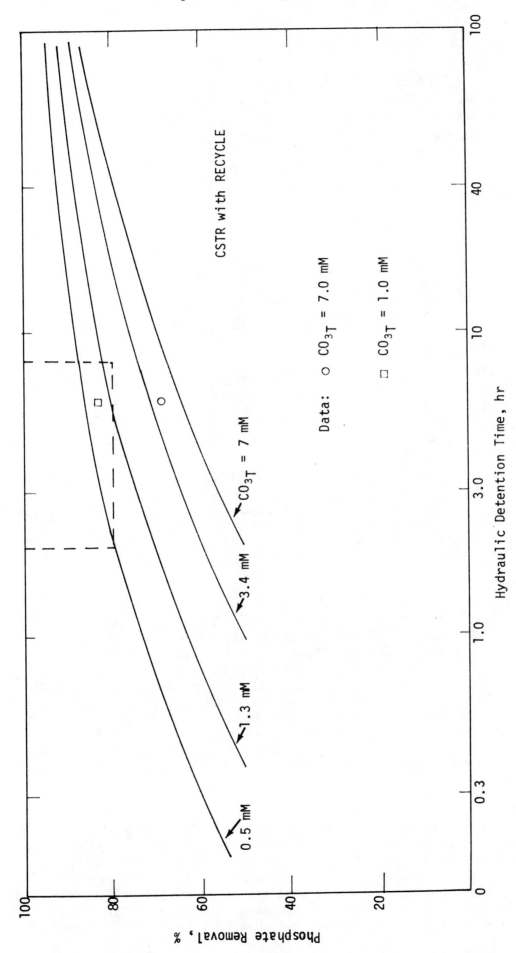

FIG. 13

Completely Stirred Tank Reactor — Effect of Bicarbonate Concentration on Predicted Phosphate Removal
(Dashed line indicates limits of practical application)

TABLE 2

Effect of Chemical and Physical Parameters on Calcium
Phosphate Precipitation at Slightly Alkaline pH

Experiment Number	1	2	3	4	5
Reactor	1 CSTR	1 CSTR	1 CSTR	1 CSTR	4 CSTR's
Solids Recycle	yes	yes	yes	yes	yes
Initial Phosphate, mM	0.29	0.30	0.30	0.30	0.30
Initial Calcium, mM	2.3	2.0	2.0	2.0	2.0
Bicarbonate, mM	1.1	1.2	1.3	7.0	6.7
pH	7.2	7.78	8.08	7.95	7.73
Detention Time, hr	7.5	6.0	6.8	5.9	6.3
Average Residual Phosphate, mM	0.086	0.061	0.053	0.099	0.060
Average Residual Calcium, mM	2.03	1.64	1.66	1.61	1.52

range 7.2-8.1 and it is likely that the pH could have been increased slightly
above 8.2 with further decreases in dissolved phosphate residual. The limiting
pH value would be that at which calcium carbonate precipitation began to occur
(with accompanying consumption of chemical) — a pH of between 8-9. Calcium
phosphate precipitation in this pH range can be superimposed on an activated
sludge flowsheet since the process can operate satisfactorily up to at least
pH 8.5.

The presence of magnesium will tend to reduce calcium phosphate precipitation at pH values below about 9, possibly by substitution in and disruption
of the calcium phosphate crystal lattice or by adsorption on to the growing
calcium phosphate crystals. The magnitude of these effects was assessed from
batch experiments on chemically defined solutions containing calcium, orthophosphate, carbonate, and magnesium (Figure 14). In the presence of 0.8 mM
magnesium (a Ca/Mg mole ratio of 2.5) phosphate removal at all times was reduced
about 20%. With 1.8 mM magnesium, phosphate removal was reduced by 80%. It
would appear, therefore, that a magnesium concentration of 1 mM (24.3 mg/ℓ as
Mg or 100 mg/ℓ as $CaCO_3$ hardness) is about the maximum that can be tolerated
without seriously reducing phosphate removal at slightly alkaline pH values.

Discussion

The two studies described in this paper show that both chemical and physical
conditions have a significant effect on the rate and extent of calcium phosphate
precipitation from wastewaters. In short-term nonrecycle calcium phosphate
precipitation processes between pH 8-11 an amorphous solid forms rapidly and
produces a dissolved phosphate residual that can be predicted well on the basis
of the solubility product of tricalcium phosphate. The precipitate separation
(by sedimentation) ultimately determines the phosphate removal efficiency of

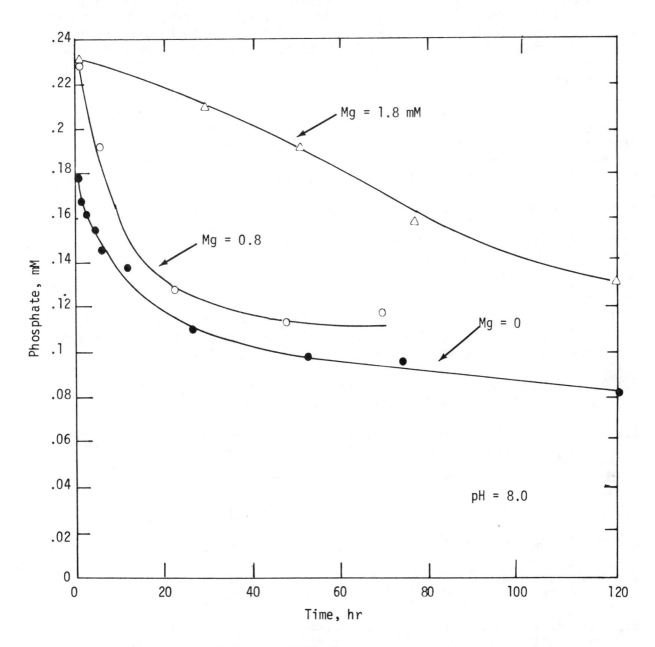

FIG. 14
Effect of Magnesium on the Rate of Calcium Phosphate Precipitation at pH 8

such a process because these precipitates dissolve very rapidly in the lower pH environments of a recarbonation basin that necessarily precedes any filtration process (designed to increase phosphate removal by increasing particle removal).

In low magnesium (24 mg Mg/ℓ), moderate alkalinity (350 mg CaCO₃/ℓ) wastewaters it should be possible to reduce dissolved phosphate levels to below 2 mg P/ℓ by calcium addition and adjustment of pH to a value below 9. Such a process offers the promise of considerable savings in chemical costs because of the low pH values at which precipitation is conducted. It can be superimposed upon the flow scheme of the conventional activated sludge processes if aeration basins of typical hydraulic residence time (~6 hr), with a considerable plug flow component (four compartments) are available. If such a process combining activated sludge treatment and low pH calcium phosphate precipitation is contemplated, sludge retention in the secondary clarifier must be kept to an absolute minimum to prevent dissolution of calcium phosphate in the low pH environment produced in unaerated concentrated activated sludge by CO_2 release from sludge respiration.

The possibility of precipitating calcium phosphate at pH values below 8 to produce orthophosphate residuals of below 1 mg/ℓ P in hydraulic residence times typical of the activated sludge process provides a rational explanation of the excess or "luxury" uptake of phosphate observed in some activated sludge plants [9,10,11]. Many of the observations associated with the phenomenon of "luxury" uptake are explained by this work. Thus, luxury uptake takes place at pH values of 8 or below; it takes place in long "plug flow" basins, but not in "completely mixed" basins; there is a release of phosphate to solution in secondary clarifier sludge blankets and when waste activated sludge is returned to primary sedimentation basins.

References

1. W. Stumm and J. O. Leckie, "Phosphate Exchange with Sediments: Its Role in the Productivity of Surface Waters," paper presented at the 5th Intern. Water Pollut. Research Conference, San Francisco, 1971.

2. A. B. Menar, "Calcium Phosphate Precipitation in Wastewater Treatment," Ph.D. thesis, University of California, Berkeley, Sept. 1972.

3. A. G. Walton, W. J. Bodin, H. Furedi, and A. Schwartz, "Nucleation of Calcium Phosphate from Solution," Canadian J. Chem., 45, 2695, 1967.

4. H. Bassett, Jr., J. Chem. Soc., 111, 620, 1917.

5. Y. Kauko and S. Eyubi, S. Tekn. Förén Finland Förhandl, 263, 1955.

6. F. G. Zharovskii and T. Komissi, Analit Khim Akad. Nauk, SSSR, 3, 301, 1951.

7. J. F. Ferguson, D. Jenkins, and W. Stumm, "Calcium Phosphate Precipitation in Wastewater Treatment," Chem. Engr. Prog. Symp. 67, 107, 279-287, 1970.

8. J. F. Ferguson, D. Jenkins and J. Eastman, "Calcium Phosphate Precipitation at Slightly Alkaline pH Values," in press, J. Water Pollut. Control Fed. Presented at 44th Ann. Conf. of Water Pollut.Control Fed., Oct. 1971.

9. A. B. Menar and D. Jenkins, "The Fate of Phosphorus in Sewage Treatment Processes, Part II - Mechanism of Enhanced Phosphate Removal by Activated Sludge," SERL Rept. 68-6, Univ. of California, Berkeley, 1968.

10. D. Vacker, C. H. Connell, and W. M. Wells, "Phosphate Removal Through Municipal Wastewater Treatment at San Antonio, Texas," J. Water Pollut. Control Fed., 39, 750, 1967.

11. R. D. Bargman, J. M. Betz, and W. F. Garber, "Nitrogen-Phosphate Relationships Obtained by Treatment Processes at the Hyperion Treatment Plant," in Advances in Water Pollution Research, 1, 1971.

LOGICAL REMOVAL OF PHOSPHORUS

Irving Yall, William H. Boughton, Frank A.
Roinestad, and Norval A. Sinclair

Department of Microbiology and Medical Technology,
The University of Arizona, Tucson, Arizona 85721

Introduction

Phosphorus is most frequently removed from waste waters by the use of activated sludge, chemical treatment, or some combination of the two. Chemical treatment usually involves precipation of the phosphorus by the use of various metal ions such as aluminum, iron, or calcium. Jenkins, Ferguson, and Menar (1), in a review paper, discuss various aspects of these chemical methods.

Activated sludge treatment alone, as it is usually practised, is frequently unable to remove sufficient phosphorus from its effluent to control algal blooms when the element is the limiting nutrient. However, sludges from plants in a number of cities throughout the United States have been reported to have high affinities for phosphorus and to remove at least 80% of this element when it occurs in normal amounts in their natural waste waters. A study of high phosphorus affinity sludges from plants located at San Antonio, Fort Worth, and Amarillo, Tex. and Baltimore, Md. has been reported by Witherow (2). Sludge from the Hyperion treatment plant at Los Angeles, Calif. has also been reported to have high phosphorus affinity (3). The ability to remove phosphorus in the field seems to be governed by the amount of sludge solids, the efficiency of aeration, the amount of time the waste water is exposed to the sludge under the most optimal conditions of aeration and agitation, and the disposal of the waste sludge which should be subjected to a minimal amount of anaerobiosis.

The mechanisms by which sludges remove amounts of phosphorus in excess of their metabolic requirements remain to be fully elucidated. Waste waters usually are relatively low in sources of carbon; therefore, microbial growth is relatively limited and slow (4). It is generally assumed that 2 to 3% P content in

sludge indicates metabolic requirements and that greater amounts
represent enhanced uptake. Sludge from the Rilling Road plant,
San Antonio, Tex. has been reported to attain total phosphate
compositions of 6 to 8% P on a dry weight basis (5).

Studies of the metabolic factors affecting enhanced phosphorus uptake by Rilling sludge were reported by Boughton et al.
in 1971 (6). In the laboratory, this sludge had the capability
of removing all of the phosphate normally found in Tucson sewage
(about 30 mg/l) in less than 3 hr. Removal was independent of
externally supplied sources of energy or ions, since added
orthophosphate and $H_3{}^{32}PO_4$ radioactivity were readily removed
from tap water, glass-distilled water, and deionized water. The
uptake had an optimum temperature range (24 to 37C) and an optimum pH range (7.7 to 9.7). It was inhibited by $HgCl_2$, iodo-
acetic acid, p-chloromercuribenzoic acid, NaN_3, and 2,4-dinitro-
phenol. Uptake was inhibited by 1% NaCl but was not affected
by 10^{-3} M ethylenediaminetetraacetic acid.

The experiments described in this report represent further
efforts to ascertain the mechanisms by which Rilling sludge removes phosphorus from its medium. The action of this sludge
will be compared to that from the plant located at Tucson, Ariz.
The latter sludge has lower phosphorus uptake capability but removes the element by biological means (7).

Methods

Sludge from the Rilling Road plant at San Antonio, Tex. was
concentrated by filtration at the plant and shipped to Tucson
overnight by surface carrier. The sludge was stored at 4C until
needed, usually within 1 week after collection. It was diluted
with tap water to the desired concentration immediately before
using. Tucson return sludge was stored in the same fashion.

In the experimental procedure, 10 ml of sludge in the
desired concentration (as determined by dry weights) was added
to 20 ml of fresh raw sewage from the primary clarifier at the
sewage treatment plant in Tucson. Inhibitors and $H_3{}^{32}PO_4$ (New
England Nuclear, Boston, Mass.)were added to the sewage, when
required, prior to the addition of the sludge. The mixtures
were contained in Kimax tubes (38 X 200 mm), aerated from the
bottom at the rate of 0.8 liter of prewet air per min and incu-
bated at 25C. At intervals any sludge adhering to the sides of
the vessel was removed with a spatula and returned to the mixture. After 3 hr, the mixtures were poured into centrifuge
tubes which were centrifuged in the cold at 27,000 X g for 10
min. The supernatant fractions were separated from the pellets
and assayed for ^{32}P radioactivity, when required, with the aid
of a model 3320 Tri-Carb liquid scintillation counting system
(Packard Instrument Co., Downers Grove, Ill.) using techniques
previously described (8) and chemically for orthophosphate (7).
All counts were corrected for decay.

Adenosine triphosphate (ATP) was extracted from the pellets
by the method of Patterson, Brezonik and Putnam (9) using tris

TABLE 1

Uptake of Phosphate by Activated Sludges

Sludge Origin	Treatment	PO$_4$ Removed (mg/l)	Uptake PO$_4$ (mg/l) per mg (dry wt) Sludge
San Antonio (Rilling)	Tucson sewage	155	1.45
	Tucson sewage + iodoacetate	40	0.37
	Tucson sewage + 2,4-dinitrophenol	dumped (+15)	dumped
Tucson	Tucson sewage	28	0.47
	Tucson sewage + iodoacetate	10	0.17
	Tucson sewage + 2,4-dinitrophenol	dumped (+10)	dumped

Approximately 107 mg (dry wt) of San Antonio sludge was added to Tucson sewage. The mixed liquor contained about 172 mg/l of phosphate. Approximately 60 mg (dry wt) of Tucson sludge was added to Tucson sewage. The mixed liquor contained about 50 mg/l of phosphate. The mixtures were aerated at 25 C, at pH 8.0, in the presence or absence of 10^{-3}M inhibitor.

buffer. The ATP content was determined in sludge extracts that did not contain radioactivity by pipetting samples into luciferin-luciferase preparations (Sigma Chemical Co., St. Louis, Mo.; reference 9). Light emissions were read with the aid of a Packard Tri-Carb liquid scintillation counting system (model 314 Ex-2). Standard solutions of ATP were used each time an unknown series was read.

The debris resulting from the tris procedure was extracted with 10 ml of 1 N HClO$_4$ at 5C for 18 hr and then with 5 ml of 1 N HClO$_4$ at 70C for 45 min. This modified procedure of Ogur-Rosen (10) extracts ribonucleic acid (RNA), the deoxyribonucleic acid (DNA) components, polyphosphates, and orthophosphates resulting from hydrolysis of nucleic acids especially DNA (11).

Radioactive organic components were separated from orthophosphate and polyphosphate in the tris extracts of sludge labelled with ^{32}P by the use of Norit A (12). The disappearance of ATP was measured at 260 nm with a Beckman DU Spectrophotometer. The ATP was eluted from the charcoal by the use of 5 ml of a mixture of ethanol (15 ml of 95%) and ammonium hydroxide (85 ml of 58%). Samples from the various fractions were applied to Whatman no. 1 chromatography paper and developed in an ascending mode in a mixture of isopropanol (75 ml), distilled water (25 ml), trichloroacetic acid (5 g) and ammonium hydroxide (0.2 ml; reference 13). By using the proper fractionation procedures, this solvent can distinguish between orthophosphate, polyphosphates, ATP, RNA and nucleotides. The positions of the ^{32}P containing compounds on the paper were determined by scanning with a Geiger-Muller counting tube. The locations of organic phosphorus containing compounds were confirmed by absorption of short wave ultraviolet light. The amount of radioactivity was measured quantitatively by eluting the spots with distilled water and then assaying with the aid of the scintillation counter.

During the course of the experiments, drops of sludge were spread thinly on standard glass microscope slides and stained for volutin by the use of Neisser's stain. The solution of methylene blue plus gentian violet stains the granules black and the chrysoidin solution stains the cells yellow (14). The smears were observed at 400 X (high dry) or 1,000 X (oil immersion) by using bright-field microscopy. Photographs were taken with a Leitz Orthomat microscope camera using Kodak Panatomic-X film.

Results

Table 1 shows the efficiency of Rilling sludge for phosphate removal from Tucson sewage as compared to Tucson sludge. The presence of 10^{-3} M iodoacetate resulted in about 76% inhibition of phosphate uptake for Rilling sludge and 64% for Tucson sludge.

Table 2 shows the ATP content of the sludges. The 3 hr sewage feeding did not result in any increase in ATP. However, treatment with antimetabolites resulted in definite decreases. On a per mg (dry wt) basis, the ATP content of both sludges were similar with Rilling having a value of 0.35 μg/mg and Tucson having a value of 0.33 μg/mg.

Table 3 shows the distribution of ^{32}P radioactivity among the various fractions extracted from the sludges. The distributions were very similar.

Table 4 shows the per cent total activity in the various products extracted from the sludges and separated by paper chromatography. There are only minor differences between the 2 sludges. The distributions are not entirely accurate since some of the orthophosphate was derived from DNA hydrolyed by the extraction method. The elution from charcoal is not quantitative. The "unrecovered" category represents radioactivity that was absorbed by the Norit and not eluted.

Studies in our laboratory using ^{33}P and autoradiographic techniques (not shown) indicated that San Antonio sludge accumulated phosphorus in distinct areas rather than uniformly throughout the floc. In subsequent experiments, both sludges were treated in various ways, spread on microscope slides, and stained for volutin.

TABLE 2

Adenosine Triphosphate (ATP) Content of Activated Sludges

Sludge Origin	Treatment	Total ATP Content (µg)
San Antonio (Rilling)	Starved	37.5
	Tucson sewage	37.8
	Tucson sewage + iodoacetate	27.6
	Tucson sewage + 2,4-dinitrophenol	15.0
Tucson	Starved	20.0
	Tucson sewage	20.0
	Tucson sewage + iodoacetate	5.0
	Tucson sewage + 2,4-dinitrophenol	5.0

Approximately 107 mg (dry wt) of San Antonio sludge or 60 mg (dry wt) of Tucson sludge were aerated at 25C, at pH 8.0, in Tucson sewage for 3 hr in the presence or absence of 10^{-3} M inhibitor.

Figure 1A shows a photomicrograph of volutin stained Rilling sludge taken (using a high-dry objective lens) 3 days after collection and prior to being exposed to a fresh source of phosphorus. The flocs have numerous dark masses. The Tucson sludge (Fig. 1B), photographed under the same conditions, also show volutin bodies. These appear to be less intense and more widely dispersed than those in the San Antonio sludge.

Figure 2 shows photomicrographs of volutin stained San Antonio sludge flocs taken using an oil immersion lens. The volutin staining material seems to be present in coccoidal-rod shaped cells which occur in pairs and appear to be 1-2 µ in length. These cells are Gram-negative. They are found in the sludge either as parts of larger flocs or as numerous sub-flocs that resemble grape clusters. These "grape clusters" are readily separated from the other components of the sludge.

Figure 2A shows details of a "grape cluster" from the previously shown slide of San Antonio sludge (Fig. 1A). Figure 2B shows the appearance of the volutin staining material after the sludge was exposed to Tucson sewage for 3 hr. The amount of material seems to have increased within the cells.

TABLE 3

Occurrence of ^{32}P Radioactivity in Various Fractions
Extracted From Activated Sludges

Sludge	Solvent	Radioactivity (counts/min)	% of Total Radioactivity
San Antonio	None	80,100,000	100
(Rilling)	Tris buffer	54,470,000	68
	1 N HClO$_4$ (cold)	28,840,000	36
	1 N HClO$_4$ (hot)	3,200,000	4
Tucson	None	49,000,000	100
	Tris buffer	32,350,000	66
	1 N HClO$_4$ (cold)	17,650,000	36
	1 N HClO$_4$ (hot)	1,960,000	4

Approximately 105 mg (dry wt) of San Antonio sludge was added to
Tucson sewage. The mixed liquor contained about 190 mg/l of
phosphate. About 67 mg of Tucson sludge was added to Tucson
sewage. The mixed liquor contained about 65 mg/l of phosphate.
Approximately 89,000,000 counts of $H_3{}^{32}PO_4$ per min were introduced per experiment. The mixtures were aerated at 25 C at
pH 8.

TABLE 4

Per Cent Total Activity in Products Resulting From Extraction
of ^{32}P Containing Materials From Activated Sludges

Sludge	Product	Activity (%)
San Antonio	Orthophosphate	40
(Rilling)	Polyphosphate	10
	ATP	18
	Nucleic Acids	10
	Unidentified	5
	Unrecovered	15
Tucson	Orthophosphate	46
	Polyphosphate	8
	ATP	10
	Nucleic acids	10
	Unidentified	17
	Unrecovered	13

See Table 3 for experimental conditions and amounts of radioactivity.

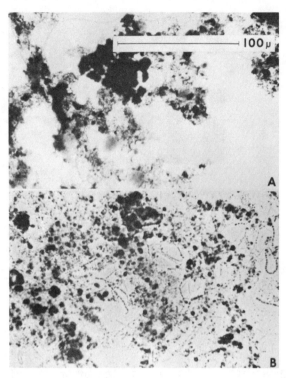

FIG. 1.

Photomicrographs of Volutin Staining Bodies in
Starved Activated Sludges (photographed with 40 X
objective). (A) San Antonio; (B) Tucson.

For the amount of phosphate taken into the sludge, see Table 1.
Figure 2C shows a "grape cluster" after treatment with 10^{-3} M
iodoacetate (Table 1). The concentration used of this antimetabolite incompletely inhibited uptake of new phosphorus.
Many of the volutin staining particles appear to be of the same
size as those in Figure 2A. Figure 2D shows volutin staining
material after treatment with 10^{-3} M 2,4-dinitrophenol. This
antimetabolite completely inhibited the uptake of new phosphorus
and caused some of the phosphorus previously stored in the sludge
to dump. The volutin staining particles appear to be markedly
smaller than those in the other photomicrographs.

     Figure 3 shows photomicrographs of volutin stained Tucson
sludge flocs taken using an oil immersion lens. In this sludge,
the volutin staining material appears to occur more often as
part of a larger floc than does the San Antonio material.
"Grape clusters" are relatively rare.

     Figure 3A shows magnified detail from the previously shown
slide of Tucson sludge (Fig. 1B). Several sizes of volutin
staining particles are present. Figure 3B shows the appearance
of a "grape cluster" after the sludge was exposed to Tucson
sewage for 3 hr (Table 1). The granules appear larger and more
deeply staining. Figure 3C shows the appearance of a "grape
cluster" after the sludge was treated with iodoacetate (Table 1).
The volutin particles are smaller than those in Figure 3B and
approximately the size of the larger ones in Figure 3A. Figure
3D shows the appearance of volutin staining material after

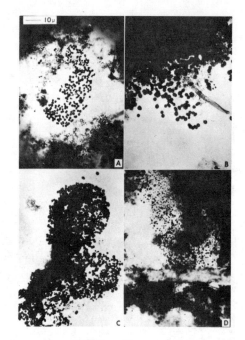

FIG. 2

Photomicrographs of Volutin Staining Bodies in San
Antonio Sludge (photographed with 100 X objective).
(A) Starved; (B) Tucson sewage, 3 hr; (C) Tucson
sewage + 10^{-3} M iodoacetate, 3 hr; (D) Tucson sewage
+ 10^{-3} M 2,4-dinitrophenol, 3 hr.

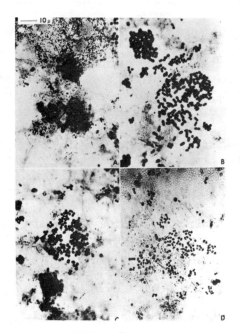

FIG. 3

Photomicrographs of Volutin Staining Bodies in
Tucson Sludge (photographed with 100 X objective).
(A) Starved; (B) Tucson sewage, 3 hr; (C) Tucson
sewage + 10^{-3} M iodoacetate, 3 hr; (D) Tucson sewage + 10^{-3} M 2,4-dinitrophenol, 3 hr.

the sludge was treated with 2,4-dinitrophenol. The volutin particles appear to have decreased in size.

A parallel series of slides were stained for inorganic
phosphate by immersing the smears in lead nitrate and then in
ammonium sulfide (15). The results (not shown) indicated that
the volutin staining material contained inorganic phosphate.

Discussion

The figures for ATP of 0.35 µg/mg for San Antonio and 0.33
ug/mg for Tucson sludge are very similar to that of 0.34 mg
ATP/g VSS reported by Weddle and Jenkins (4) for their initial
sludge. If 2 µg/mg of cell material is assumed to be the value
representing the amount of ATP present in living material, then
the sludges contain about 17.5% of viable cell mass. It has
been reported that only from 15-20% of the MLVSS of sludge from
the University of Florida contact stabilization plant may be
active biomass (9).

The higher phosphorus affinity of San Antonio sludge as
compared to that from Tucson cannot be accounted for on the
basis of the amount of ATP present or incorporation into RNA
and DNA (11). The difference seems to lie in the number of
volutin forming organisms in the two sludges. The photomicrographs in this report indicate that there is a relationship
between the amount of phosphorus taken up by the sludge flocs
and the density of the volutin staining material. According
to Widra (16), volutin in Aerobacter aerogenes is composed of
polyphosphates in association with lipoprotein, RNA and Mg++.

Volutin seems to be formed in microorganisms when they
are subjected to some form of nutritional imbalance in the
medium such as deprivation of nitrogen or sulfur (17) or
phosphorus (8). What the deprivation is in sewage remains to
be ascertained. A logical pathway to account for the uptake of
phosphorus in sludge as suggested by our observations and those
of Harold for some bacteria (18) is:

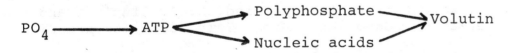

These reactions are subject to reversal as conditions change.

Conclusions

The uptake of phosphorus, in the laboratory, by sludge from
San Antonio, Tex. and Tucson, Ariz. is inhibited by antimetabolites and, therefore, is biological in nature.

According to ATP measurements, both sludges are composed of
about 17.5% active biomass. The difference in phosphorus uptake
between the two sludges cannot be accounted for on the basis of
differences in ^{32}P radioactivity distribution among various
chemical components.

The phosphorus is incorporated into volutin staining bodies in the sludge that frequently take the form of grape clusters. The volutin staining bodies appear to increase in size under conditions when the sludge accumulates phosphorus and decrease when the sludge releases phosphorus. The San Antonio sludge seems to have a greater number of these bodies than does Tucson sludge.

References

1. D. Jenkins, J. F. Ferguson and A. B. Menar, Water Research, 5, 369 (1971).

2. J. L. Witherow, Proc. 24th Ind. Waste Conf. p. 1169. Purdue University, Lafayette, Ind. (1969).

3. R. D. Bargman, J. M. Betz and W. F. Garber, Water-1970. Chem. Eng. Prog. Symp. Ser. 67, 117 (1971).

4. C. L. Weddle and D. Jenkins, Water Research, 5, 621 (1971).

5. W. N. Wells, J. Water Pollut. Contr. Fed. 41, 765 (1969).

6. W. H. Boughton, R. J. Gottfried, N. A. Sinclair and I. Yall, Appl. Microbiol. 22, 571 (1971).

7. I. Yall, W. H. Boughton, R. C. Knudsen and N. A. Sinclair, Appl. Microbiol. 20, 145 (1970).

8. F. A. Roinestad and I. Yall, Appl. Microbiol. 19, 973, (1970).

9. J. W. Patterson, P. L. Brezonik, and H. D. Putnam, Environ. Sci. Technol. 4, 569 (1970).

10. M. Ogur and G. Rosen, Arch. Biochem. 25, 262 (1950).

11. W. H. Boughton, R. J. Gottfried, N. A. Sinclair and I. Yall, Water-1971. Chem. Eng. Prog. Symp. Ser. (in press).

12. R. K. Crane, Science 127, 285 (1958).

13. Y. Volmar, J. P. Ebel and B. Jacob, Compt rend. de l'Acad. des Sciences 235, 372 (1952).

14. H. J. Conn, J. W. Bartholomew and M. W. Jennison, Manual of Microbiological Methods, p. 10. McGraw-Hill, Inc. N. Y. (1957).

15. J. P. Ebel, J. Colas and S. Muller, Exp. Cell. Res. 15, 28 (1958).

16. A. Widra, J. Bacteriol. 78, 664 (1959).

17. I. W. Smith, J. F. Wilkinson and J. P. Duguid, J. Bacteriol. <u>68</u>, 450 (1954).

18. F. M. Harold, Bacteriol. Rev. <u>30</u>, 772 (1966).

ACKNOWLEDGMENT

We are grateful to W. N. Wells of San Antonio, Tex. for providing samples of Rilling sludge.

Applications of New Concepts of Physical-Chemical
Wastewater Treatment
Sept.18-22, 1972

PHOSPHOROUS REMOVAL BY CHEMICAL AND
BIOLOGICAL MECHANISMS

William F. Garber
Chief Engineer
Sewage Treatment Division - Bureau of Sanitation
City of Los Angeles

Introduction

Professionals working in the field of wastewater treatment and
disposal are well aquainted with both the positive and negative
effects of the so called "Environmental Revolution". As workers
who have dedicated their careers to environmental improvement it
is at times exhilarating for them to be at the center of a problem of national interest and concern; but it is also difficult
to be subjected to the apparent necessity for finding scapegoats
and for apply tremendous publicity, which often results in more
heat than light, to subjects of concern. To point up the problem, two instances are cited. One involved the Specialty Conference of the American Society of Civil Engineers held in Los
Angeles on March 21-24, 1971 and entitled "Clean Water For Our
Future Environment". Here the executive secretary of one of the
important conservation groups said, after hearing the concern of
the engineers for the environment and learning of what they had
and were doing to balance the needs of our civilization against
ecological needs, that perhaps it was not the engineers who were
responsible for present environmental degradation but someone
was and organizations of his type were out to find out who. (1)
The second was a statement in "Highlights", a publication of the
Water Pollution Control Federation, in which the officials of
one of the control agencies were quoted as not being interested
in engineers and scientists but rather in generalists who could
make decisions even though information was not available. (2)
Since the amounts of capital and operation and maintenance funds
involved in most treatment and disposal problems are very high
and since environmental impact may not be known, this approach
appears to be unsatisfactory. A decision may be necessary

before all facts are known; but certain basic facts are required
and research must be continued until these are apparent. It is
the writer's belief that the Lake Tahoe installation in his state
is a prime example of leaping before looking. A great deal of
money and effort has gone into an installation to prevent the discharge of nitrogen and phosphorous to the lake when at least one
prominent limnologist - ecologist has pointed out that these elements
are not the problem at all. (3) The whole question of phosphorous
removal has been clouded by this same type of confusion. Phosphorous, nitrogen, vitamin B12, silica and certain other nutrients
most probably should not be removed from wastewaters entering
the ocean. The surface waters of the ocean are deficient in
nutrients and the fertility of an area is dependent upon a sufficient nutrient content. For example, in Southern California all
streams have been dammed and/or otherwise controlled to short
annual flows because of ground water removal. The coastal waters
are now dependent upon wastewater discharges for their nutrients
and there is a large body of evidence which indicates that the
amount now received is about equal to that formerly carried in
when streams were perennial* There is no evidence of an overall
phosphorous problem, but rather specific phosphorous problems
at specific locations based upon specific environmental conditions.
It does not make sense to always design to remove phosphorous unless it is certain that the effect will be measureably beneficial.

In addition to the fact it would appear to be necessary to determine whether phosphorous removal is indicated before requiring
its removal, is the fact that we do not at this time really know
the best method of removal given present information. The literature contains data showing successful removal using both chemical and biological methods. (4,5,6,7) Biological methods are
considered by many to be unreliable since the exact mechanism
is probably not known and some workers still believe the phenomenon to be based upon stoichiometric factors. Chemical methods
are very high energy methods; and, based upon operation of a
10 MGD chemical treatment plant in Los Angeles for three years,
(8) difficult to control. It is the writer's opinion that basic
research covering the biological method is badly needed. As a
general rule the biological processes have the least energy use
and thus the least environmental impact and pollution potential.
Inasmuch as the biological method is known to be successful in
a number of operating facilities, it would appear to be important
to determine the basic mechanisms so that the engineer would
have another process to consider. The bulk of the paper will
then consider phosphorous removal at Hyperion with factors considered to be important.

*Imel, C.E. "Nutrient Assessment Littoral Waters, Southern
 California". U. S. Naval Research Laboratory Port Hueneme,
 California, (Personal Communication) 1972.

Discussion

A. Hyperion Results

As noted above the literature includes reports of several biological treatment (activated sludge) facilities, processing
large flows, in which high phosphorous removals are obtained
(4) (5) (6). However, the exact mechanism of such removal is
not clear and it is often stated that the phenomenon is sporadic
and not subject to the type of control needed to assure reliable
treatment. There also appears to be a strong difference of
opinion between workers who believe that the activated sludge
process simply changes parameters such as pH thus allowing a
stoichiometric precipitation; and investigators who believe that
under certain conditions there is an accumulation of phosphorous within the bacterial cells themselves. Work done by
the City of Los Angeles does not allow such questions to be
completely resolved; but it has been clearly demonstrated in
the activated sludge portion of the Hyperion Treatment Plant
that reductions of phosphorous from about 10 mg/liter to
$<$ 2 mg/liter are routinely achieved. As will be developed, a
biological process as opposed to chemical treatment probably
uses the least energy and results in the least environmental
impact. It therefore appears logical to expend research effort
to determine the mechanism of removal so that the environmentally
most desirable alternate can be considered for any location.
From the viewpoint of at least some operators the present pressure to adopt only chemical processes is an attempt to limit
the amount of capital costs which may be partly reimbursed by
Federal or State grants by passing such costs on to local
authorities in the form of high operational expenses. The
apparent reluctance to even consider the obvious success of the
low energy biological process at a number of locations makes
such a conclusion one obvious alternate.

Table 1 is a brief summary of data obtained on 100 MGD of flow
over a 16 month period in 1969-1970. One half of this flow
received air in a plug flow activated sludge configuration at
a rate of about 1.8 cubic feet per gallon with waste sludge
discharged directly to digestion and with an effluent phosphorous of 0.5 mg/liter. The other half received air at about
1.4 cubic feet per gallon with the waste sludge being returned
to primary sedimentation where a lower pH resolublized some of
the phosphorous and resulted in a recirculation load. Effluent
phosphorous was about 3.2 mg/liter. High removals on one half
of the flow were desirable at that time because of pilot scale
tests on water reclamation. Since then sludge return to primary sedimentation has continued and no consistent effort to
keep phosphorous removals high has been undertaken. Effluent
levels of the order of 3 mg/liter have been maintained. It is
possible to immediately return to the very high removal levels
by increasing the air and removing the recirculating load.

Ofcourse the phosphorous must then be precipitated from the
sludge; but the problem has been changed from one involving
100 MGD to one involving high phosphorous levels in about 400,000
gallons of sludge. A paper outlining some of the problems resulting from high phosphorous discharge to anaerobic digestion
and confirming the potential of recovery as magnesium ammonium
phosphate has recently been published (9).

It will be noted in Table 1 that overall removals, considering
the 50 MGD with removals to 0.5 mg/liter and the 50 MGD with
removals to 3.2 mg/liter, were about 77%. Removals of 95% were
obtained in the 50 MGD with higher air addition and no return of
phosphorous containing waste sludge to the system. If stoiche-
metric precipitation of the metallic phosphates indicated by the
change in the concentration of such metals through the activated
sludge step actually occurs, then the removals indicated in
Table 1 can be readily explained. However since Volutin granules high in phosphorous do occur within the activated sludge
cells, questions remain as to whether the important mechanism
is inorganic precipitation outside of the cells with subsequent
surface attachment or precipitation within the cells in either
an inorganic or polymeric form. At Hyperion some work was done
in attempting to wash the cells free of any inorganic precipitates and then of lysing them to see what phosphorous had
accumulated inside. This amounted to 6.5 - 7.5% by weight. Because of this we tend to believe that the uptake is sludge
specific and to be either chemical with participation of living
organisms or purely biological. Just how and why cells carry
such a heavy phosphorous load needs to be determined. Work
by Harold (10) and by Yall et al (11)(12) appears to advance
the levels of knowledge and to indicate some interesting factors
covering the uptake of phosphorous in biological systems. Thus:

1. Pure polymeric phosphorous is acid-soluble regardless of
    chain length.   It will combine with and precipitate
    positively charged macro-molecules such as proteins and can
    exist in living cells as a polymeric phosphorous complex
    with proteins or nucleic acids. The complex is apparently
    dependent upon Mg++ serving as a bridge.

2. Polymeric phosphorous is deposited in the cells in volutin
    granules which possibly include RNA, lipids, protein and
    Mg++.

3. The reaction is thought to be

$$\text{ATP} + n(P) \xleftarrow{\text{Polyphosphate Kinase}} \text{ADP} + (P)\ n+1$$

    ATP = Adenosine Tri Phosphate
    ADP = Adenosine Di Phosphate
      P = Inorganic Phosphorous

TABLE 1 *

Phosphorous Removals April 1969 Through July 1970

Constitutent	Primary effluent concentration mg/liter	Secondary effluent concentration mg/liter	Removal through secondaries mg/liter	100 million gal./day flow — Removal through secondaries mm/liter	Removal through secondaries mm/liter	Calculated removal as P mg/liter
Calcium	63	58	5	0.125		3.88
Aluminum	0.86	0.18	0.68	0.125		0.78
Zinc	0.55	0.20	0.35	0.0054	0.0036	0.11
Iron	0.87	0.20	0.67	0.012	0.012	0.37
Phosphorous	7.90	1.80	6.10			

Calculated removal, mg/liter 5.14
Observed removal, mg/liter 6.10
Difference (metabolic?), mg/liter 0.96

*To obtain a stoichiometric balance, it was assumed
that phosphorous reacted with calcium, aluminum, zinc,
and iron ions in the following molar ratios:

```
Ca: P = 1:1
Al: P = 1:1
Zn: P = 1.5:1
Fe: P = 1:1
```

4. ADP inhibits the bio accumulation and ATP encourages it so the ATP/ADP ratio is very important.

5. The polymeric phosphate content of the cells is low during rapid growth and increases under conditions of nutritional imbalance unfavorable for growth. Addition of inorganic phosphorous to cells previously subjected to phosphorous starvation induces rapid and extensive accumulation of polymeric phosphorous.

6. In addition to the activity of polyphosphate kinase and the availability of ATP, polymeric phosphorous synthesis is proportional to the inhibition of concurrent nucleic acid synthesis. If nucleic acid snythesis ceases because of the absence of an essential nutrient, polymeric phosphorous degradation is inhibited and accumulation occurs within the cells.

7. The absence of Ca++ inhibited polymeric phosphorous synthesis whereas nucleic acid synthesis continued and the ATP/ADP ratio was lowered. Such cells then did not store phosphorous.

Given the fact that a stoichiometric balance appears to present with the probability that the phenomenon is also dependent upon biological activity, how much does the Hyperion experience contribute towards a solution of the problem of mechanism? Table 2 illustrates a number of long term ratios of phosphorous to elements which might be of importance in the uptake mechanisms in Hyperion wastewaters.

Table 3 shows some operational parameters found when phosphorous removals were very high.

A plug flow configuration was used, much of the BOD or COD had been utilized, nitrification was well underway, and the organisms were in the so called endogenous respiration stage of growth. Phosphorous uptake was very rapid at the beginning of the plug flow tank and some leak back occured at the end in tanks with long sludge ages. These operational conditions would then appear to point towards some of the growth considerations pointed out by Harold (10) and Yall et al (11) (12) and already noted. The organisms have been carried into the endogenous respiration phase where the substrate has been stripped of most of its nutrient value and the bacteria are working on their own body protoplasm. They then are settled in secondary sedimentation tanks where any possible further nutrients are removed and sustinence growth must continue to come from cell tissues themselves. They are then returned to a rich substrate at the head of the aeration system in a starved condition. Rapid accumulation of phosphorous occurs.

TABLE 2

Ratios of Selected Elements to Phosphorous

Sample	Carbon	Organic N	Ammonia N	Phosphorous*
Raw influent	28	7.4	1.9	1
Primary effluent	18	2.6	1.7	1
Secondary effluent	9.5	1.4	1	1

Sample	Calcium	Magnesium	Phosphorous*
Primary effluent	8		
* Primary efflunet		2	1

Sample	Phosphorous values
Raw influent	10 mg/liter
Primary effluent	9.5 mg/liter
Secondary effluent (avg. for 100 million gal./day)	1.8 mg/liter

TABLE 3

Values Of Selected Operational Parameters For The Activated Sludge Process

Showing High P Removal

Parameter	Value
Process loading factor (14)	~ 0.6 lb. COD applied/lb. M L V S S
Net growth rate (14)	~ 0.1 days^{-1}
Air-wastewater ratio	> 1.6 cu. ft./gal.
S. D. I.	> 1.0
Secondary effluent suspended solids	< 10 mg/liter
Aeration tank pH	> 7.4
M L V S S	< 70%
Insoluble phosphate return sludge	> 5%
Effluent nitrate - N	Nutrification well underway
Sludge Age	9 - 10 days
Air rate, cu. ft./lb. BOD Removed	1220

If this model is true, it is obvious that both process design (e.g. plug flow) and process operation are important.

In line with the argument that biological phosphorous removal is erratic and that chemical treatment must be used to assure continuing removal at all locations is the fact that, although the biological process worked at Hyperion, it did not work at an upstream plant in the Los Angeles system nor in plants of the Los Angeles County Sanitation Districts. A laboratory activated sludge facility was constructed in the Hyperion laboratory to further investigate this phenomenon. Wastewater was brought on a daily basis from the Whittier Narrows plants of the Los Angeles County Sanitation Districts and the laboratory installation operated on this substrate alone. The following was found:

1. Operation using the parameters of Table 3 gave no phosphorous removal.

2. Addition of Hyperion activated sludge initially showed high phosphorous removal followed by a drop back to none.

3. Pilot plant activated sludge built from Whittier Narrows substrate and added to a similar pilot plant using Hyperion substrate did not remove phosphorous until it had been diluted in operation by the sludge built from Hyperion substrate.

4. An adjustment to the Ca: P ratio of 8:1 found at Hyperion was made with no phosphorous removal. This was then allowed to return back to the ratio normal to Whittier Narrows.

5. A similar Mg: P ratio adjustment to the 2:1 found at Hyperion was made with no phosphorous removal.

6. An adjustment of the C:P ratio of 18:1 found at Hyperion was made and immediate removal of phosphorous to levels similar to those at Hyperion was found.

The above appeared to indicate that perhaps plug flow, operation and a proper C:P ratio was needed and that a biological reaction was definitely present. The necessity for a proper C:P ratio is suggested by the work of Harold (10), Yall et al (11)(12), and Sekikawa et al (13) as well as others. In the pilot plant this was accomplished with glucose. In a large plant a less expensive source such as methanol might be satisfactory. The Hyperion work has not progressed beyond this point; but it would appear that the possibility for a viable controllable biological process is indicated and that the necessity for limiting environmental impact by using low energy biological methods would necessitate further research by the control agencies.

B. Energy Considerations

During the period 1954 through 1958 the City of Los Angeles
operated its Valley Settling Basin facility as a chemical treatment plant. Flows averaged 10 MGD with peaks to 20 MGD.
Tables 4, 5 and 6 summarize operational data and costs as of
1955. It will be noted that with suspended solids removals of
90% only 65% of the BOD was removed and coagulant dosages of
350 mg/liter of aluminum sulphate and 2-3 mg/liter of polymer
were required. In addition it was found necessary to continuously add chemicals at a rate sufficient to meet the highest
demand expected since there was no reliable way to match demand
and dosage in an operating plant. This is difficult enough to
do in a water treatment plant where changes in raw water composition are small. In a wastewater facility where changes in
composition are large the matching of demand and dosage is
almost impossible. It will also be observed that effluent
suspended solids averaged more than 40 mg/liter. The Los Angeles
experience with chemical treatment can be summarized by the
following:

1. Very high operating cost.

2. Unreliable in that coagulant matching to demand was very
    difficult.

3. Effluent substantially less satisfactory than biological
    facility of the same size.

4. Degraded effluent in that the total dissolved solids were
    increased by the conditioning chemicals and most commercial
    conditioning chemicals contain amounts of heavy metals
    which may substantially increase toxicant loads.

5. Very high energy usage process.

6. Unreliability of chemical supply. The 10 MGD plant was
    using all of the aluminum sulphate produced west of the
    Rocky Mountains except for the small amounts previously
    committed to municipal water treatment operations. Sub-
    stitution of ferric chloride was not possible because it
    would have reduced effluent pH to about 2.

Table 7 is a summary of tests made in 1972 of extended primary
treatment using inorganic coagulants and organic polymer
flocculants. It will be noted that required dosages were
similar to those found necessary at the Valley Settling Basin
during its period of operation (Table 4). The high alkalinity
of the Hyperion influent, 350-400 mg/liter, means that a high
inorganic coagulant dosage must be added. Comparison of the
California dishcarge requirements (Table 7) with the results
obtained show the activated sludge effluent to be at least
equal in quality in terms of the heavy metals and organics
and with a lower total dissolved solids. Use of commerical
grade ferric

TABLE 4
Basic Design and Operating Data

Item	Design	Operation
Peak flow rate (c.f.s.)	30	45
Detention at maximum flow (hr.)	1	0.67
Suspended solids (p.p.m.)	430	516
Removal (%)	80–90	90
B.O.D. (p.p.m.)	250	435
Removal (%)	75–80	65
Chemical dosage (p.p.m.):		
Ferric Sulfate	200	400
Aluminum Sulfate	–	350
Synthetic coagulant*	–	2–3
Prechlorination	5	0
Postchlorination	25	18

*Dow polymer Separan 2610

TABLE 5

Operating Data for the Valley Settling Basin,
Los Angeles, California, for the Fiscal Year
1954-55 and the Period July 1, 1955 to
December 31, 1955

Item	Fiscal Year 1965-55	Six months July-Dec., 1955
Flow (m.g.d.)	3.5	7.2
Suspended solids:		
Influent (p.p.m.)	595	516
Effluent (p.p.m.)	42	49
Removal (%)	92.7	90.5
B.O.D.:		
Influent (p.p.m.)	334	435
Effluent (p.p.m.)	121	155
Removal (%)	63.8	64.3
Chemicals used:		
Aluminum sulfate (lb./mil. gal)	2,940	2,810
Chlorine (lb./mil. gal.)	139	147

TABLE 6

Operating Costs for the Valley Settling
Basin, Los Angeles, California, for the
Fiscal Year 1954-55 and the Period July 1,
1955 to December 31, 1955

Item	Fiscal Year 1954-55		Six Months July-Dec., 1955	
	Total ($)	Per Mil Gal ($)	Total ($)	Per Mil Gal ($)
Salaries	42,947	37	29,943	22
Chemicals	81,997	70	94,187	71
Miscellaneous	6,491	6	2,750	2
Total	131,435	113	126,880	95

TABLE 7

1972 Tests Extended Primary Treatment – Hyperion Treatment Plant

Item	Primary Effluent	Activated Sludge Effluent	Fe Cl3 +A 23	Al2(SO4)3 +A 23	Lime +A 23	California Discharge Standards
Inorg. coag. mg/1	-	-	300	200	300	-
Polymer Floc. mg/1	-	-	1.5	7.2	1.2	-
BOD mg/1	170	11	145	131	124	≥ 85% Rem.
Sett. Sol. mg/1	0.6	< 0.1	< 0.1	< 0.1	< 0.1	< 0.1
Susp. Sol. mg/1	103	11	26	19	24	< 50
Turbidity JTU	60	2	11	11	8	< 50
pH	7.4	7.6	6.3	7.1	9.4	6-9
Ether Sol. mg/1	39	< 0.1	0.8	< 0.5	< 0.5	< 10
Arsenic mg/1	0.018	0.002	0.002	0.052	< 0.001	< 0.01
Cadmium mg/1	< 0.001	< 0.001	< 0.001	< 0.001	< 0.001	< 0.02
Chrome, Total mg/1	0.50	0.04	< 0.01	0.03	0.10	< 0.005
Copper mg/1	0.22	0.07	0.04	0.02	0.02	< 0.2
Lead mg/1	0.09	0.05	0.07	0.15	< 0.01	< 0.1
Mercury mg/1	0.012	< 0.002	0.007	0.006	0.002	< 0.001
Nickel (cannot meet)	0.25	0.22	0.31	0.26	0.23	< 0.1
Silver mg/1	0.002	0.001	0.002	0.001	0.001	< 0.02
Zinc mg/1	0.59	0.33	0.14	0.15	0.03	< 0.3
Cyanide mg/1	0.10	0.05	0.17	0.39	0.67	< 0.1
Phenols mg/1	0.20	0.006	0.12	0.48	0.12	< 0.5
Chlor. Hydra mg/1	?	0.010	?	?	?	< 0.002
TDS mg/1	888	820	894	862	910	--
NH3 mg/1	20	9	20	20	20	< 40
Phosphorous	7.9	0.4	1.1	0.1	4.4	-

chloride, aluminum sulphate, or lime would undoubtedly have increased the heavy metals more than the laboratory tests showed.
Phosphorous results were spotty. The activated sludge and alum
gave very good removals whereas the ferric chloride and lime
left fairly good residuals. It is believed that in the case of
these two coagulants the phosphorous and total dissolved solids
were increased by precipitate material which would not either
settle or filter.

Tables 4, 5, 6 and 7 provide data upon which energy costs at
Hyperion for chemical treatment can be estimated. The present
Hyperion energy usage for 100 MGD of activated sludge treatment
and 240 MGD of primary treatment is shown in Table 8. Table 8
also shows an estimate of the total energy need if the full
340 MGD is converted to secondary with operation such that
phosphorous removal to less than 0.5 mg/liter can be obtained.

Energy use for chemical treatment has not usually been considered. Because the three coagulants tested showed that similar
tonnages would be needed in each case and because lime treatment
would in any case be the least expensive, the energy estimate
of Table 9 was therefore prepared using lime.

It is realized that the values of Table 9 are rough estimates
and that the problem of organic sludge solids removal is not
considered in either of Tables 8 or 9. In addition it is not
intended that all uses of energy be included. However an
attempt was made to cover the major ones. It can be seen that
for 340 MGD of activated sludge at the high air rate of 3 cubic
feet per gallon there is an estimated energy expenditure of
about 900,000 KWH per day. At least 60% of this energy would
come from the anaerobic digestion of the sludge solids. For lime
treatment, which in Los Angeles City experience is a less reliable
process, energy use is also about 900,000 KWH per day and all must
come from fossil fuel sources since most of it is used in transportation and processing and the lime sludge is not suitable for
digestion. Also the finished effluent from the process shows
less efficiency in removing phosphorous, still contains 35% of
the original BOD, has a high pH, and has increased total dissolved solids. A careful study of alternate costs for chemical
versus activated sludge treatment made by the City Bureau of
Engineering for the Los Angeles system is shown in Table 10.
The table is based upon 6% interest and the assumption that
operation and maintenance costs would be continuous from now on.

Table 10 makes it clear that, although chemical treatment saves
on initial capital costs, it does so at the expense of operation
and maintenance costs. In the event chemical treatment is
directed as the least costly (initial capital) satisfactory
method; it in effect passes on to the operator and local authority very high costs which, when capitalized, show a much more

TABLE 8

Energy Use at Hyperion

SUPPLY

Digester Gas 10^6 Cu. Ft./Day	4.678
10^6 BTU/Day	2760
Diesel Fuel Gallons/Day	664
10^6 BTU/Day	89
Total (In House) 10^6 BTU/Day	2,849
Purchased KWH/Day	27,500
10^6 BTU/Day	292

USAGE

Digester Gas 10^6 Cu. Ft./Day	2.620
10^6 BTU/Day	1,546
Diesel Fuel Gallons/Day	664
10^6 BTU/Day	89

SOLD

Digester Gas

10^6 Cu. Ft./Day	2.137
10^6 BTU/Day	1,261

WASTE

Digester Gas

10^6 Cu.Ft./Day	0.045
10^6 BTU/Day	27

Extra Energy Needed to Treat Full 340 MGD at 3 Cu. Ft./Gallon

KWH/Day	370,000
10^6 BTU/Day	2050

Estimated Total for 340 MGD Activated Sludge

KWH/Day	894,000
10^6 BTU/Day	5,265

Slight Difference in Some Totals Arise From Instrument Errors.

TABLE 9

Estimated Energy Use for Lime Treatment of 340 MGD Flow at Hyperion (14) (15)

1. Lime dosage @ 200 mg/liter = 283.5 Tons/Day

2. Recalcining or incineration illegal in Los Angeles so 283.5 Tons/Day trucked in and out.

Mining, Crushing, Transport, Etc. 283.5 Tons/Day	=	44,000 KWH/Day
Calcining, Processing, Etc. 283.5 Tons/Day	=	765,000 KWH/Day
Transporting 283.5 Tons/Day IN*	=	48,000 KWH/Day
Transporting** 1130 Tons/Day OUT***	=	78,000 KWH/Day
Primary Treatment 340 MGD	=	2,000 KWH/Day
Total	=	937,000 KWH/Day

*   Assume 250 mi, train, 200 mi truck
**   Assume 75% moisture (lime wt. only)
***   Assume 200 mi truck. Recalcining or incineration illegal in much of California.

TABLE 10

Hyperion Treatment Plant - 340 MGD Flow Incremental Costs Over Present Treatment

	Construction	Operations & Maintenance	Capitalized
Activated Sludge	$ 70,000,000	$ 700,000/year	$ 82,000,000
Chemical Treatment	$ 13,000,000	$12,000,000/year	$213,000,000

costly method.

A problem which is usually not considered; but which must be carefully evaluated is that of supply and of transport logistics. As already noted, the 10 MGD Valley Settling Basin chemical treatment facility in Los Angeles previously used essentially all of the aluminum sulphate production available west of the Rocky Mountains. Supply was always tenuous with many instances of remaining chemicals on hand for only a few hours of operation. One of the vivid memories of the period was that of constant desperate searches for further supplies. In addition, when lime was added to aid fertilizer filtration during one period, the logistics of constantly bringing in and removing 5 to 10 tons a day became a major problem. Fortunately the fertilizer with lime was unusable on western soils so that phase ended. However constantly bringing in and removing over 280 tons per day of a chemical is not a prospect we relish.

Summary and Conclusions

At the present time there are two known methods for the removal of phosphorous. One is based upon stoichiometric precipitation of phosphates using coagulants such as lime, aluminum sulphate or ferric chloride aided by organic polymer flocculants. The other is based upon the demonstrated ability of activated sludge treatment in a number of facilities to biologically or chemically remove phosphorous. Recent research has concentrated upon stiochiometric processes on the basis of its presumed efficiency, the "biological" process being charged with an unknown mechanism and erratic behavior. It is strongly recommended that equal effort be given to the "biological" process for the following reasons:

1. It has demonstrated its ability and reliability on both small and large flows, and there is an increasing body of evidence which will clarify the mechanism of removal.

2. Chemical wastewater treatment processes are inherently less environmentally acceptable. High energy use from fossil fuels, high TDS and increased trace metals concentrations of the effluent and consequently restricted potential for reclamation and utilization of the effluent water and sludge are among the factors involved.

3. Chemical treatment facilities have substantially lower initial capital costs to both local and State or Federal grant authorities but pass on to the operating authorities and local governments very much higher operational and maintenance costs. When these are capitalized they show the chemical processes to be much more expensive.

4. Experience at Los Angeles with both systems show the bio-
    logical method to be substantially more controllable and
    reliable. Matching chemical dosage to wastewater conditions
    is very difficult and expensive.

5. The logistics and reliability of chemical supply in a large
    facility, leaving out factors such as increased air pollution
    from fuel burning and increased traffic congestion, is a
    major concern which has not been considered before and which
    most facilities have not previously dealt with.

6. The chemical processes are incomplete processes in that
    about 35% of the BOD remains and the water is degraded in
    terms of factors such as heavy metals and total dissolved
    solids. Adding activated carbon or similar processes in
    series would provide improvement, but at the cost of con-
    siderably greater energy expenditure and environmental
    impact.

In summary, it appears to the writer that environmental impact
and energy use considerations make it imperative that the
inherently lower energy biological processes be thoroughly
evaluated prior to selecting only chemical treatment for the
removal of phosphorous. Overriding all other considerations
is the important one of determining whether the basic factor
of the actual cost and benefits of phosphorous removal has been
factually evaluated.

References

(1) "Planning for Clean Water in Terms of Ecology and the Social Environment", Clean Water For Our Future Environment, Sanitary Engineering Division, American Society of Civil Engineers, National Specialty Conference, Los Angeles, California, pp 35-42 (March 1971).

(2) Federal Agencies Defend Environmental Policies, Highlights - Water Pollution Control Federation, 9, No. 6, H-4 (June 1972).

(3) Goldman, C. R., "Effects of Molybdenum on a Food Chain", Man's Chemical Invasion of the Ocean, An Inquiry, Scripps Institution of Oceanography (February 1969).

(4) Bargman, R. D., Betz, J. M. and Garber, W. F., "Nitrogen-Phosphorous Relationships and Removals Obtained by Treatment Processes at the Hyperion Treatment Plant, Proceedings 5th International Conference on Water Pollution Research, San Francisco, California, Paper I-14 (August 1970).

(5) "Phosphate Study at the Baltimore Back River Wastewater Treatment Plant", Environmental Protection Agency, Publication 17010 DFV09/70 (September 1970).

(6) Priesing, C. P., Witherow, J. L., Lively, L. D., Scalf, M. R., DePrater, B. L., and Myers, L. H., "Phosphate Removal by Activated Sludge", U. S. Department of the Interior, Federal Water Pollution Control Administration (November 1966).

(7) Black and Veatch Consulting Engineers, "Process Design Manual for Phosphorous Removal", Environmental Protection Agency, Technology Transfer (October 1971).

(8) Betz, M. M., "The Valley Settling Basin Facilities, Los Angeles, California", Sewage and Industrial Wastes, Vol. 29 No. 6 667-671 (June 1957).

(9) Borgerding, J., "Phosphate Deposits in Digestion Systems", Journal WPCF, Vol. 44, No. 5, 813-19 (May 1972).

References

(10)   Harold, F. M., "Inorganic Polyphosphates in Biology:
       Structure, Metabolism, and Function", <u>Bacteriological
       Review</u>, Vol. 30, No. 1, 772-794 (December 1966).

(11)   Yall, I., Boughton, W. H., Knudsen, R. C., and Sinclair,
       N. A., "Biological Uptake of Phosphorous by Activated
       Sludge", <u>Applied Microbiology</u>, Vol. 20, No. 1, 145-150
       (July 1970).

(12)   Roinestad, F. A., and Yall, I., "Volutin Granules in
       Zoogloea ramigera", <u>Applied Microbiology</u>, Vol. 19,
       No. 6, 973-979 (June 1970).

(13)   Sekikawa, Y., Nishikawa, S., Okazaki, M., and Kato, K.,
       "Release of Soluble Ortho-Phosphate in the Activated
       Sludge Process", <u>Proceedings Third International
       Conference on Water Pollution Research</u>, Paper II-12
       (August 1966).

(14)   Makhijani, A. B. and Lichtenberg, A. J., "Energy and
       Well Being", Environment, Vol. 14, No. 5, pp 12-18
       (June 1972).

(15)   Perry, J. H., "Lime Calcining", <u>Chemical Engineers'
       Handbook</u>, Third Edition, McGraw-Hill (1950).

USE OF SURFACE STIRRERS FOR AMMONIA DESORPTION FROM PONDS

A.M. Wachs*, Y. Folkman** and D. Shemesh***

* Environmental Engineering, Technion - Israel Institute of Technology
** Dan Region Reclamation Project, Tahal Ltd., Tel Aviv, Israel
*** Ministry of Health, Haifa, Israel

INTRODUCTION

The removal of nitrogen from wastewater by converting most of the ammonia ions into dissolved free ammonia that can then be transferred from the liquid phase to the atmosphere has been the subject of theoretical and experimental work and has been applied in treatment plants (1, 2, 3, 4, 5).

In a recently published paper (6) the authors described experimental work carried out by treating with lime the effluents of stabilization ponds, settling the resulting solids and keeping or flowing the supernatant in tanks or ponds, used singly or in series. In the research project hereby described, surface stirring devices were used to promote the transfer of ammonia to the atmosphere. Some of the experiments were performed in a large wind tunnel that made possible to study the influence of wind velocity on the rate of transfer. Other experiments were performed in a small pilot plant using the effluent of a stabilization pond model. The devices used for stirring the liquid surface had been designed as surface aerators for biological treatment of wastewater.

EXPERIMENTAL WORK

a. Experiments in wind tunnel, with ammonia solutions.

In order to study the effect of wind velocities a series of experiments were carried out in a wind tunnel where the wind velocity could be regulated by the use of screens and measured by a network of thermo-anemometers.

In all runs of this series a cilindrical tank of d = 1.42 m was used. Two thousand liters of ammonium chloride solution were introduced in the tank, the resulting liquid depth being 70 cm. Sufficient lime slurry was then added to reach a pH of 11. The ammonia concentration was 80 mg/l.

The surface stirring devices used in these experiments had been originally designed and used as surface aerators in research on biological treatment of wastewater. They were of the turbine type, made of a disk of perspex to which curved blades were attached. The one designed with the number 1 had a d = 16 cm, the one designed with number 2 had a diameter of 24 cm.

Experimental conditions for the four runs in this series have been summarized in Table 1. In each run samples were taken hourly, both just under the surface of the liquid and at a point at about one half of the liquid depth. The tank installed in the wind tunnel can be seen in Fig. 1.

TABLE 1

Experimental Conditions for Work Carried Out in Wind Tunnel, Using Surface Stirrers

Run number	Wind velocity km/hr	Water temp. °C	Surface stirring device no.	r.p.m.	Watts supplied to motor	Initial ammonia concentration mg/l
1	13.8	11.8	2	180	160	80
2	2.0	14.0	2	180	160	80
3	0	16.8	2	180	160	80
4	14.4	14.5	1	165	160	50

FIG. 1
Tank with Surface Stirrer Installed in Wind Tunnel

b. Experimental work using effluent of stabilization pond.

For these investigations the ammonia-containing liquid was the effluent of a stabilization pond model, consisting of five cells operated in series. It had been found previously that efficient lime treatment for separation of algae and other objectionable susbstances from the pond effluent required a pH close to 11 (7, 8) and therefore the lime slurry dosage was adjusted to obtain this pH value, which was also desirable from the point of view of converting ammonia ions into ammonia. The lime treatment was performed in a pilot-plant-sized clarifier and the supernatant conveyed to the tank that had been used for the wind tunnel experiments. The stirrer used was the one identified as No. 2 and was operated at 160 r.p.m. Wind velocity was recorded and air and liquid temperatures were measured. Samples were taken both just below the surface and at about half the liquid depth. Immediately after collection the samples were examined for ammonia, pH, alkalinity and calcium ion.

DISCUSSION OF RESULTS

a. Experimental work in wind tunnel.

In Fig. 2 the results of four runs carried out in the wind tunnel are presented in terms of relative remaining concentrations C_t/C_o plotted against time, C_o representing the initial concentration and C_t the concentration at time t. The C_t/C_o values are plotted on logarithmic ordinates, and the exponential function corresponding to the lines in the graph can be expressed as

$$\frac{C_t}{C_o} = e^{-K \frac{A}{V} t} \qquad\qquad /1/$$

In this equation K represents the overall transfer coefficient, V is the volume of the liquid and A the surface area of the air-water interface. When the surface of the liquid in the tank is undisturbed, it can be assumed that

$$\frac{A}{V} = \frac{1}{h} \qquad\qquad /2/$$

where h is the depth of the liquid. In their previous work on ammonia release from ponds (1) this assumption permitted the authors to write equation /1/ as

$$\frac{C_t}{C_o} = e^{-K \frac{t}{h}} \qquad\qquad /3/$$

where K was considered to be the product of k_1 and k_u, coefficients that represented respectively temperature effects and wind velocity effects. When the surface of the liquid is intensively disturbed, as it is the case when a surface stirrer is used, the value of A can not be measured and therefore equation /3/ can not be used. This makes difficult to calculate the values of the components of K. However, since the main objective of this work was to compare the rates of ammonia desorption obtained by using surface stirring devices (i.e. surface aerators) with those that had been previously obtained without using them, such comparison is possible in terms of, for instance, the values of the slopes of the lines C_t/C_o corresponding to experimental work carried out with and without the use of surface aerators, or the times required to achieve a given value of C_t/C_o.

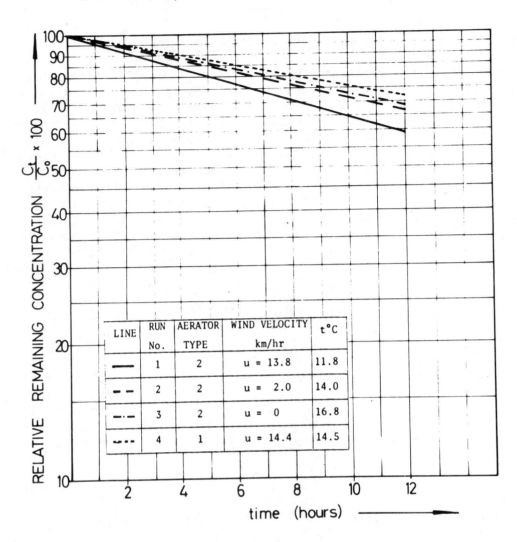

FIG. 2
Results of Runs Carried out in Wind Tunnel, using Surface Stirrers

Results of runs in the wind tunnel with and without surface stirring, are shown in Fig. 3. All runs correspond to an initial liquid depth of 70 cm. The upper shaded area represents runs without stirrer for an air velocity range of 0 to 11.9 km/hr, while the lower shaded area represents results for runs with surface stirrer, for an air velocity range of 0 to 13.8 km/hr. It must be understood that average temperatures were different for the two series of runs, the one for "no stirrer" being 25°C and the one "with stirrer" being 14.5°C. If the "no stirrer" runs would be corrected to the temperature of the "with stirrer" runs, the difference between the respective ammonia desorption rates would be greater than that indicated by the graph. But even without that correction the significant improvement resulting from the use of surface stirrers is quite obvious.

The higher desorption rates attained when the surface stirrers are used can be attributed to two main factors. One is the increase in the interfacial surface area due to the formation of ripples and droplets. The other is that under intensive surface stirring conditions, a rapid renewal of the interface occurs. It is of interest to note that even when surface stirrers are used, wind velocity still has an appreciable effect on the desorption rates. Since the increase in A due to wind effect (at the range of wind velocities used in

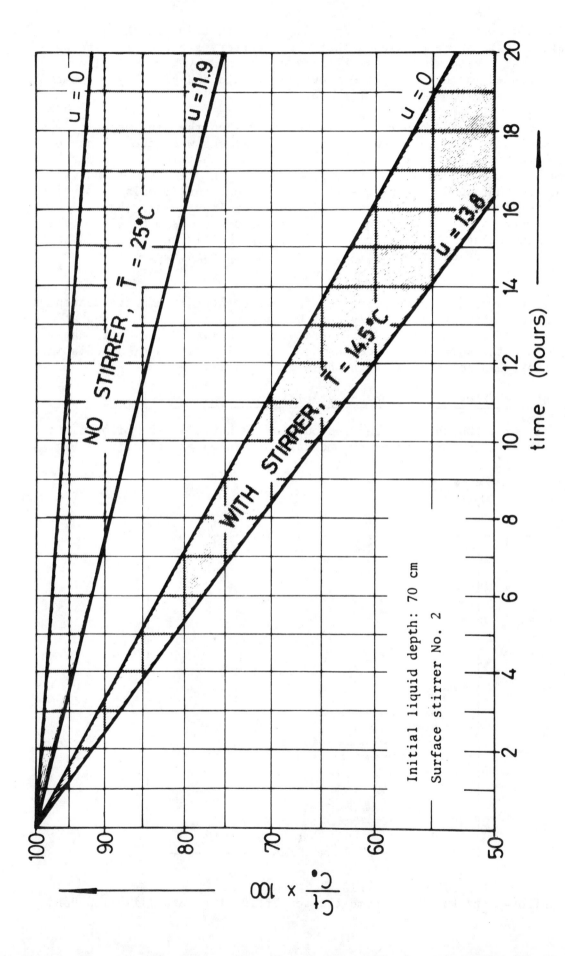

FIG. 3

Results of Runs in Wind Tunnel, with and without Surface Stirrers

the experimental runs) is negligible in relation to that due to the surface stirring effect, it can be concluded that when stirrers are used the increase in desorption rates with wind velocity is mainly caused by the removal of ammonia accumulated in the atmosphere above the liquid and possibly to some breakdown of the gas film at the interface.

It is important to compare results obtained when the wind velocity was zero. In field ponds this would correspond to situations when the air above the pond is practically stagnant. In Fig. 3 the lines that form the upper limit of each of the shaded areas correspond to runs in which u = 0 and it can be observed that when no surface stirrer was used, the ammonia desorption rate was extremely low, the slope value of the upper line being 0.43×10^{-2} hr^{-1}. When the stirrer was used, the desorption rate for u = 0 was much higher, the line slope value being 1.18×10^{-2} hr^{-1}. It is clear that in the case of ammonia desorption ponds, surface stirring is of decisive importance at zero or very low wind velocities.

In all runs with surface stirring, the differences in ammonia concentrations in samples taken simultaneously just below the liquid surface and at approximately half of the liquid depth were nil or negligible, a fact that proves that when the surface stirrer was used, the diffusivity of ammonia in the liquid phase had no bearing on the desorption rates.

b. Experimental work using effluent of stabilization pond.

Results of a "batch" run, carried out for 12 hours in the pilot plant set-up described in the preceeding chapter are presented graphically in Fig. 4.

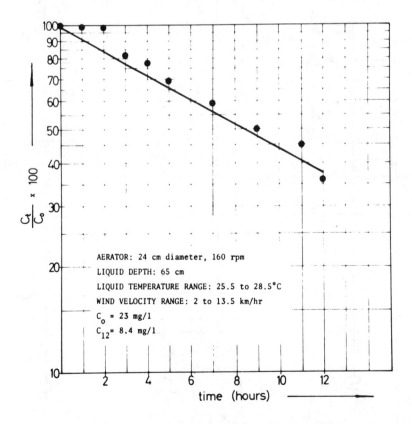

FIG. 4
Results of Pilot-plant Run Using Lime-treated Effluent of Stabilization Pond

The fact that little change of ammonia concentration occured during the first hour of the run can be attributed to the fact that the tank was filled one day before the start of the experiment, this allowing some stratification to occur in the liquid. This assumption is born out by the measured increases in pH, alkalinity and calcium concentration in samples taken at a point just under the liquid surface one hour after the surface stirrer was started, as shown in Fig. 5.

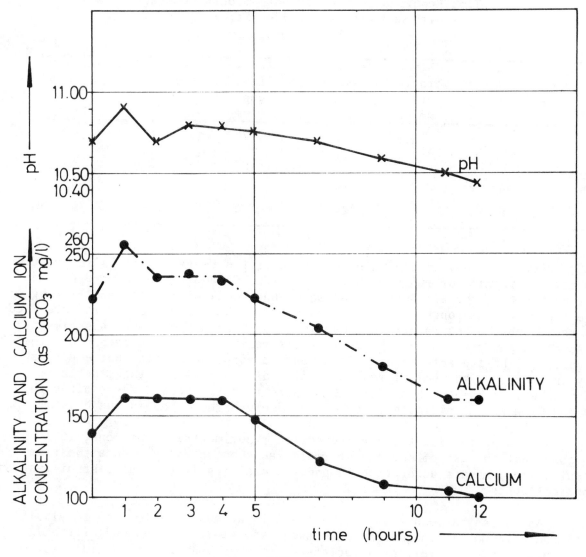

FIG. 5
Changes in pH, Alkalinity and Calcium ion in Run 5

It can be observed from the results of this run, which will henceforth be designed run 5, that the rate of ammonia desorption was significantly higher than those obtained in the wind tunnel runs using ammonia solutions in distilled water. Table 2 permits the comparison of results of run 5 with those of run 1, the one that produced the best results in the wind tunnel series, and with runs C_2 and C_3, which were continuous flow runs carried out without surface stirrers.

Again, results of the runs 1 and 5 cannot be directly compared since they were carried out under different conditions of temperature and wind velocity. The rate of desorption in run 5 was approximately twice as fast as that in run 1, which can mostly be accounted by a) the higher temperature, b) the fact that the water contained impurities that decreased the solubility of ammonia. On the other hand the wind velocity was lower in run 5 than that at run 1.

TABLE 2
Time Required for 50 percent Ammonia Removal
with and without Surface Stirrers

Run number	Stirrer No.	Wind Velocity Average km/hr	Temperature Average °C	Initial liquid depth cm	Slope of line hr^{-1}	Time for $C_t/C_o=0.50$ hr
1	2	13.8	11.8	70	1.78×10^{-2}	16
5	2	9.7	26.7	65	3.60×10^{-2}	9
C_2	none	8.1	14.0	96		105
C_3	none	14.4	21.0	96		130

Returning now to the main objective of the research, it is interesting to compare results of run 5 with those of runs C_2 and C_3 which were carried out by treating the effluent of the same pond model with lime, and flowing the supernatant through a series of open tanks. No surface stirrers were used in these runs. As explained in a previous publication (6), when a series of desorption ponds are used, the shorter the residence time in each pond, the higher will be the rate of overall ammonia desorption a fact that explains why in run C_2, with nominal residence time of 1.5 days per tank, better results were obtained than in run C_3, where that time was 2 days, although in run C_3 wind velocities and temperature were higher than in run C_2.

Allowing for all the differences in experimental conditions, it is obvious that the use of surface stirrers can substantially reduce the time required for ammonia desorption in ponds containing lime-treated stabilization pond effluents.

As it can be seen from Fig. 5 there is a considerable drop in alkalinity and calcium ion concentration between the start and the end of the 12 hour "batch" run, as well as a decrease of the pH value. The initial rise of all those values can be attributed to the reason given above: stratification of the liquid in the tank before the start of the run. Therefore,the values obtained after the first hour of the run should be considered as the ones representing the initial conditions. The differences between initial and final values are then: for calcium ion 160 - 100 = 60 mg/1 (as $CaCO_3$), for alkalinity 258 - 160 = 98 mg/1 (as $CaCO_3$), while for the pH value the drop was from 10.90 to 10.45. This means in effect that while ammonia desorption proceeds, and as air is introduced into the liquid phase, insoluble calcium compounds are formed and precipitated. In asbestos-cement tanks that were used for long periods for ammonia desorption (without stirrers) from similarly treated pond effluent, a dense but thin layer of calcium carbonate was formed on walls and floors.

CONCLUSIONS

The main purpose of the research work presented in this paper was to investigate whether the use of surface stirrers would result in a significant saving in area when compared with that required for non-surface-stirred ponds intended for ammonia desorption in a wastewater renovation project.

As described elsewhere (7) a particular feature of that project is the availability of large tracts of fiscal, relatively low-priced land, a feature which explains why stabilization ponds were selected for the biological part of the treatment, throughout the first and second stage of the project development. At the time of the original planning, enough land was available to allow the use of simple holding ponds to remove most of the ammonia-nitrogen after lime treatment of the stabilization ponds effluents. New conditions make highly desirable to reduce the area allocated for the ammonia desorption ponds, and the result of this research indicate that such reduction is feasible when surface stirring is used.

To obtain more precise design data, continuous flow experiments are being carried out at present, but on the basis of a conservative evaluation of the data already available it seems that the use of surface stirrers would permit to obtain the requied degree of ammonia desorption in from one third to one fourth of the area originally assigned to that purpose, even under the most unfavourable conditions of wind velocity and temperature.

REFERENCES

1. A.F. Slechta and G.L. Culp, Water Reclamation Studies at the South Lake Tahoe Public Utility District, J.W.P.C.F. 39(5), p.787 (1967).

2. J.B. Farrell, Physical-Chemical Methods for Nitrogen Removal, Sympozium on Nutrient Removal and Advanced Waste Treatment, Cincinnati, Ohio (1969).

3. J.B. Farrell, Ammonia Nitrogen Removal by Stripping with Air. Nitrogen Removal from Wastewater, Federal Water Quality Administration, Cincinnati (1970).

4. J.F. Roessler, R. Smith and R.G. Eilers, Simulation of Ammonia Stripping from Wastewater, J. Sanitary Engineering Division, ASCE, 97(SA3), p.269, (June 1971).

5. Advanced Wastewater Treatment as Practiced at South Tahoe, Water Pollution Control Research Series, E.P.A. Water Quality Office (August 1971).

6. Y. Folkman and A.M. Wachs, Nitrogen Removal Through Ammonia Release from Holding Ponds, Sixth Conference of the IWPRA, Jerusalem (1972).

7. A. Wachs, The Outlook for Wastewater Utilization in Israel, Developments in Water Quality Research, Ann Arbor-Humphrey Science Publishers, Ann Arbor (1970).

8. Y. Folkman, Reclamation of Stabilization Pond Effluents by Chemical Treatment, Thesis for degree of D.Sc., Sanitary Engineering Laboratories, Technion, Haifa (1970).

Applications of New Concepts of Physical-Chemical
Wastewater Treatment
Sept.18-22, 1972

DEWATERING PHYSICAL-CHEMICAL SLUDGES

Donald Dean Adrian, Ph.D.
Associate Professor of Civil Engineering, University of Massachusetts,
Amherst, Massachusetts

James E. Smith, Jr., Sc.D.
Research Sanitary Engineer, Ultimate Disposal Research Program,
Advanced Waste Treatment Research Laboratory, National Environmental Research
Center, Environmental Protection Agency, Cincinnati, Ohio

Abstract

Wastewater treatment plants upgrading their existing solids removal
processes by polymer or chemical addition, and plants precipitating phosphorous increase the amount of sludge to be handled. The increased amount of
sludge may be approximated by referring to pilot plant and field data and
by a calucational approach. The characteristics of the sludge have an
impact on the method selected for dewatering although physical-chemical
processes produce sludges whose dewatering properties are not yet well understood. The cost of dewatering these sludges may be estimated by referring to
the costs of sludge dewatering at conventional activated sludge plants and
making appropriate adjustments for scale economies brought about by the
greater amounts of sludge to be handled.

Introduction

The dewatering characteristics of conventional primary and secondary sludges
have become rather well known in practice and are reasonably well discussed
in the literature. Likewise the dewatering techniques and the methodology
for applying them on these sludges are known. Although a number of physical-
chemical treatment processes predate conventional secondary treatment
processes (1), their sludge production rates dewatering characteristics and
dewatering methodologies are poorly understood.

Numerous physical-chemical treatment processes are presently being considered and employed for use with wastewater. It appears that the principal
source of new sludge solids to be handled will be from chemical precipitation
processes. This paper will emphasize the production rates, dewatering
characteristics, dewatering methodologies and the costs of treating sludges
produced by the addition of chemicals to primary, secondary, and tertiary
systems for the removal of phosphorus or to upgrade present treatment.

Upgrading present treatment efficiencies can be accomplished either by
design changes or by adding chemicals (alum, lime, ferric chloride or

polyelectrolytes) to enhance principally the removal of suspended solids, BOD, and phosphorous in primary and secondary clarifiers (2). Increasing the removal of suspended solids in the primary clarifier increases primary sludge production and decreases waste activated sludge production. The combined sludge stream produced from chemical upgrading may be more amenable to dewatering because primary sludge is more readily dewatered than waste activated sludge. However, the total mass of sludge solids to be dewatered will increase due to greater solids removal, precipitating dissolved materials and the weight of chemical additives.

Volume and Mass of Sludge

Upgrading Treatment

After a treatment process is upgraded the sludge production may be estimated from suspended solids removal. For example, if a primary clarifier is producing 600 lb of sludge solids per million gallons (lb/MG) while operating at a 50% efficiency and the efficiency is increased to 60% by adding polymers, the sludge production will be 720 lb/MG. Polymer addition would likely increase the sludge solids contents although it is difficult to predict the effect on sludge volume. If one assumed a 5% solids content before and a 6% solids content after polymer addition, the sludge volume would remain at 1,440 gal/million gallons. Operating results for two locations in which a polyelectrolyte was added to the plant influent are shown in Table 1.

Analysis of the results shown in Table 1 illustrates the divergent effects which may be obtained when adding polyelectrolytes to upgrade treatment. The change in sludge volume and mass that occurs with polymer addition is a function of the prior efficiency of the clarifier. The volume of primary sludge after upgrading increased by 180% in the Painesville primary plant and 27% in the Cleveland Easterly plant. The volume of secondary sludge decreased by 26% at the Cleveland plant. The overall sludge volume decreased by 10% at Cleveland, while the secondary sludge mass decreased by 57% at Cleveland shifting the weight ratio of waste activated sludge to primary sludge from 0.68 before upgrading to 0.19 after chemical treatment. From a dewatering viewpoint this represented a relative shift from a preponderance of difficult to dewater secondary sludge to a preponderance of more readily dewaterable primary sludge. The two examples cited were for upgrading using polyelectrolyte addition; other coagulants lime, ferric chloride or alum would increase the sludge mass from primary clarifiers and accomplish phosphate removal.

Phosphate Removal

A calculation approach may be taken to obtain a first estimate of the additional sludge mass to be produced by chemical addition to various unit processes in the treatment of wastewater for phosphorus removal. This has been illustrated for the case of iron addition to raw wastewater by the Environmental Protection Agency (4). In addition to using their raw wastewater characteristics and basic assumptions, we assumed the following:
- Influent BOD = 230 mg/l, influent SS = 300 mg/l, influent P = 10 mg/l
- 0.3 mg/l polymer added in every case
- 0.5 lb VSS is produced for every lb BOD removed in biological system
- Cation/P dose = 1.5 moles to aerator, = 1.75 moles to primary and/or final
- Each lb Al added = 3.9 and 3.8 lb Al sludge for the 1.5 and 1.75 mole doses

- Each lb Fe added = 2.4 and 2.3 lb Fe sludge for the 1.5 and 1.75 mole does
- Conventional sludge solids are 70% volatile.

One can develop estimates of sludge mass production for the cases of: a) Iron addition to the primary, b) Iron addition to the aerator, c) Aluminum addition to the aerator, and d) Aluminum addition to the final clarifier for a standard rate trickling filter. These estimates are shown in Table 2. They differ in the case of phosphorus removal by iron addition to the raw wastewater from the estimates given in the design manual (4). This difference is due to the fact that here consideration was given to the degradation of solids to gas, water and other soluble products in the aerator. It may readily be seen from Table 2 that any phosphorus removal process by chemical addition will cause a significant increase in the mass of sludge to be handled.

TABLE 1 - Sludge Production for Upgrading Treatment (2,3)

Plant and Process	Dose mg/l	Solids %	Sludge Volume gal/MG	Sludge Mass lb/MG
Painesville, Ohio	0	5 (est.)	2,280	950
Primary Clarifier	0.4 (5T-269)	5.3	6,400	2,800
Cleveland, Ohio				
Easterly Plant				
Primary Clarifier	0	4.1	670*	440
Secondary Sed.	0	2.4	1,640*	300
Primary Clarifier	0.21(Purifloc A-23)	4.3	850*	670
Secondary Sed.	0.21(Purifloc A-23)	2.0	1,220*	130
Overall Performance	-	-	2,310*	740
Overall Performance	0.21(Purifloc A-23)	-	2,070*	800

*Based on an assumed design overflow rate of 1,000 gpd/sq. ft.

Aluminum addition caused the smallest mass increase of sludge.

The available data on sludges from phosphorus removal systems is very limited. However, data was obtained from the review of 13 case histories, which included both pilot and plant scale studies (3,5,6,7,8,9,10,11,12,13, 14). Table 3 covers the processes of phosphorus removal from raw wastewater. Table 4 by mineral addition to the aerator, and Table 5 by chemical treatment of secondary effluent. From these tables it may be noted that the greatest increase in sludge mass occurred when aluminum was added to secondary effluent, and the minimum increase in sludge mass occurred when lime was added at a level of from 800-1600 mg/l to raw sewage. In terms of additional sludge volume to be handled, the maximum amount would be expected with iron addition to the secondary effluent and the minimum with lime addition to raw sewage. The above comparisons are on the basis of the mass and volume of additional sludge produced per pound of chemical added.

TABLE 2 - Calculated Sludge Mass Production (lb/MG)

Process	Type Waste Solids	Conventional Case		Fe to Primary		Fe to Aerator		Al to Aerator		Al to Trickling Filter Final	
		TSS	VSS	TSS	VSS	TSS	VSS	TSS	VSS	TSS	VSS
Primary	Primary	1250	875	1875	1314	1250	875	1250	875	1250	875
	Fe Solids	--	--	605	--	--	--	--	--	--	--
	Al Solids	--	--	--	--	--	--	--	--	--	--
	SubTotal	1250	875	2480	1314	1250	875	1250	875	1250	875
Activated Sludge	Fe Solids	--	--	--	--	541	--	--	--	--	--
	Al Solids	--	--	--	--	--	--	425	--	--	--
	Secondary Solids	715	500	536	375	804	563	804	563	563	563
	SubTotal	715	500	536	375	1345	563	1229	563	563	563
Trickling Filter	Al Solids	--	--	--	--	--	--	--	--	483	--
	Secondary Solids	656	459	--	--	--	--	--	--	745	521
	SubTotal	656	459	--	--	--	--	--	--	1228	521
Totals											
Case I-Combined Primary and Activated Sludge Solids		1965	1375	3016	1689	2595	1438	2479	1438	--	--
Case II-Combined Primary and Trickling Filter Solids		1906	1334	--	--	--	--	--	--	2478	1396

TABLE 3 - Additional Sludge to be Handled with Chemical Treatment Systems: Primary Treatment for Removal of Phosphorus

Sludge Production Parameter		Conventional Primary	Lime Addition to Primary Influent	Lime Addition to Primary Influent	Aluminum Addition to Primary Influent	Iron Addition to Primary Influent
Level of Chemical Addition (mg/l)		0	350-500	800-1600	13-22.7	25.80
% Sludge Solids	Mean	5.25	11.1	4.4	1.2	2.25
	Range	5.0-5.5	3.0-19.5	2.1-5.5	0.4-2.0	1.0-4.5
lb/MG	Mean	788	5,630	9,567	1,323	2,775
	Range	600-950	2500-8000	4,700-15,000	1200-1545	1400-4500
gal/MG	Mean	4465	8924	28,254	23,000	21,922
	Range	3600-5000	4663-18,000	16,787-38,000	10,000-36,000	9,000-38,000

TABLE 4 – Additional Sludge to be Handled with Chemical Treatment Systems: Phosphorus Removal by Mineral Addition to Aerator

Sludge Production Parameter		Al^{+++} Addition to Aerator		Fe^{+++} Addition to Aerator	
		Conventional Secondary	With Al^{+++} Addition	Conventional Secondary	With Fe^{+++} Addition
Level of Chemical Addition (mg/l)		0	9.4-23	0	10-30
% Sludge Solids	Mean	0.91	1.12	1.2	1.3
	Range	0.58-1.4	0.75-2.0	1.0-1.4	1.0-2.2
lb/MG	Mean	672	1180	1059	1705
	Range	384-820	744-1462	918-1200	1100-2035
gal/MG	Mean	9100	13,477	10,650	18,650
	Range	7250-12,300	7,360-20,000	10,300-11,000	6,000-24,400

TABLE 5 - Additional Sludge to be Handled with Chemical Treatment Systems: Phosphorus Removal by Mineral Addition to Secondary Effluent

Sludge Production Parameters		Lime Addition	Alum Addition	Iron Addition
Level of Chemical Addition (mg/l)		268-450	16	10-30
% Sludge Solids	Mean	1.1	2.0	0.29
	Range	0.6-1.72	-	-
lb/MG	Mean	4,650	2,000	507
	Range	3100-6800	-	175-781
gal/MG	Mean	53,400	12,000	22,066
	Range	50,000-63,000	-	6,000-36,000

Characteristics of Sludge

Experience with primary and secondary sludges has shown how they differ in their dewatering properties, primary sludges being considerably easier to dewater than secondary sludges. In a qualitative manner this variation in dewaterability may be described by the secondary sludge's gelatinous structure, its dilute concentration, the degree of agglomeration of the particles, the presence or absence of binding material, and the amount of bound water. In a quantitative manner the variation in dewaterability of sludges may be expressed in terms of the Buchner funnel filtration parameters specific resistance and coefficient of compressibility or the filter leaf parameter specific filter yield. Typical values are shown in Table 6 which also illustrates how the filtration properties can be made to vary through the addition of chemical conditions.

TABLE 6 - Comparative Values of Sludge Filtration Properties

Sludge Type & Chemical Treatment & Reference	Chemical Dose	Solids Content	Coefficient of Compressibility σ	Specific Resistance @ 38.1 cm Hg
	%	%	Dimensionless	sec^2/gm
Raw(16)	0	NA	0.54	4.7×10^9
Amherst, MA Digested Primary(17)	0	9.5	0.67	26.8×10^9
Calgon ST 260	0.1	9.5	0.75	18.0×10^9
" " "	0.2	9.5	0.84	14.4×10^9
" " "	0.3	9.5	0.90	11.6×10^9
" " "	0.4	9.5	0.88	10.0×10^9
" " "	0.5	9.5	0.93	6.6×10^9
Pittsfield, MA Digested Mixed Primary & Trickling Filter (17)	0	4.9	0.64	65.0×10^9
Calgon ST 260	0.2	4.9	0.65	52.5×10^9
" " "	0.4	4.9	0.66	41.8×10^9
" " "	0.6	4.9	0.65	36.0×10^9
" " "	0.8	4.9	0.68	33.2×10^9
" " "	1.0	4.9	0.70	30.5×10^9
Franklin, Tenn. Digested Mixed Primary & Activated Sludge (18)	0	3.7	0.63	48.0×10^9

Interest in physical-chemical treatment processes has been focused on the development of the major components of the processes with less attention being given to sludge handling and disposal. As a result, few quantitative data are available in the literature on physical-chemical sludge properties; furthermore there is some divergence of evidence on how well these sludges will dewater. For example, most Dow Chemical Co. reports (19,20,21,22) have reported filter yields would not decrease when dewatering physical-chemical sludges precipitated with iron, although one case of vacuum filter blinding was reported (23). Other investigators (8) have found a filter yield of 1.0 lb/ft^2-hr for an iron sludge. Table 7 presents filter yields for iron, aluminum and lime precipitation of phosphorus (6). Lime filter yields were substantially higher than the yields for iron or aluminum sludges.

TABLE 7 - Vacuum Filtration Leaf Tests for Salt Lake City Raw Wastewater(6). Series A tests were for Ferric chloride does of 99-145 mg/l and 0-1.5 mg/l polymer. Series B tests were for alum does of 143 mg/l. All filter yields were for 33% submergence, a 3/16 inch cake, a 20 in Hg vacuum and a 0.8 scale up factor.

Chemical Treatment	Test	Feed Solids	Cycle Time	Filter Yield	Cake Solids
		%	min.	lb/hr-ft^2	%
$FeCl_3$ + Polymer	A1	2.5	16	0.37	15
" "	A2	3.3	20	0.32	17
" "	A3	1.6	25	0.27	20
" " + 24% $Ca(OH)_2$	A4	1.6	5	1.0	18
" " + 47% $Ca(OH)_2$	A5	1.6	5.3	1.0	17
$Al(SO_4)_3 \cdot 14H_2O$ + .43% Anionic polymer	B1	0.70	8	0.38	15
$Al_2(SO_4)_3 \cdot 14H_2O$ + 0.70% Cationic polymer	B2	0.36	23	0.14	23
$Ca(OH)_2$	C1	17.9	1.8	12.8	33
"	C2	9.4	3.0	4.5	28

The dewatering properties of tertiary sludges may be approximated by looking at the filtration properties of water treatment sludges. The American Water Works Association Research Foundation (24) has discussed the water treatment industries' sludge handling practices which employ dewatering methods similar to those employed in wastewater treatment. Water treatment sludge characteristics are presented in Table 8. A comparison of these values for specific resistance with the typical values for wastewater sludges from Table 6 suggests that tertiary sludges may filter easier than secondary sludges.

TABLE 8 - Water Treatment Sludge Properties

Sludge Type Chemical Treatment & Reference	Chemical Dose		Solids Content	Coefficient of Compressibility σ	Specific Resistance @ 38.1 cm Hg
	To raw Water	wt.Chem x 100 / wt.Sludge Solids			
	mg/l	%	%	Dimensionless	sec^2/gm
Lime added	0	0	1.0	NA	$1.85-10^9$
to raw	2.0	0.05	1.0	NA	$1.29-10^9$
water (25)	3.0	0.08	1.0	NA	$0.10-10^9$
Alum added to raw water					
Albany,N.Y. (26)	NA	NA	1.86	0.49	$8.00-10^9$
Amesbury,N.Y. (27)	NA	NA	2.06	0.80	$1.04-10^9$
Billerica,MA (27)	NA	NA	4.65	1.21	$2.49-10^9$
Lawrence,MA (27)	NA	NA	0.94	1.02	$10.4-10^9$
Lowell,MA (27)	NA	NA	3.81	0.89	$4.23-10^9$

Dewatering Methods

Sludges produced by upgrading existing treatment plants are likely to continue to be dewatered by the processes and facilities previously in use in the plant. Jenkins, et al (15) pointed out that a partial alternative to sludge disposal is precipitate recovery, although this is most likely to occur with lime processes. Thus centrifuges, vacuum filters, sand beds and filter presses will continue to be used. As previously pointed out, upgrading through polymer addition can double or triple the amount of primary sludge while halving the production of secondary sludge. As a result, it is likely that existing dewatering facilities will have to be enlarged when upgrading primary treatment plants by polymer addition. However, for upgrading secondary treatment plants the combination of an increase in more readily dewatered primary sludges and a decrease in more difficult to dewater secondary sludges may not require expansion of the existing dewatering facilities.

Physical-chemical treatment for phosphorous removal has been shown in Tables 2-5 to increase the mass of the sludge to be handled. Calculations suggest that the volume of sludge may increase, but in Dow Chemical Co. studies (19,20,21,22) an increase in sludge volume has not been observed.

Sand dewatering and drying beds have been used successfully by certain water utilities to dewater sludges composed of precipitated iron with alum as a coagulant (27) alum sludge (26), and have been reported to work satisfactorily with tertiary sludges (28). The tertiary sludge consisted of a mixture of primary and trickling filter sludge and alum sludge generated by treating the secondary effluent. In this case, Richardson, Texas, the sand beds performed better after the addition of the tertiary sludge so that bed expansion to handle the increased amount of sludge was unnecessary. Lo (29) has addressed himself to the problem of sizing sand beds taking into account climatological variables, sludge properties and economics.

Centrifuges have been popular in dewatering lime and lime-soda softening sludges (24) and tertiary lime sludges (11), especially for plants practicing lime reclamation. One can anticipate that they will be installed in some plants which use lime to precipitate phosphates.

Dewatering Costs

The method chosen to express dewatering costs has traditionally been on a dollars per ton of dry solids basis. As long as the major treatment alternatives consisted of primary treatment and secondary treatment this method of expressing costs was satisfactory, although it may be argued that one implicitly assumed the sludge production rates were the same from similar plants. For example, if one compared two primary plants of the same capacity, but which operated at different suspended solids removal efficiencies, then the plant removing the most solids would achieve certain scale economies so that its dewatering costs per ton of dry solids would be lower, although its absolute expenditures as a cost per million gallons would be higher. Similar expressing costs on a dollars per ton of dry solids basis may mask the cost impact of chemical conditioners and additives if they increase the total weight of dry solids to be handled. Physical-chemical treatment enlarges the number of process alternatives so that one should reexamine whether cost comparisons are meaningful if expressed in the traditional manner. Expressing costs as a cost per million gallons of plant influent assists the decision maker in assessing the merits of various treatment process alternatives, i.e., whether to coagulate with alum, ferric chloride or lime. Once the treatment process has been established and the engineer is comparing alternative methods of dewatering the sludge produced then it may be meaningful to express the cost of each sludge dewatering alternative on a cost per ton basis.

Dewatering costs for upgrading existing treatment facilities is influenced by the larger quantities of sludge bringing about scale economies and the shift in composition of the sludge with the greater emphasis on primary sludges. Unit costs of dewatering sludge when expressed on a dollars per ton of dry solids basis would be expected to decrease. Unit costs of dewatering sludge when expressed on a dollars per million gallons of plant influent basis would likely increase, although there may be special cases in which the shift from biological sludges to primary sludges would allow the dewatering cost per million gallons of plant influent to stay constant.

From recent work by Black and Veatch (30) and Smith and Eilers (31,32) the sludge handling process cost estimates shown in Table 9 were obtained. To obtain an estimate of the impact of phosphorus removal by chemical precipitation, two processing alternatives and the change in sludge mass to be handled were considered. Alternative No. 1 considered a 10 mgd plant with the following sludge handling facilities: gravity thickening of the combined primary and waste activated sludges, anaerobic digestion, and sand drying beds. While Alternative No. 2 considered a 100 mgd plant with the following sludge facilities: gravity thickening of primary sludge, air flotation thickening of waste activated sludge, vacuum filtration of the combined thickened sludges, and multiple hearth incinerators. Smith's (31) computer program assumed that the mass of primary sludge to be handled is 833 lb/MG and the mass of waste activated sludge to be handled is 893 lb/MG of wastewater treated.

TABLE 9 - Total Cost in Cents per 1000 Gallons of Wastewater Processed for Indicated Sludge Handling Processes*

Process	Plant Size		
	1 MGD	10 MGD	100 MGD
- Gravity thickening of primary and waste activated sludge	1.61	0.31	0.13
- Gravity thickening of primary sludge above	1.47	0.22	0.08
- Air flotation thickening of waste activated sludge above	2.28	1.01	0.75
- Anaerobic digestion of combined primary & waste activated sludge	6.09	2.09	1.89
- Dewatering of digested sludge on sand beds	2.20	1.64	NA
- Dewatering thickened raw sludges on rotary vacuum filters	8.39	5.15	3.70
- Multiple hearth incineration of filter cake	13.53	5.02	1.16

*Costs included are capital, debt and operation and maintenance
NA = not applicable

For each treatment alternative the minimum sludge production and the maximum sludge production in lb/MG were determined from Tables 3, 4 and 5 and entered into Table   For example, when considering conventional primary treatment followed by aluminum addition to the aerator the sludge mass was calculated to be 833 lb/MG from the primary clarifier, plus an additional 508 lb/MG (1180 lb/MG-672 lb/MG) from the aerator (see Table 4) which when added to the 893 lb/MG estimated by Smith (31) from a secondary clarifier would give a total sludge production of 2,234 lb/MG. Costs for the two alternatives were estimated from the cost factors for each process, making appropriate adjustments for the amount of sludge to be handled and scale economies achieved by handling greater amounts of sludge. Adding aluminum to the aerator produced 130% (2234 x 100/1726) as much sludge as a conventional secondary plant would produce. Scale economies are realized in processing larger quantities of solids so that the unit cost for Process I and II was read by interpolating for an equivalent 13 MGD or 130 MGD plant, respectively. These figures were then multiplied by the 130% increase in solids to be handled to give the final cost estimates of 5.1¢/1000 gal for Process I and 7.0¢/1000 gal for Process II. In a similar fashion the other cost estimates in Table 10 were developed.

The limitations of the resulting cost estimates should be noted. Sludge filtration characteristics did not enter directly into the calculations. Therefore if physical-chemical sludges filter more readily than secondary sludges the cost estimates are high, and vice versa. The ranking of treatment alternatives would change if one type of sludge filtered much more easily than secondary sludge (lowering the cost) and another type was much more difficult to filter than sludge from secondary treatment facilities (increasing the cost).

TABLE 10- Cost of Sludge Handling for Various Phosphorus Removal Methods

Treatment Alternative	Process I 10 MGD			Process II 100 MGD		
	Low	Mean	High	Low	Mean	High
Conv.Prim. + Conv. Sec.						
lb/MG		1726			1726	
¢/1000 gal		4.2			5.7	
Al to Prim. + Conv. Sec.						
lb/MG	2261			2261		
¢/1000 gal	5.1			7.0		
Lime to Prim. + Conv. Sec.						
lb/MG			10,505			10,505
¢/1000 gal			23.9			27.5
Conv.Prim. + Al to Aerator						
lb/MG	2234			2234		
¢/1000 gal	5.1			7.0		
Conv.Prim. + Fe to Aerator						
lb/MG			2,372			2,372
¢/1000 gal			5.2			7.3
Sec.Eff. + Fe						
lb/MG	2233			2233		
¢/1000 gal	5.1			7.0		
Sec.Eff. + Lime						
lb/MG			6,376			6,376
¢/1000 gal			14.1			14.8

Conclusions

Upgrading primary treatment facilities increases sludge quantities. Upgrading secondary treatment facilities may produce results ranging from little, if any, sludge increase to large sludge increases. Secondary facilities will experience an increase in the ratio of primary to waste activated sludge.

There is a diversity of opinion on how physical-chemical sludges differ in their filtration properties from conventional secondary sludge. Lime sludges are expected to filter well although high lime treatment produces sludges with poor filtration characteristics.

The cost comparisons presented here were for two plant sizes and for two sludge handling systems which might be employed. Cost calcuations show an increase in the cost of handling physical-chemical sludges as compared to secondary treatment with the cost being dependent upon the type of chemical and the process to which it was added. The costs may be reduced if the sludges are released from the clarifier at high solids content. Existing plants may find their actual costs somewhat different from the costs calculated here due to factors such as previously existing excess sludge handling capacity.

Acknowledgments

The assistance of S. W. Hathaway, J. J. Westrick, J. B. Farrell, S. Bennett, O. L. Grant and H. E. Thomas of the Advanced Waste Treatment Laboratory, National Environmental Research Center, Cincinnati is gratefully acknowledged. The senior author acknowledges the support of Environmental Protection Agency research grant 17070-DZS and Office of Water Resources Research grant WR-B011-MASS.

References

1. Committee on Sewage Disposal, Chemical Treatment of Sewage, American Public Health Association, Public Health Engineering Section; Sew. Works J., **7**, 1 (1935).

2. P. Krishnan and C. M. Mangan, Process Design Manual for Upgrading Existing Wastewater Treatment Plants, p. 1-1. Environmental Protection Agency, Technology Transfer Office, Washington, D.C. (1971).

3. J. J. Westrick, Environmental Protection Agency, Advanced Waste Treatment Research Laboratory, Physical and Chemical Treatment Program, Cincinnati, Ohio, (1971). Personal Communication.

4. Black and Veatch, Process Design Manual for Phosphorous Removal, p. 4-13, Environmental Protection Agency, Technology Transfer Office, Washington, D.C. (1971).

5. University of Texas Medical Branch, Galveston, Texas, Phosphorous Removal and Disposal from Municipal Wastewater, Water Pollution Control Research Series 17010DYB 02/71, p. 1, Government Printing Office, Washington, D.C. (1971).

6. D. E. Burns and G. L. Shell, Physical-Chemical Treatment of a Municipal Wastewater Using Powdered Carbon, Water Pollution Control Research Series 17020EFB, p. 57, Government Printing Office, Washington, D.C. (1972).

7. O. E. Albertson and L. T. Sheker, Sludge Handling and Disposal, p. 1, Presented at the 42nd Annual Conference of Water Pollution Control Association of Pennsylvania, University Park, Pennsylvania (Aug. 7, 1970).

8. J. E. Smith, S. W. Hathaway, J. B. Farrell, and R. B. Dean, Lime Stabilization of Chemical Primary Sludges at 1.1 MGD, p. 1, Presented at the Water Pollution Control Federation 45th Annual Conference, Atlanta, Georgia (Oct. 9, 1972).

9. K. Wuhrmann, Objectives, Technology and Results of Nitrogen and Phosphorous Removal Processes, in Advances in Water Quality Improvement, E. Gloyna and W. W. Eckenfelder, Editors, University of Texas Press, Austin, Texas (1968).

10. A. A. Kalinske and G. L. Shell, Phosphate Removal from Waste Effluents and Raw Wastes Using Chemical Treatment, Presented at Phosphorous Removal Conference, Federal Water Pollution Control Administration, Chicago, Illinois (June 26-27, 1968).

11. South Tahoe Public Utility District, South Tahoe, Calif. Advanced Wastewater Treatment as Practiced at South Tahoe, Water Pollution Control Research Series 17010ELQ 08/71, Government Printing Office, Washington, D.C. (1971).

12. D. R. Zenz and J. R. Pivnicka, Effective Phosphorus Removal by the Addition of Alum to the Activated Sludge Process, Proc. of the 24th Ann. Purdue Ind. Waste Conf., **24**, 273 (1969).

13. M. C. Mulbarger, E. Grossman, III, R. B. Dean and O. L. Grant, Lime Clarification, Recovery, Reuse and Sludge Dewatering Characteristics, J. Water Poll. Cont. Fed. 41, 2070 (1969).

14. M. C. Mulbarger and D. G. Shifflett, Combined Biological and Chemical Treatment for Phosphorus Removal, Water-1970, Chemical Engineering Progress Symposium Series No. 107, 67, 107 (1970).

15. D. Jenkins, J. F. Ferguson and A. B. Menar, Chemical Processes for Phosphate Removal, Water Research, 5, 369 (1971).

16. W. W. Eckenfelder and D. J. O'Connor, Biological Waste Treatment, p. 277, Pergamon Press, New York (1961).

17. J. H. Nebiker, D. D. Adrian and K. M. Lo, Evaluation of Chemical Conditioning for Gravity Dewatering of Wastewater Sludge, Presented at the American Chemical Society Symposium on Colloid and Surface Chemistry in Air and Water Pollution, Atlantic City, New Jersey (Sept. 10, 1968).

18. J. H. Nebiker, T. G. Sanders and D. D. Adrian, An Investigation of Sludge Dewatering Rates, J. Water Poll. Control Fed. 41, R255 (1969).

19. Anon., Application of Chemical Precipitation, Phosphorous Removal at the Cleveland Westerly Wastewater Treatment Plant, Dow Chemical Co., Midland, Mich. (April 1970).

20. A. L. Cherry and R. G. Schuessler, Private Company Improves Municipal Waste Facility, Water and Wastes Engineering, 8, 32 (1971).

21. Anon., Phosphorus Removal Trial, Fond du Lac, Wisconsin, Dow Chemical Co., Midland, Mich. (July 1971).

22. J. Hennessey, R. Jelinski, J. H. Beeghly and T. J. Pawlak, Phosphorous Removal at Pontiac, Michigan, Presented at the Michigan State Water Pollution Control Association Meeting, Boyne Mountain, Mich. (June 1970).

23. W. J. Stonebrook, V. Dykhuzen, J. H. Beeghly and T. J. Pawlak, Phosphorus Removal at a Trickling Filter Plant, Wyoming, Michigan, Dow Chemical Co., Midland, Michigan (1970).

24. American Water Works Association Research Foundation, Disposal of Wastes from Water Treatment Plants, J. Am. Water Works Assoc., Part I, 61, 541 (Oct. 1969); Part II, 61, 619 (Nov. 1969); Part III, 61, 681 (Dec. 1969); Part IV, 62, 60 (Jan. 1971).

25. O'Brien and Gere Consulting Engineers, Waste Alum Sludge Characteristics and Treatment, p. 59, Research Report No. 15, New York State Health Department, Albany, New York (Dec. 1966).

26. E. E. Clark, Water Treatment Sludge Drying and Drainage on Sand Beds, Report EVE 24-70-4, Environmental Engineering Program, University of Massachusetts, Amherst, Mass. (Aug. 1970).

27. D. D. Adrian, P. A. Lutin and J. H. Nebiker, Source Control of Water Treatment Waste Solids, Report EVE 7-68-1, Environmental Engineering Program, University of Massachusetts, Amherst, Mass. (April 1968).

28. R. Brenner, Personal Communication about A Demonstration of Enhancement of Effluent from a Trickling Filter Plant, Grant 11010EGL, (July 1972).

29. K. M. Lo, Digital Computer Simulation of Water and Wastewater Sludge Dewatering on Sand Beds, Report EVE 26-71-1, Environmental Engineering Program, University of Massachusetts, Amherst, Mass. (July 1971).

30. W. L. Patterson and R. F. Banker, Estimating Costs and Manpower Requirements for Conventional Wastewater Treatment Facilities, Water Pollution Control Research Series 17090DAN 10/71, Government Printing Office, Washington, D.C. (1971).

31. R. Smith, and R. G. Eilers, Personal Communication (May 31, 1972).

32. R. G. Eilers and R. Smith, Executive Digital Computer Program for Preliminary Design of Wastewater Treatment Systems, U.S. Dept. of Interior, Federal Water Quality Administration, Advanced Waste Treatment Branch, Division of Research, Cincinnati, Ohio (Nov. 1970).

DISPOSAL AND RECOVERY OF SLUDGES FROM PHYSICAL-CHEMICAL PROCESSES

Dr. Carl E. Adams, Jr., President
Associated Water and Air Resources Engineers, Inc.
Nashville, Tennessee

Introduction

The experience to date in sludge handling with conventional waste treatment schemes has been concerned with the processing and disposal of
primary and secondary organic sludges. In advanced wastewater treatment
schemes, the use of chemical coagulating agents generates tremendous
quantities of sludge of a different nature than these organic sludges.
However, the disposal and recovery of chemical sludges is not a new or
novel experience. Many industries and water treatment plants have been
dealing with similar sludges for years and much of the present experience
was gained from these facilities.   Two combined, biological-physical/
chemical, treatment are in operation at South Tahoe, Nevada and Colorado
Springs, Colorado. The first complete physical/chemical facility in the
United States is being constructed by Graver at Rosemont, Minnesota.

Origin of Physical-Chemical Sludges

Excluding spent activated carbon, the two major sources of sludges originating from physical-chemical treatment systems are those generated
during chemical addition to primary clarification facilities and those
sludges which result from tertiary addition of chemicals to reduce effluent
suspended solids and phosphorus levels. The major difference between these
two sludges, assuming the same chemicals are utilized, is the organic
content in the primary sludges which result from influent and colloidal
suspended materials removed by the chemical coagulation. The addition of
chemicals directly to the aeration basin has been suggested as a possible
method of phosphate removal; however, the resulting sludge will not be
discussed as a physical-chemical sludge since it behaves and requires
similar treatment as a normal waste activated sludge.

The most commonly employed coagulants are the salts of calcium, aluminum,
and iron. Generally, calcium is the form used exclusively when clarification

and upwards pH adjustment are needed for subsequent processing such as
ammonia stripping, phosphorus removal, etc. Aluminum, usually in the
form of alum (aluminum sulfate), is used for clarification and phosphorus
removal at lower pH ranges, i.e., 5 to 7. The resulting sludges settle
slower and are more difficult to dewater than lime sludges. Iron, generally in the form of ferric chloride, has been used similarly for both
coagulation and phosphorus removal. Normally, the iron requirements for
phosphorus removal are considerably higher than alum and will many times
impart an undesirable yellowish color to the treated water. Disposal and
recovery applications with iron sludges are very limited and will not be
discussed in detail herein.

In considering the various alternatives of sludge handling, the most
single important factor involves the ability of the sludge to be dewatered.
The drier sludges, particularly lime, are more amenable to recovery due to
the economy of energy input. Many times the difficulty in dewatering one
type of sludge will generate high costs for a dry method of recovery,
thereby, encouraging the use of another coagulant, such as alum, which
can be recovered by wet processing. With advanced waste treatment schemes
oriented toward recovery, some form of dewatering is essential and must
be considered.

Ocean barging has not been included as a feasible alternative due to its
applicability only to coastal populations and the present hesitancy of
federal authorities to commit on the future status of this method for final
sludge disposal.

Disposal and Recovery of Alum Sludges

The use of alum in waste treatment has been primarily for phosphorus
removal and final polishing ahead of filtration. Alum is an excellent
coagulant for many wastes and will not only clarify turbidity and remove
suspended materials, but forms a stable precipitate with phosphorus. Much
more alum is required for phosphorus removal than for clarification and
the resulting sludges are different in nature and dewater differently.

The sludges generated from the use of alum are gelatinous in structure
with a feathery, bulky nature. The moisture content is approximately
98.5 to 99 percent and, although it settles fairly readily, it is extremely
difficult to dewater. The economics of dewatering alum and alum-sewage
sludges often prohibit its use in wastewater treatment. The most feasible
method of handling alum sludges include land disposal and recovery by
alkaline and acidic methods.

Lagooning and Land Disposal of Alum Sludges

Lagooning has been the most popular and acceptable method of disposal of
alum water treatment sludges for many years. The sludges are added continuously or intermittently until the lagoon is filled whereupon the lagoon
may be abandoned or cleaned for reuse. Frequently, the lagoon is not the
final treatment process but is only an intermediate means for dewatering,
thickening, and providing temporary storage prior to final land disposal.

Although the operating costs of sludge lagooning are quite low, influences,
such as climate, intermittent or continuous operation, percent influent
solids, and availability of one or more alternate lagoons, will decide the

land requirements. At least two lagoons are necessary. It has been well documented that alum sludges are difficult to dewater by lagooning methods. Some studies have indicated that alum sludges would not thicken to greater than 9 percent after years of settling (12). Neubauer (15) performed thorough investigations on lagooning of alum sludges with the results shown in Figure 1.

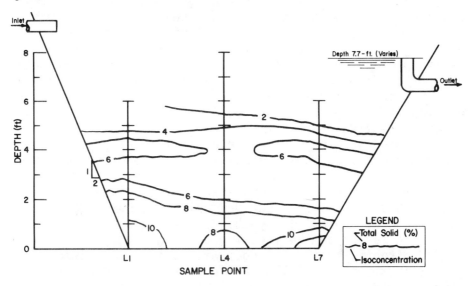

FIG. 1
Iso-concentration Profile for Lagooning of Alum Sludge (Ref 6)

The lagoon, which was 400 ft long by 320 ft wide was in operation for 3 years with an average sludge depth of 7 ft. An average solids level of 4.3 percent persisted and the majority concentration was less than 10 percent in the lagoon. It was concluded that a lagoon of this nature would not produce a sludge suitable for landfill without additional dewatering.

In lagooning operations, the water removal is by decantation, evaporation, and a little drainage or percolation. Evaporation may produce a hard crust on top but the remaining depth of alum sludge is thixotrophic and turns to viscous liquid upon agitation with near zero shear resistance under a static load (1). In many cases, the sludge is removed from the lagoons by drag line or clam shell operations and allowed to dry on the lagoon banks prior to subsequent removal and disposal by land. Overall, lagooning must be considered an inefficient process. Nonetheless, where land is cheap and available, it is difficult to justify any other disposal method.

Sankey (19) showed that alum sludge could be successfully applied directly to land at rates of one inch per year with no effect on vegetation. A method was proposed which involved removing the top soil, spreading dried alum sludge, and resoiling. Other investigators (6, 9) reported significant clogging of the soil by aluminum hydroxide so that the resulting material was unfit for vegetation. However, after the soil had frozen and thawed, the sludge structure was altered and vegetation thrived. It appears that land disposal is satisfactory if the chemical structure can be destroyed to release the water of hydration in order to prevent significant soil clogging. Generally, it is assumed that a solids content of 20 percent or more is necessary for landfill applications (11, 15).

Natural Freezing of Aluminum Sludges

Although not a direct disposal process, the application of freezing techniques has increased the potential of lagooning as a disposal method in colder climates due to the increased solids content which can be obtained. The freezing and thawing of aluminum sulfate sludges has been shown to significantly increase the rates of settling and filtration by conversion of the hydrogel structure into a suspension of granular solids. The process of freezing removes the water of hydration from the gelatinous aluminum hydroxide sludge. Upon thawing the sludge does not revert to its original stucture, but ends up as clear water and small granular particles resembling grains of sand with the volume reduced approximately one-sixth of the original. The use of artificial or mechanical freezing equipment has not been shown to be economical and is not discussed herein.

Farrell et al (10) conducted comprehensive investigations on natural freezing of aluminum sludges and demonstrated that aluminum hydroxide sludges freeze at approximately the same rate as water. Therefore, existing information on the freezing of water, which is plentiful, can be applied directly to alum sludges with a small provision of safety factor. In summary, the following conclusions were drawn for natural freezing techniques (10):

1. Aluminum hydroxide sludges will freeze at rates similar to those of water. These sludges may require 10 percent longer to freeze due to the reduction in natural convection during the time the liquid is being cooled to the freezing point.

2. The presence of snow on aluminum sludges can severely retard the rate of freezing, even in very cold conditions. Operational techniques can minimize the effects of snow cover by freezing in thin layers, by ceasing sludge application when snow is expected, by melting the snow previous to sludge application by a water flush or by layering out more sludge, and by plowing off the snow prior to sludge application.

3. The beneficial effects of freezing, such as dewaterability and solids content, are dramatically increased by a slow freezing and thawing, provided that freezing is complete. Partial freezing will also increase the dewaterability and solids content of the sludge but to a much smaller degree than complete freezing.

4. Freezing in a mild climate can be successful only if thin layers of sludge, i.e., less than one inch, are applied in each application. This will increase management costs significantly.

5. The presence of significant quantities of aluminum phosphate will not significantly effect the freezing of aluminum sludges.

Alkaline Recovery of Alum

The alkaline methods of alum recovery have derived from the original work

conducted by Lea and Rohlich (13). Slectha and Culp (21) evaluated this
method on a laboratory method at South Tahoe, Nevada, and concluded that
the economics were not favorable based on chemical costs alone.

Basically, the alkaline recovery method consists of adding sodium hydroxide to raise the pH above 12 and dissolve the aluminum hydroxide sludge.
The aluminum hydroxide is converted to sodium aluminate at the high pH
and the precipitated phosphate ions are released back into solution.
Calcium chloride is then supplemented to precipitate the phosphate as
insoluble calcium phosphate (hydroxyapatite). The following reaction
describes the process:

$$Al(OH)_3 \cdot PO_4 + 4NaOH \qquad NaAlO_2 + Na_3PO_4 + 2H_2O + 3OH^-$$

$$NaAlO_2 + 2Na_3PO_4 + 3CaCl_2 \qquad NaAlO_2 + Ca_3(PO_4)_2 + 6NaCl$$

The sodium aluminate, which is thus generated, is not as effective a coagulant as aluminum sulfate and sufficient sulfuric acid must be added to
adjust the pH downward and reform aluminum sulfate. Figure 2 indicates

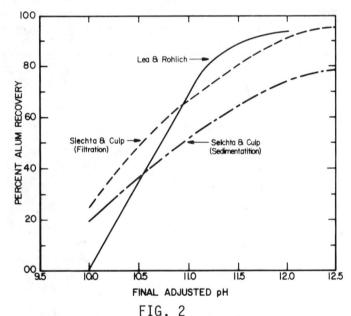

FIG. 2
Percent Alum Recovery by the Alkaline Method (Ref 6)

that the efficiency of alum recovery may be as high as 90 to 94 percent (4) depending on whether sedimentation or filtration is used to
recover the alum. The pH must be adjusted to greater than 12 initially
or side reactions will occur with the calcium chloride to precipitate
the aluminum ion. Slechta and Culp (21) obtained 78 percent recovery by
sedimentation and required 50 percent makeup alum to achieve the same
turbidity removals as fresh alum. Therefore, the process was deemed uneconomical and was not pursued further at South Tahoe.

Another method of alkaline recovery of alum, which is reported by Slechta
and Culp (21), employs lime for both the pH adjustment and phosphorus precipitation. Lime is much cheaper than sodium hydroxide and facilities would
only be needed for one chemical. The recovery efficiency was low at

about 35 percent; however, the overall costs were lower at $60.90 per ton of reclaimed alum than the alkaline method. Fresh alum can be purchased for $57.50 per ton. Thereby, on a strict chemical cost the recovery method is uneconomical. If the economics of sludge handling are included, these recovery techniques will become much more feasible.

Acidic Method of Alum Recovery

Generation of aluminum sulfate from the addition of sulfuric acid to aluminum hydroxide or hydrous aluminum oxide sludges has been practiced at several water treatment plants throughout the United States and Great Britain. Extensive studies by Roberts and Roddy (18) at the Tampa, Florida, water treatment plant paved the way for successful demonstration of reuse and recovery of alum. The equations of the process are:

$$2Al(OH)_3 + 3H_2SO_4 \qquad Al_2(SO_4)_3 + 6H_2O$$

Approximately 1.9 lb of sulfuric acid are required per lb of aluminum hydroxide recovered. Slechta and Culp (21) found greater than 98 percent recovery at pH levels of 1 to 2 as shown in Figure 3. More refined alum can

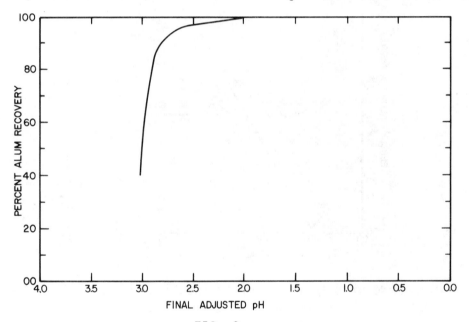

FIG. 3
Percent Alum Recovery by the Acidic Method (Ref 6)

be formed by drying the sludge to form aluminum oxide prior to the sulfuric acid addition. The subsequent reaction is:

$$Al_2O_3 + 11H_2O + 3H_2SO_4 \qquad Al_2(SO_4)_3 + 14H_2O$$

In experimental studies (18), an air dried sludge contained 25 to 30 percent aluminum oxide while an oven dried sludge at 105°C contained about 45 to 50 percent which is similar to low grade bauxite. Roberts and Roddy decided that the wet method of treating thickened, clarified sludge involved the least amount of equipment and was cheaper than subsequent dewatering and drying methods. Culp (4) estimated that the cost of the acidic recovery method is approximately one-third the cost of buying fresh alum.

The major disadvantage of the acidic recovery method of alum is the dissolution of phosphorus, which returns to the supernatant of the sludge handling system, at the low pH range. Phosphate removal from the supernatant by ion exchange and activated silica was examined and found to be economically prohibitive (4). Therefore, the acidic method of alum recovery can be very effective, but only if phosphate removal is not required.

Disposal and Recovery of Calcium Sludges

The cost of lime has favored use of this chemical in wastewater treatment for coagulation of suspended and colloidal organic materials and also precipitation of phosphate compounds. At the proper pH, insoluble calcium carbonate is formed as shown below:

$$Ca(OH)_2 + Ca(HCO_3)_2 \qquad 2\ CaCO_3 + 2H_2O$$

Calcium carbonate floc entraps suspended organics and is useful for primary clarification. The sludge which is produced has a very high lime and organic content. Calcium phosphate or hydroxyapatite is formed by the reaction of lime and phosphate as follows:

$$5CA^{++} + 4OH^- + 3HPO_4^- \qquad Ca_5OH(PO_4)_3 + 3H_2O$$

The most applicable techniques of lime disposal and reuse include lagooning and recovery by recalcining.

Lagooning of Calcium Sludges

The use of lagoons for water treatment plant softening sludges is quite common. Moisture contents of 50 percent can be obtained by lagooning and an applied sludge with an initial moisture content of 90 percent will concentrate to approximately 16 percent of the original volume. Depths from 3 to 10 ft are employed and generally the supernatant is returned to treatment processes or allowed to percolate

Where sludge is settled through ponded water, one can expect the solids content to only reach a 40-percent maximum and in many cases only 20 to 30 percent have been observed. If the lagoon is allowed to drain so that the supernatant water is removed without ponding, the upper limit of 50 percent dry solids may be obtained.

Odors from lime sludge are rare, and when they occur they are usually only temporary and insignificant. It is possible to recover the land after the lagoons have filled if cleaning and additional disposal to other land sites is not required. The low land around the Miami, Florida treatment plant was filled with lime sludge, allowed to thicken, and covered with one to two inches of soil and sown with grass (8). Dean (5) indicates that fills may be made by continuously applying sludge without pre-drying but that heavy loads should not be applied for several years after the land is reclaimed.

Primary, lime-organic sludges can be disposed directly onto land at pH values greater than 11.5 (2). The sludge has been placed in trenches at loading rates as high as 240 tons per acre per year. However, preliminary indications implied that there is excessive nitrogen for agricultural uses and that this nitrogen can enter the water table. Therefore, nutrient loading may control the rate of sludge application rather than moisture content or pH.

Lime Recovery and Reuse

As primary coagulation and primary and tertiary phosphate removal systems become more prevalent in waste treatment, the recovery and reuse of lime will undoubtedly receive considerable attention. The most commonly used method of reclaiming lime is by the process of recalcination where lime sludges are heated to regenerate calcium oxide. Basically, the process of calcination involves incineration of sludges to convert carbonates and hydroxides to oxides. The re-calcination of lime sludges consists of burning the dewatered sludge in the temperature range of 1800 to 2000°F in order to drive off water and carbon dioxide leaving the calcium oxide as shown in the following equation:

$$CaCO_3 \quad \xrightarrow{\quad 2000°F \quad} \quad CaO + CO_2$$

The CO_2 which is released is generally reclaimed and employed for recarbonation and pH adjustment. Major processes which have found the most frequent application in recalcining line sludges are the rotary kiln, the pellet-seeding calciner, and the multiple hearth furnace. Each of these processes will be briefly described below.

Rotary Kiln Process

A typical rotary kiln process is illustrated in Figure 4 (3).

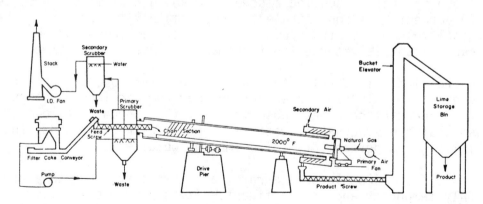

FIG. 4
Rotary Kiln Process for Lime Sludge Calcining (Ref 2)

Generally, the lime sludge is dewatered to approximately 60 to 65 percent solids by a vacuum filter or centrifuge. The feed end (chain section) of the kiln contains a heat-recuperating chain in order to promote efficient exchange of heat between the hot exhaust gases and the incoming sludges. The net effect of the chain is to reduce the fuel required in the process, e.g., five gallons of fuel oil per ton of lime produced (22).

The kiln shell consists of a refractory lining and rotates at approximately one revolution per minute although slower speeds are not uncommon. The long kiln is inclined at about 1/2 in. per ft to allow proper traveling of the dried sludge to the firing end and to allow an approximate retention time of 1.5 hours. At the firing or calcining end of the kiln, the sludge is nodulized and converted to calcium oxide. The product is then discharged into an integral tube cooler where it is cooled to 600°F by secondary combustion air, thereby, effecting another transfer of heat to improve the overall efficiency of the operation.

Primary combustion air is supplied to a gas or oil burner for flame control. Secondary combustion air is drawn through the cooler cylinders and the temperature at the firing end of the kiln is maintained at 1800-2200°F. The kiln exhaust gas is approximately 400°F at the feed end of the housing. In Miami (22) a draft control is provided by a butterfly damper in the exit duct located beyond the tower scrubber and ahead of the induced draft fan. It has been found that a residual of 2 percent oxygen in these gases at the duct will result in the maximum fuel (oil) efficiency without the danger of producing noxious stack fumes. Therefore, damper control on a kiln is all important.

Fluidized Bed Process

The term "fluidized" is used because a dewatered sludge is fed into a bed of fluidized particles, such as sand, supported by upward moving air which is supplied at pressures of 3.5 to 5.0 psig. The sludge particles are kept suspended by the moving gas stream which causes the gas-particle mixture to behave as a liquid. In recalcining calcium carbonate sludge, the sludge is fed to the fluidized bed reactor where it is instantly transformed into calcium oxide. The calcium oxide adheres to previously calcined and crushed "seed" pellets which are also fed into the calciner. The schematic illustration in Figure 5 describes the fluidized process for a water treatment sludge at Lansing, Michigan (1). The filter cake from the dewatering process (vacuum filter or centrifuge) is mixed with previously dried sludge in a paddle mixture and discharged to the cage mill and flash drying system which employs reactor off-gases for the drying operation. The dried sludge is removed in a centrifugal cyclone and discharged to storage bins prior to feeding to the reactor.

Crushed lime pellets are added as active seeds on which the dried lime is deposited and calcined. Pellet growth is promoted by the addition of soda ash to the dewatered sludge just prior to flash drying. The soda ash fuses at a calcining temperature of approximately 1600°F, thereby, forming a molten nucleus to which the calcium oxide particles and seed pellets can adhere. The pellets continuously grow in size until they can no longer be supported by the gas velocity in the reactor. The heavy pellets are then withdrawn through a differential pressure transfer device to a cooling compartment where the product is cooled and the influent

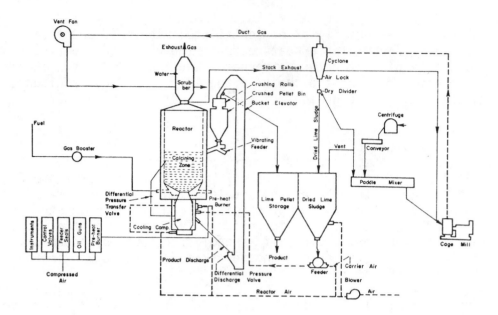

FIG. 5
Fluidized Bed Process for Lime Sludge Calcining (Ref 2)

air preheated. The finished product is carried to a storage bin by bucket elevator where a small amount is diverted by crushing to form reseeding particles.

Multiple Hearth Furnace

The multiple hearth furnace is a simple device consisting of a steel shell lined with a refractory material. The furnace is divided into separate compartments, called hearths, with alternate holes at the periphery to allow feed solids to drop through each succeeding hearth. The center shaft is driven by a variable speed motor which rotates the "rabble" arms on each hearth to move the sludge inwards and outwards on alternate hearths.

The multiple hearth system has three distinct operating zones as shown in Figure 6. The top hearths, or dry zones, dry the incoming sludge

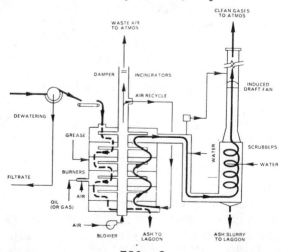

FIG. 6
Multiple Hearth Furnace for Sludge Disposal (Ref 12)

by utilizing heat from the lower hearths. Some organic burning and
carbon dioxide distillation occur in this top zone. It is very important not to let the temperature of the top hearth become too low
in order to avoid clinker formation due to the slow drying and balling
of the lime cake. The middle hearth, or recalcining zone, produces
the calcining reaction at temperatures in the range of 1800 to 1900°F.
Finally, the lower hearths, or cooling zone, cool the calcines and give
up heat to the incoming combustion air. A typical profile across a
six hearth furnace (20) is shown below:

| | Temperature | |
Hearth	(°F)	(°C)
1	770	410
2	1490	810
3	1670	910
4	1940	1060
5	1900	1038
6	800	427

The hot calcined lime exits the furnace at approximately 800°F and
passes through a lump breaker and into a cooler where the lime is
cooled to about 200°F by indirect water cooling. The recalcined
lime is then sent to storage prior to reuse.

Summary of Recalcining Processes

These previous studies have indicated that recalcining plants with
capacities as low as 6 tons per day may be feasible where sludge
disposal is critical and an expensive problem. The product costs
for the recalcined lime range from 7.5 to $30.00 per ton in water
and wastewater recovery facilities. The lower costs are generally
with water treatment plants because they are reclaiming more lime
from the softening operation for resale. The use of the rotary kiln
is limited to large operations greater than 50 tons per day while the
fluidized beds are more useful for small installations. The rotary
kiln requires a greater land area and dissipates more heat because it
is less compact. The multiple hearth furnace satisfies intermediate
needs between the small and large operations. A summary of existing
data from various water and wastewater recalcining plants in the
United States is shown in Table I (7).

General Considerations for Lime Recycle

Probably the most important consideration in reclaiming lime is the
quantity of inert materials which accumulate through the continuous
recycling. The wastage rate and the degree of lime makeup required
are controlled by the quantity of inerts which must be removed from
the system. With primary sludges a tremendous quantity of inert
materials accumulate from the residue of burning organics. This
large quantity of inert residue generally will render the recycle of lime
infeasible. However, two centrifuges in series can be employed to
correct the situation. The first centrifuge is used to classify the
solids by separating the more dense calcium carbonate solids into the
cake. This calcium carbonate sludge then goes to incineration for
recalcining and reclaiming. The centrate from the first centrifuge,
consisting of organic and phosphate solids, is passed to the second

TABLE I

List of Recalcining Plants in United States

Location	Approximate Construction Year	Capacity Product Output tons/day	Type of Dewatering Equipment*	Type of Recalciner**	Requirement mil Btu/Ton of Output	CO$_2$ Recovered	Lime Plant Cost $	Water Plant Rated Capacity mgd	Lime Fed lb/mg	Sludge/Lime Ratio
Marshall, Iowa	-	4.2	C	FC	9.66	-	-	3.5	-	-
Pontiac, Mich.	-	-	C	FC	-	-	241,000	10	2,200	2.50
Miami, Fla.	1948	80	C	RK	8.5	Yes	856,000	180	1,800	2.20
Lansing, Mich.	1954	30	C	FB	8.0	-	-	20	2,200	2.27
Salina, Kan.	1957	-	C	FC	-	-	-	-	-	-
Dayton, Ohio	1960	150	C	RK	9.0	Yes	1,500,000	96	2,140	2.47
San Diego, Calif.	1961	25	C	RK	9.0	-	534,000	-	-	-
S. D. Warren Co.+	1963	70	VF	FB	7.2	-	-	-	-	-
Merida, Yucatan	1965	40	C	RK	-	Yes	-	24	-	-
Ann Arbor, Mich.	1968	24	C	FB	-	Yes	-	-	-	-
Lake Tahoe, Nev.++	1968	10.8	C	MH	7.0	Yes	516,000	7.5	3,300	-
St. Paul Minn.§	1969	50	C	FB	8.2	Yes	1,750,000	120§§	990	2.40

*   C - centrifuge; VF - vacuum filter.
** RK - rotary kiln; FB - fluidized bed; FC - flash calcination; MH - Multiple hearth.
+   Paper Mill owned by Scott Paper Company.
++ Activated sludge disposal plant.
§   Now under construction.
§§ Lime recalcining plant designed to accommodate future water treatment plant capacity of 170 mgd.

centrifuge for concentration and eventual burning in the sewage sludge
incinerator.

In both primary and tertiary applications of lime, the reaction pH of
9.5 results in phosphorus removals on the order of 85 to 95 percent
but is accomplished by high effluent turbidity due to colloidal and
pinpoint particles. As the pH is increased above 9.5, magnesium
hydroxide will precipitate until the reaction is essentially complete
at a pH of 10.5 to 11.0. The magnesium hydroxide precipitate will
remove the majority of the colloidal materials resulting in a clear
supernatant; however, the hydrous nature of this precipitate has adverse
effects on subsequent thickening and dewatering processes. Recarbonation
of the resulting sludges to a pH of 9.5 will redissolve much of the
magnesium hydroxide and precipitate the remaining free calcium as calcium
carbonate.

Three investigations have studied the characteristics of lime with regard
to reuse and waste treatment systems (14, 17, 21). Rand and Nemerow (17)
performed laboratory studies with raw wastewater and found no change in
the ability to remove phosphates with recovered lime and measured an
average makeup demand of 13 percent. Slectha and Culp (21) worked
with secondary effluent and recovered lime through 11 cycles. The
results indicated an average makeup of 36 percent. The accumulation
of inert materials, such as hydroxyapatite, in the recalcined ash resulted in a decrease of the percent calcium oxide in the ash as shown
in Figure 7. The phosphorus removal was not hindered by returning

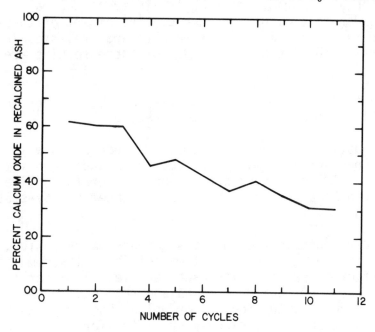

FIG. 7
Effect of Lime Recycle on Phosphate Removal (Ref 6)

reclaimed lime as illustrated in Figure 8.

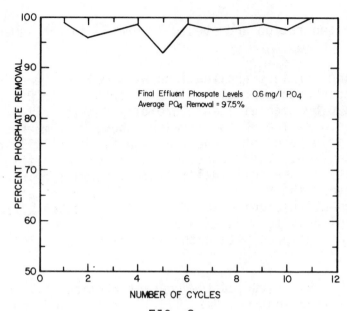

FIG. 8
Characteristics of Recalcined Lime Sludge (Ref 6)

Mulbarger et al (14) observed that recycled lime sludges dewatered better than raw sludges, probably due to the accumulation of inert materials. In the same studies the characteristics of lime sludges, both before and after three recycles, were examined in soft and hard waters. The phosphate content and the alkalinity of the raw waste were varied in order to examine the accumulative effect of these qualities on the recycled lime. These results are presented in Figure 9 and several conclusions derived from these results are summarized below:

1. The quantity of sludge generated is directly proportional to the lime dosage, as would be expected.

2. The combustion loss and calcium oxide production will remain constant with multiple lime reuse under consistent water quality and system performance. Therefore, lime makeup can be expected to be constant under steady-state conditions.

3. Also under consistent conditions, magnesium hydroxide, hydroxyapatite and other inert materials will accumulate at constant increments with each recycle. In order to control the inert buildup, a wastage rate must be established for each situation, after which the sludge mass and composition can be closely predicted for any number of recycles.

4. Carbonation of the sludge will dissolve the magnesium hydroxide resulting in slightly less sludge production.

5. In hard waters with a low phosphorus content, lime recovery may exceed lime usage. Consequently, lime wastage will control the wastage of other inert constituents.

6. In wastewaters of low alkalinity and high phosphate content, hydroxyapatite accumulates rapidly and may reach levels found in low grade phosphate rock.

The hydroxyapatite levels in recalcined lime ash have been found to have neglible solubility in distilled water (14). Therefore, disposal in landfill sites should not cause a pollutional discharge due to phosphorus runoff. The citrate solubility, which measures phosphate availability to plants, in the recalcined ash was measured to be in the range of 80 to 94 percent which is the range of good phosphate fertilizers. Thus, the sale of sludges for fertilizer value might be feasible in some locations.

References

1. Adrian, D. D. and J. H. Nebiker, "Disposal of Wastes from Water Treatment Plants," AWWA Research Foundation Report, J.AWAA, v. 61 p. 682 (Dec 1969).

2. Bishop, D. F., Environmental Protection Agency, Washington, D. C., Personal Communication (1972).

3. Crow, W. B., "Techniques and Economics of Calcining Softening Sludges: Calcination Techniques," J.AWWA, v. 52, p. 332 (Mar 1960).

4. Culp, R. L. and G. L. Culp, Advanced Wastewater Treatment, Van Nostrand Reinhold Co. (1971).

5. Dean, R. B., "Disposal of Wastes from Filter Plants and Coagulation Basins," J.AWWA, v. 45, p. 1226 (Nov 1953).

6. Dean, R. B., "Waste Disposal from Water and Wastewater Treatment Processes," Proc. 10th San. Engr. Conf., Uni. of Ill., Urbana, Ill. (1968).

7. Dick, R. I. and R. B. Dean, "Disposal of Wastes from Water Treatment Plants," AWWA Research Foundation Report, J.AWWA, v.61, p. 543 (Oct. 1969).

8. Dittoe, W. A., "Disposal of Sludge at Water Purification and Softening Works of the Mahoning Valley Sanitary District," J.AWWA, v. 25, p. 1523 (Nov 1933).

9. Doe, P. W., "The Treatment and Disposal of Washwater Sludge," J. Inst. Water Engrs., v. 12, p. 409 (1958).

10. Farrell, J. B., J. E. Smith Jr., R. B. Dean, E. Grossman III, and O. L. Grant, "Natural Freezing for Dewatering of Aluminum Hydroxide Sludges," J.AWWA, v. 62, p. 786 (Oct 1970).

11. Krasauskas, J. W., "Review of Sludge Disposal Practices," J.AWWA, v. 61, p. 225 (May 1969).

12. Krasauskas, J. W. and L. Streicher, "Disposal of Wastes from Water Treatment Plants," AWWA Research Foundation Report, J.AWWA, v. 61, p. 621 (Nov 1969).

13. Lea, W. L., G. A. Rohlich, and W. J. Kalz, "Removal of Phosphates from Treated Sewage," Sew. and Ind. Wastes, v. 26, p. 261 (Mar 1954).

14. Mulbarger, M.C., E. Grossman III, and R. B. Dean, "Lime Clarification, Recovery and Reuse for Wastewater Treatment," Project for U. S. Dept. of Interior, FWPCA, Cincinnati, Ohio (Jun 1968).

15. Neubauer, W. K., " Waste Alum Sludge Treatment," J.AWWA, v. 60, p. 819 (Jul 1968).

16. Proudfit, D. P., "Selection of Disposal Methods for Water Treatment Plant Wastes," J.AWWA, v. 60, p. 674 (Jun 1968).

17. Rand, M.C. and N. L. Nemerow, "Removal of Algal Nutrients from Domestic Wastewaters," Rept. No. 9, Dept. of Civil Engr., Syracuse Uni. Research Inst. (1965).

18. Roberts, J. M. and Roddy, C. P., "Recovery and Reuse of Alum Sludge at Tampa," J.AWWA, v. 52, p. 857 (Jul 1960).

19. Sankey, K. A., "The Problem of Sludge Disposal at the Arnfield Treatment Plant," J. Inst. of Water Engrs, v. 21, p. 367 (1967).

20. Sebastian, F.P., "Advances in Incineration and Thermal Processes," The Theory and Design of Advanced Waste Treatment Processes, Short Course, Uni. of Calif., Berkeley, University Extension (Sept 1971).

21. Slechta, A. F. and G. L. Culp, "Water Reclamation Studies at the South Tahoe Public Utility District," J.WPCF, v. 39, p. 787 (May 1967).

22. Wertz, C. F., "Techniques and Economics of Calcining Softening Sludges: Miami Lime Recovery Plant," J.AWWA, v. 52, p. 326 (May 1960).

23. Young, E.F., "Water Treatment Plant Sludge Disposal Practices in the United Kingdom," J.AWWA, v. 60, p. 717 (Jun 1968).

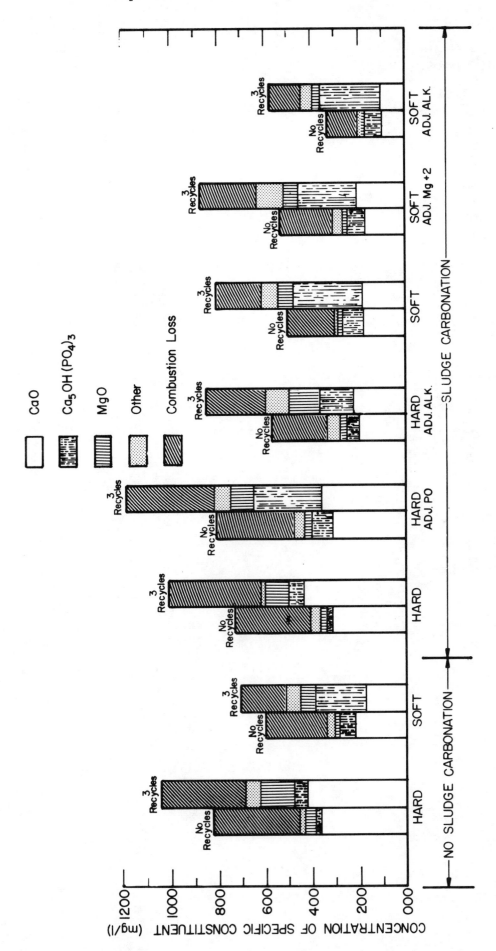

FIG. 9

Characteristics of Recalcined Lime Sludge (Ref 20)

Applications of New Concepts of Physical-Chemical
Wastewater Treatment
Sept.18-22, 1972

THE ROLE OF FREEZING PROCESSES IN WASTEWATER TREATMENT

James H. Fraser and Wallace E. Johnson
Avco Systems Division
Wilmington, MA   01887

The primary role of freezing is to provide a means to concentrate and at the same time recover water for reuse from wastes containing dissolved solids.

The idea of a freezing process to extract pure water from a solution has been under active investigation for about 20 years, primarily as a means of converting saline water to fresh water. Several forms of freezing processes have evolved. (1, 2) All of the processes are based on the fact that when ice crystal forms in an aqueous solution the crystal formed is pure water excluding all impurities found in the original solution. These processes all have inherent advantages that have given them potential economic attractiveness:

1. Low energy consumption
2. Low capital cost
3. No pre-treatment required
4. Universal applicability to a wide variety of waters

Low energy consumption results from direct contact heat transfer, resulting in small temperature differences relative to the theoretical temperature difference and thus low energy consumption. (The low energy consumption is not related to the fact that the latent heat of fusion is 1/7 that of the latent heat of evaporation. This is compensated for by the fact that the freezing point depression is approximately seven times greater than the boiling point evaluation, so that theoretically the energy requirements are nearly identical to that of distillation. (2)) Also where heat transfer surfaces are involved the optimum temperature differences are lower than in other desalination processes because of the use of lower cost materials of construction.

Low capital cost results from: 1) direct contact heat exchange in the
crystallizer and thus the elimination of heat transfer surfaces and possible
scaling and/or fouling thereof and 2) low cost materials of construction
made possible by the lack of corrosion in the low temperature, oxygen
free environment found in the process. This permits use of mild steel,
plastics and other inexpensive materials.

Pre-treatment of the feed water is not required, also because of direct
contact heat transfer in the crystallizer. Because there are no inert transfer surfaces there are no concentration gradients near surfaces to cause
scaling and if the solubility limit of some material in the feed is exceeded
in the bulk, precipitates if formed, leave the process with the brine. (3, 4)

The universality of the process results from the above advantages in that
since scaling and fouling are not problems and no pre-treatment is required, almost any liquid containing dissolved solids can be treated. Systems are being developed (5) whereby precipitates can be separated from
ice thus lending the process suitable for by-product recovery. Solutions
of organics in water can be handled as well as inorganics thus making the
process suitable for the concentration of a variety of industrial wastes as
well as desalination.

Process Descriptions

General Process Description

For all direct contact processes a general process description can be applied, Figure 1. The feed water is passed through a heat exchanger and
is cooled to within a few degrees of its freezing point by the fresh water
and brine streams leaving the process. The feed enters a crystallizer
where the refrigerant (either water vapor or immiscible refrigerant) is
evaporated thus removing heat from the feed water and forming ice. The
refrigerant vapor is pumped out of the crystallizer by a compressor and
is recontacted with the ice, which has been washed, gives up its latent
heat by condensing thus causing the ice to melt. The melted ice becomes
the fresh water. The ice is pumped out of the crystallizer to the counterwasher as a slurry with the concentrated feed water.

The counterwasher is a simple vertical vessel with a screened outlet
located midway between top and bottom. The slurry enters the bottom and
forms a porous plug, the majority of the brine leaving the plug through
the screen, and out of the process through the heat exchanger. The ice
plug is propelled through the column by the force created by the pressure
drop caused by the flow of brine through the ice plug, the velocity of the
brine being much greater than that of the ice. A portion of the product
water is applied to the top of the ice plug and flows down counter-currently
to the ice. Ideally, the velocity of the wash water and the ice are numeri-
cally equal, but opposite in sign, so that theoretically no net downward
flow of wash water is required, the fresh water displacing the brine by

ideal plug flow. Because of the low velocities involved the theoretical case
is closely approximated with a net loss of less than 5% of the product
resulting.

Because of inefficiencies in the process, a greater amount of refrigerant
vapor is generated than can be condensed by the ice. A heat removal system is, therefore, required to maintain thermal equilibrium. This consists
of a system which removes the excess vapor by condensing it or raising it
to a pressure such that it will condense on ambient cooling water.

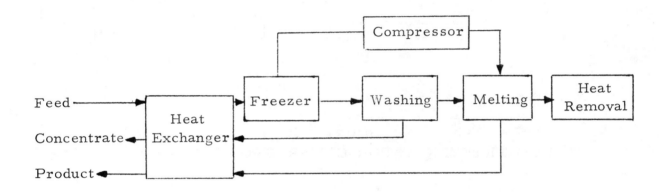

FIGURE 1
GENERAL PROCESS SCHEMATIC

Three of the most distinctive versions of freezing processes are described briefly in the following paragraphs.

Vacuum-Freezing Vapor-Compression Process

This process (6) has the least elements and most closely resembles the
general case. As shown in Figure 2, water vapor is evaporated in the
crystallizer, compressed and condensed in the melter. The heat removal
system consists of a refrigerated heat transfer surface, located in the
melter, on which the excess vapor is condensed. Because of the large
volumes of vapor that would have to be handled in the air purge if the air
in the feed were not removed a deaerator is required. The primary advantage of this system is its relative simplicity. Disadvantages are its
large size and low efficiency, both caused by the large volumes of low
pressure vapor which must be handled, i. e., the crystallizer operates at
a pressure of about 0. 08 psia where water vapor has a specific volume of
4, 600 cubic feet per pound.

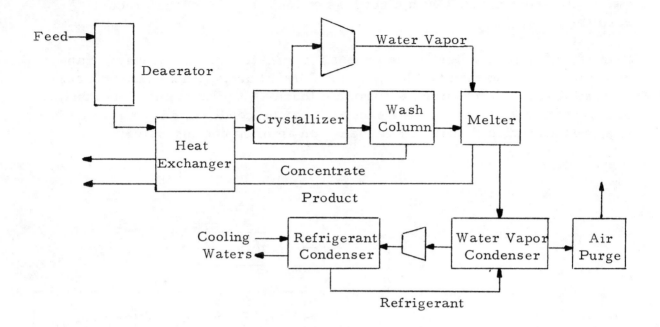

FIGURE 2

VACUUM-FREEZING VAPOR-COMPRESSION PROCESS

Secondary Refrigerant Process

This process, (3, 7, 8, 9) Figure 3, overcomes the disadvantages of low
pressure by introducing an immiscible refrigerant into the system, rather
than using water vapor as a refrigerant.

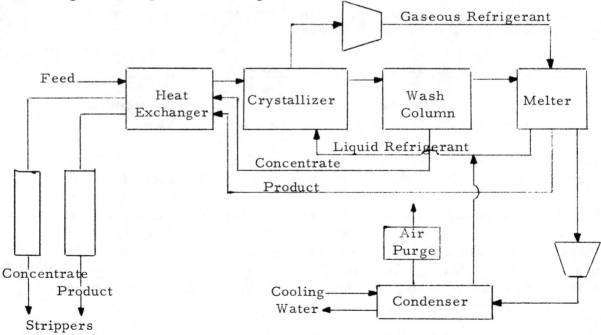

FIGURE 3

SECONDARY REFRIGERANT PROCESS

Several refrigerants have been used including butane and R-114. The volume of vapor that has to be handled is about 1/100 that of water vapor. Because of the higher pressure operation it is economical to purge the air dissolved in the feed in the air purge system and eliminate the deaerator. Also because of the higher pressures it is economical to compress the excess refrigerant vapor in the melter to a pressure high enough so that it will condense on cooling water. This eliminates the energy loss associated with condensing the excess water vapor on a refrigerated surface as in the vacuum freezing process. The primary disadvantage of the process is that additional equipment must be added to strip the refrigerant from the effluent streams.

Vacuum-Freezing Ejector Absorbtion Process

This process (10), Figure 4, attempts to overcome the size limitations and efficiency disadvantage of the mechanical compressors in the vacuum-freezing vapor-compression (VFVC) by combining an ejector and absorbtion system. An ejector powered by a portion of the steam produced in the regenerator replaces the main (mechanical) compressor used in the VFVC process. The absorbtion system and condenser also replaces the mechanical refrigeration system. The system is more complex than the previous systems described, however, it is less mechanical in that both the primary and heat removal compressors have been eliminated.

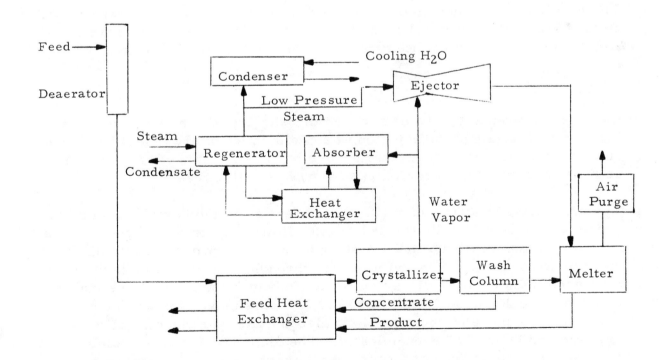

FIGURE 4
VACUUM-FREEZING EJECTOR ABSORBTION PROCESS

The system still retains the physical size disadvantage associated with low pressure operation. In addition, it has the disadvantage that the absorbent being used, sodium hydroxide, is corrosive and more expensive materials of construction must be used in parts of the system. The use of steam as the primary source of energy is an advantage in locations where low cost steam is readily available, however, if steam generation equipment must be installed the additional capital equipment makes this process expensive and more complex.

Another absorbtion process was proposed by Carrier (11) in which the main compressor was eliminated by absorbing all of the water vapor and then melting the ice with hot water. Heat removal was accomplished by conventional mechanical refrigeration, however.

The Crystalex^{tm} Process

The CRYSTALEX process (7, 8, 12) incorporates several features which differ significantly from previous freezing processes. They are:

1. A pressurized wash column
2. A small high capacity crystallizer
3. Indirect contact melting
4. Use of a non-toxic, non-flammable halocarbon refrigerant

The combination of these items permit development of what we believe will be an advanced commercially viable process. The schematic shown in Figure 5 shows how these items will be combined in this process.

Wash Column

Contrary to several published articles the washing of the brine from the ice by means of a hydraulic piston wash column (counterwasher) has been quite successful. (6, 9) Several groups have demonstrated this equipment as being reliable. However, it has been physically large.

The pressurized wash column (13, 14) differs from those utilized in other processes by being totally flooded and not limited by gravity forces. This makes the wash column a much more responsive part of the system. Because it is not limited to gravity heads, throughput of the column can be increased many fold. A unit throughput of about 30 times that obtainable in a gravity column has been realized. (7) This makes the wash column a much smaller piece of equipment. It also means that much smaller ice crystals can be washed economically. The diameter of a wash column, for a given set of operating conditions, varies inversely as the square of the crystal diameter. In the past, major development efforts have been expended to obtain large ice crystals so as to reduce the size of the wash column. This no longer is a prime requisite as changes in operating pressures can compensate for changes in crystal size. This also makes

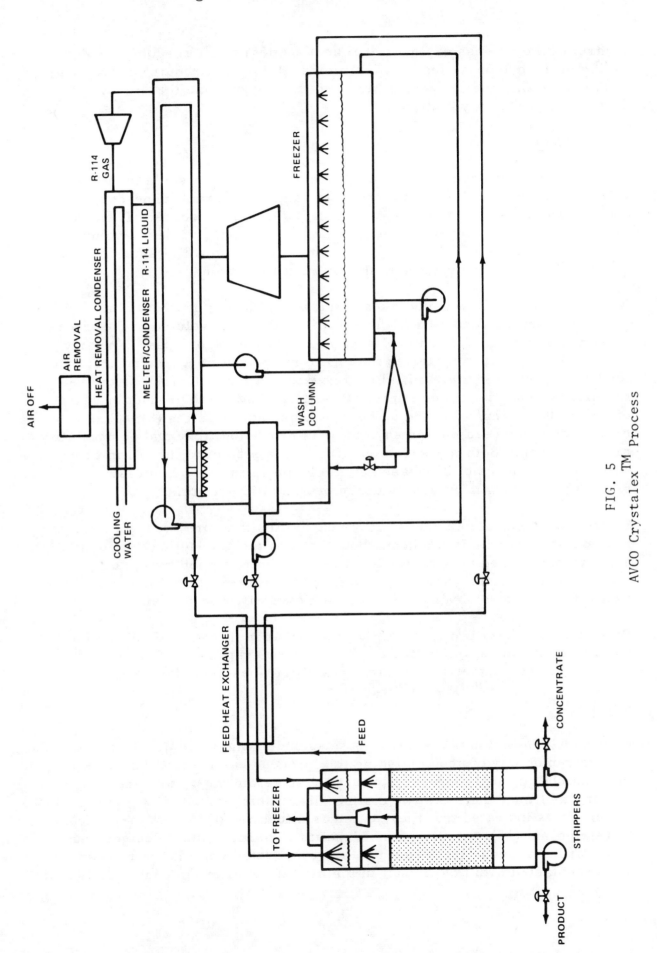

FIG. 5

AVCO Crystalex^{TM} Process

control of the process easier in that another degree of freedom is added.
Sizing of the column is less critical in that it does not require overly conservative sizing due to uncertainty as to the crystal size that might result
from an unproven crystallizer.

Crystallizer

The crystallizers used in previous processes have been some form of
stirred tank device, i.e., the feed water and refrigerant (if needed) are
introduced into an agitated tank where crystallization occurs. These devices had relatively large retention times and did not utilize the volume of
the vessel very well. A more compact higher unit capacity spray crystallizer has been incorporated into the CRYSTALEX process.

In a spray freezer, freezing takes place within falling droplets of a mixture
of refrigerant and feed water. This utilizes the total volume of the freezer
rather than just an agitated pool accompanied by a vapor space of nearly
equal volume which contributes nothing to the production. Refrigerant and
feed water are intimately mixed by a nozzle and ice is formed as the refrigerant evaporates in the spray. Results indicate that production rates
of about 700 lbs/hr/ft^3 based on spray volume are possible (8). The
crystal size has been measured both by microphotography and permeability
pressure drop. Both measurements indicate a size of about 120 microns
with a driving force of 2°F based on the brine salinity. The crystals are
smooth platelets without any evidence of dendritic growth.

Stirred tanks can also achieve higher capacities than previous pilot plants
have demonstrated. It has been shown (8) that the production rate obtainable from a stirred tank varies inversely as the retention time, for a
given degree of turbulence and a fixed driving force. Thus higher production rates are possible from stirred tank crystallizers as well.

By combining these two types of freezers, that is introduction of the feed
and refrigerant in a spray, followed by further evaporation in a pool, a
compact, high volumetric capacity crystallizer has been developed.

Melter

Melting has been a unit operation that has undergone little development. The
most common way of accomplishing this process has been to dump the ice
in a chamber and allow the compressed refrigerant vapor to contact it
causing melting. This has two distinct disadvantages: 1) development of
scaling laws indicated that they scaled as a function of the ice area exposed
to refrigerant (15), which does not lend itself to large plant designs and
2) volatile vapors in the feed which are evaporated with the refrigerant
are recontacted with the product in the melter resulting in contamination of
the product.

An indirect contact melter overcomes these disadvantages. A slurry of product ice is passed through the tubes of a heat exchanger while the refrigerant is condensed on the outside of the tubes. Thus the scaling of the unit is the same as any heat exchanger and the volatiles are kept separated from the product water by the heat transfer surface. This latter feature is especially important in waste treatment applications where volatiles are common. This overcomes a problem common in evaporators where volatiles in the feed contaminate the condensate. In a pressurized column the washed ice is removed as a slurry by the nature of the operation, thus fluidization of the ice for melting is achieved at no penalty.

Refrigerant

All of the previous secondary refrigerant processes which progressed to the pilot plant stage used n-butane as a refrigerant. This led to the use of expensive explosion proof equipment and exacting stripping requirements due to toxicity and flammability hazards. To eliminate these hazards and reduce equipment costs a non-toxic halocarbon refrigerant, R-114, has been chosen. R-114 was chosen because it offers both the lowest cost of the suitable halocarbon refrigerants and reasonable operating pressures, thus permitting less expensive process vessels and compressors. Although it is more costly than butane, it's use does not alter the economics of the process significantly, because it is necessary to contain the refrigerant effectively and strip the effluent streams in any event. This refrigerant overcomes many of the objections encountered with butane and offers the advantage of higher density which makes the separation of refrigerant from the brine less difficult.

Influent Affects

Up to this point only saline water has been discussed as a working fluid in the process. In handling industrial wastes the properties and conditions of the fluid may vary considerably from those of saline water. The two properties that have the most effect are freezing point depression and viscosity. In addition, the temperature of the fluid and suspended particles in the fluid may affect the process. The following discussion is based on the CRYSTALEX process, but in general applies to all freezing processes.

Temperature

Temperature affects the process in two ways: 1) the feed temperature affects the size and approach temperature of the feed heat exchanger and 2) the cooling water temperature affects the size and approach temperature of the heat removal condenser. In desalting applications sea water is used for both feed and heat removal, while in some industrial applications this may not be desirable as the process from which the waste is taken may operate significantly above ambient temperatures and would make an uneconomical heat removal solution. The resultant of both of these temperature effects is usually quite small, however, tradeoff studies of equipment cost vs. energy cost should be made for each case. If the two effects are combined the resulting effect on cost of treatment is 2 to 3¢ per 1,000 gallons of feed per 10°F temperature rise, while if only feed temperature rises the effect is less than 1¢ per 10°F temperature rise.

Freezing Point

The energy consumption of the process is fairly sensitive to the temperature at which the freezer operates, which is a direct function of the freezing point of the effluent concentrate. If the freezing point of the solution is not known an approximate value can be calculated based on Raoult's Law. From this it has been shown that the freezing point of a water based solution is depressed 3.35°F (1.86°C) for a one molar solution. The following table gives some typical values for 5% (wt) solutions of several common chemicals.

TABLE I

Freezing Point Depression of 5% Solutions

Dextrose	31 °F
Ferric Chloride	28.6
Sulfuric Acid	28.4
Calcium Chloride	27.8
Methyl Alcohol	26.5
Sodium Chloride	26.5

The combined effect of increased energy consumption and capital cost increases due to lowering the freezing point amount to about 2¢/°F per 1,000 gallons of product water.

Viscosity

Increased viscosity, compared to water, has several effects on the process. These are usually minor, however, they should be considered for each case. Heat transfer in the feed heat exchanger varies inversely as the square root of viscosity and the pressure drop varies as the 2/3 power of viscosity. A new optimum heat exchanger should be calculated to take these variations into account. Likewise pressure drops in piping will be increased and would be compensated for by larger pipe sizes. The other effect is on heat transfer in the crystallizer and this has not been determined. Agitator power in the crystallizer is unaffected as long as the viscosity change is not sufficient to prevent turbulent flow. Heat transfer is expected to reduce somewhat, however. In this case the trade off to be made is between compressor power, agitator power and crystallizer size, with the tendency toward increasing agitator power to compensate for reduced heat transfer co-efficient. A quantative evaluation of these effects has not been made, but should be on the order of a few cents per 1,000 gallons of water.

Particulate Matter

Suspended solids in the feed stream to the process are not particularly troublesome. Filtration of the feed with a 20 mesh strainer is sufficient to allow normal operation of the process. Particulate matter finer than this will pass through the process. Those particles that are extremely fine (less than 10-50 microns) will pass out of the process with the brine while those above this will leave the process with the product. In many water reuse applications suspended solids of this type in the water are not of concern. If they are, they can be removed either before entering the freezing process or from the product, whichever is easier.

Materials

The low temperature operation of the process results in less expensive materials of construction. For fairly neutral solutions mild steel and aluminum have proven to be adequate. For more corrosive solutions less expensive materials than normal can be specified. 304 stainless is adequate for most solutions, where 316, alloy 20, or copper nickel might normally be specified for processes operating at higher temperatures.

Industrial Waste Applications

The CRYSTALEX process has virtually unlimited applications to the treatment of industrial wastes. Some of the possibilities are paper mills, metal finishing, chemical wastes, acid mine water, tanneries and textiles. Three general categories of use are: 1) concentration for by-product recovery, 2) concentration for reuse in a process and 3) concentration for disposal. In all cases the recovered fresh water is reused within the system. With discharge standards being increasingly more stringent this latter point becomes more important for, to turn a phrase - recycled water meets any conceivable stream standards.

Metal Finishing Wastes

The application described here is for recovery of chemicals from a plating bath for reuse. The cost of the recovery system is largely dependent on the quantity of liquid to be treated, not the concentration of the waste. Therefore, the use of countercurrent rinsing and other methods to reduce flow are an important part of the entire system. A schematic diagram showing the incorporation of a recovery system into the plating process is shown in Figure 6 (16).

The overflow from the first rinse bath is fed to the freezing recovery unit. In addition to the freezing unit for separation of the fresh water there may also be included a cation exchange unit for elimination of heavy metal impurities from a chromium rinse solution, or a filter for removal of solids which may be precipitated as a result of the cooling. The recovered water is returned to the last rinse tank while the concentrated solution is returned to the plating bath.

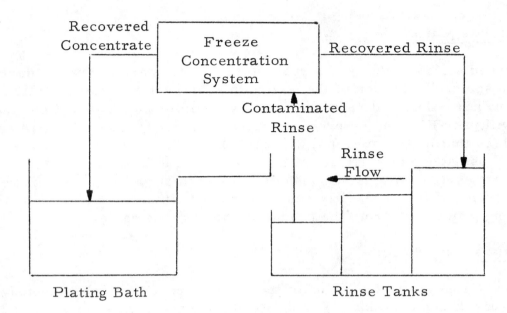

FIGURE 6
INDUSTRIAL WASTE APPLICATION

Since the rinse water eventually makes its way to the plating bath it is essential that make-up be of such purity that the plating bath will not become contaminated. This may include deionization and filtration, but since it involves only make-up water the total quantity will be low.

The basic freezing process is insensitive to the nature of the dissolved contaminants except as they affect freezing point and viscosity. It is, therefore, applicable to wastes from chromium, cyanide and other plating baths, ion exchange concentrates, and regenerant wastes. In each case the chemical nature of the concentrate is unchanged. The remaining contaminants in the recovered water will exist in the same ratio to each other as they exist in the concentrate.

The freezing process recovers water from contaminated feed with the concentration of all contaminants being reduced in the same ratio. This is in contrast with many systems which effectively remove a single constituent such as chromium or cyanide, while leaving other contaminants in solution. Reuse of water containing such contaminants will eventually result in concentrations which are detrimental to the process.

A pilot plant with a capacity of 2,500 gallons per day is currently in operation. Testing has been conducted in that plant to establish the applicability of this process to the recovery of water from metal finishing waste.

A solution was prepared containing nickel, cadmium, chromium and zinc. The recovered water contained less than 0.5 milligram per liter of any metal ion, yielding removal of more than 99.5 percent of the contaminants.

Pulp and Paper Wastes

Freezing process can be applied to several streams in pulp and paper mills. One of the most likely candidates is the concentration of liquors from sulfite pulp mills. This is a large source of pollution and when concentrated has significant heating value which can be utilized to produce steam used in the mill. A typical installation is shown in Figure 7.

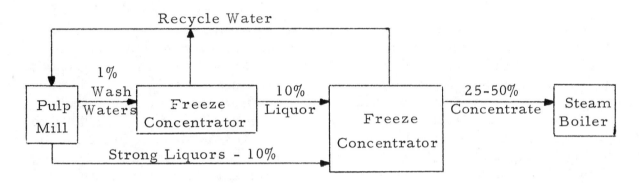

FIGURE 7
PULP LIQUOR CONCENTRATION

The two streams involved in this application are the strong liquor from the digestor which contains 7-10% total dissolved solids and wash water which has about 1% total dissolved solids. The wash water stream is the larger stream, being about five times greater than the strong liquor stream. Two steps of concentration are shown because of the differences in concentration of the two streams. In a small mill this would not be economical and a single concentrator would be specified. The degree of concentration required from the final stage is that required to support combustion, about 35-50% in most cases. Other uses for the concentrate have been proposed including chemical recovery and use as a binder in animal feed.

The CRYSTALEX process is especially suited to paper wastes because of its insensitivity to scaling and there is no volatile contamination of the water extract as there is with evaporators. The freezing point depression with increasing concentration is moderate, resulting in low energy requirements for a high degree of concentration.

The CRYSTALEX process is equally well suited to replace evaporators in kraft processes. This provides a lower energy consumption device and at the same time eliminates the condensate pollution problem.

Economics

The cost of treatment is quite variable depending on the cost of power, cost of money and type of labor that is available. Power consumption of the process can vary from 30 to 60 kwhr per 1,000 gallons of water extracted, so with power costs varying from .005 to 2¢ per kwhr a range of $.15 to

$1.20 per 1,000 gallons for this portion of the cost is possible.

Capital costs can vary from $1 to $3 per gallon of daily capacity, depending
on plant size and materials of construction. Converting this to treatment
cost requires that a fixed charge rate be established which takes into
account interest rate, life of the equipment, depreciation, taxes and insurance. The general range for this is 7 to 14% per year. This then will
contribute $.21 to $1.27 per 1,000 gallons to the treatment cost.

Labor costs are even more variable in that the plants are fully automated
and do not require constant operator attention. This means that labor costs
should be quite minimal and shared time labor be used in most installations.
Operating labor should run about two hours per unit per shift. This is about
equivalent to one man year per unit, which must be ratioed to the output
of the unit. Maintenance labor is about 1/2 man per year for a 100,000 gpd
plant and a man year for a 1 mgd plant. Using $13,500 a year for labor in-
cluding fringes and G & A, labor costs amount to 61¢/1,000 gallons for a
.1 mgd plant and 8¢/1,000 gallons for a 1 mgd plant. This range can be
further increased in cases where shared time labor is not available and
more personnel are added because few industries will allow a piece of
machinery of this type to operate totally unattended.

Other costs such as maintenance materials and make up refrigerant add
another 4 to 8¢ per 1,000 gallons.

In summary, the economics are quite variable and it is misleading to quote
a total cost of treatment. The value of by products recovered or chemicals
reused, the saving in make up water costs and sewerage charge savings
must also be considered before a total evaluation can be made.

Conclusion

It is clear that the thrust of current and future legislation and the increasing
consciousness of the impact of industrial effluents on our environment is
resulting in ever more stringent discharge standards. As a consequence of
this thrust, the direction of total recycling of industrial effluents would
appear to be industry's most responsive answer. As noted previously,
total recycle will meet any stream discharge standards, and often can show
an economic payout through recovery of valuable constituents. We believe
that freeze crystallization will play a significant role in the field of water
reuse because of its inherent advantages of low energy, no scaling, low
corrosion and universality of application.

REFERENCES

1. A. J. Barduhn, "The State of the Crystallization Processes for De-salting Saline Waters", Desalination, 5, pgs. 173-184 (1968).

2. P. L. T. Brian, "Engineering for Pure Water: Freezing", Mechanical Engineering, 19, pgs. 42-50 (February 1968).

3. Carrier Corporation, "Experimental Investigation of Direct Freeze Separation Process Using R-C318", OSW R & D Report No. 256 (April 1967).

4. J. H. Fraser and M. L. Jhawar, "Vacuum-Freezing Vapor Compression Process: Evaluation on Brackish Water", OSW R & D Report No. 541 (March 1970).

5. Avco Systems Division, "Eutectic Freezing to Obtain Dry Salts", OSW Contract No. 14-30-2945 (September 1971).

6. J. H. Fraser and D. K. Emmermann, "Vacuum-Freezing Vapor Compression Process: 60,000 GPD Pilot Plant Evaluation", OSW R & D Report No. 573 (July 1970).

7. W. E. Gibson, D. K. Emmermann, G. Grossman, A. P. Modica and A. Pallone, "Developments in Secondary Refrigeration Desalination", Presented at the Desalting Section, AIChE Meeting, Dallas, Texas (February 1972).

8. J. H. Fraser and W. E. Gibson, "A New Look at Secondary Refrigerant Desalination", Presented at the XIIIth International Congress of Refrigeration, Washington, DC (August 1971).

9. N. Ganiaris, J. Lambiris and R. Glasser, "Secondary Refrigerant Freezing Desalting Process: Operation of a 15,000 GPD Pilot Plant", OSW R & D Report No. 416 (March 1969).

10. Water Desalination Report, VII, 34 (August 1971).

11. W. Hahn, R. C. Burns, R. S. Fullerton and D. J. Sandell, "Development of the Direct Freeze Separation Process", OSW R & D Report No. 113 (June 1964).

12. W. E. Johnson and J. H. Fraser, "Water Purification System", U. S. Patent Pending.

13. R. F. Probstein and J. Shwartz, "Method of Separating Solid Particles from a Slurry with a Wash Column Separator", U. S. Patent 3587859 (1971).

14. J. Shwartz and R. F. Probstein, "Experimental Study of Slurry Separators for Use in Desalination", Desalination, 6, pgs. 239-266 (1969).

15. P. L. T. Brian, K. A. Smith and L. W. Petri, "Vapor Flow Limitations in a Melter Condenser", OSW R & D Report No. 269 (September 1969).

16. R. J. Campbell and D. K. Emmermann, "Recycling of Water from Metal Finishing Wastes by Freezing Processes", Presented at the Reuse and Treatment of Wastewater Symposium, ASME/EPA/AIChE, New Orleans, LA (March 1972).

Applications of New Concepts of Physical-Chemical
Wastewater Treatment
Sept.18-22, 1972

WASTE WATER TREATMENT
THROUGH ELECTROCHEMISTRY

C. Edmond Smith, CE., PE.
Vice President in Charge
of Research and Development
Pollution Engineering
International, Inc.

Introduction

Was the flush toilet really a great invention? In 1870
a fellow by the name of Thomas Crapper invented the first practical flush toilet. He was hailed far and wide as a savior of
the environment and mankind. Thanks to him, the smell of human
waste began to disappear from the streets and infectious diseases
rapidly declined.

Now, 100 years later, the water toilet is not looking so
good. Environmental engineers see the folly of mixing one part
of human waste to 99 parts of water to make sewage. Billions of
gallons of this noxious fluid are churned out every day because
we are so prolific in our use of water to dilute toilet waste.

Sewage is truly a problem fluid. In addition to human
waste, it also contains industrial chemicals and high powered
cleaners that are being poured down the connecting drains. This
exotic muck that gets to the treatment plant is loaded with toxic
elements of many types. These toxic elements that we are referring to are bacteria, BOD, ABS and LAS, phosphates, cyanides, and
heavy metals like lead and mercury.

The growing problem of treating waste waters to prevent
pollution of the ground streams and lakes, as well as part of the
ocean, can be divided into three main interest areas: the removal
of toxic materials, odor, and taste. The preservation and restoration of an adequate supply of pure water free from industrial,
municipal, and natural contaminants is absolutely necessary for
mankind survival. There is, therefore, a critical need to investigate new and potentially superior waste water treatment methods
and systems which will remove the alarming growth and variety of
industrial and municipal pollutants as well as natural
contaminants.

Birth of "ELECHEM" System

The important aspect of waste water purification and clarification is the removal of suspended colloidal material and as much of the solutes as is possible. These constraints would eliminate most chemical treatment systems. To remove colloidal material a floc forming chemical is needed. Most floc forming chemicals are tied up with peripheral chemicals which cause the resulting effluent to have a greater total dissolved solids (TDS) content. If this water is being reused the increased TDS increases the final water usage cost.

What if the floc chemical could be put into the contaminated water in its ionic form, without adding the peripheral chemicals. This should be ideal--the colloidal contaminants and soluble contaminants that would react with the floc chemical could be removed. The results would be a clarified water with a reduced TDS. Considerable research was undertaken to evaluate methods by which these objectives could be obtained. We feel that waste water treatment through electrochemistry most nearly approaches the desired water qualities and still be within an economical framework.

Very simply, electrochemistry is defined as the use of direct current to cause sacrificial electrode ions to move into an electrolyte and remove undesirable contaminants either by chemical reaction and precipitation or by causing colloidal materials to collesce and then be removed by electrolytic flotation. The electrochemical system has proven to be able to cope with a variety of waste waters. These waters are paper pulp mill waste, metal plating, tanneries, canning factories, steel mill effluent, slaughter houses, chromate, lead and mercury-laden effluents as well as domestic sewage. These waste waters will be reduced to clear, clean, odorless and reusable water. In most cases, more especially domestic sewage, the treated water effluent will be better than the raw water from which it had originated.

What does the basic sewage treatment system look like? The system shown in Figure No. 1 will clarify raw sewage and reduce it to a clean, clear, pure water, ready for reuse. It will remove the BOD, ABS and LAS (detergents), phosphates, silicates, sulfates, and other toxic materials to as low a figure as is required. Normally, this system will consist of the following components:

1) A Primary Treatment System

2) An Electroflotation System

3) A Disinfecting System

"ELECHEM PROCESS"

Patent No. 3,664,951

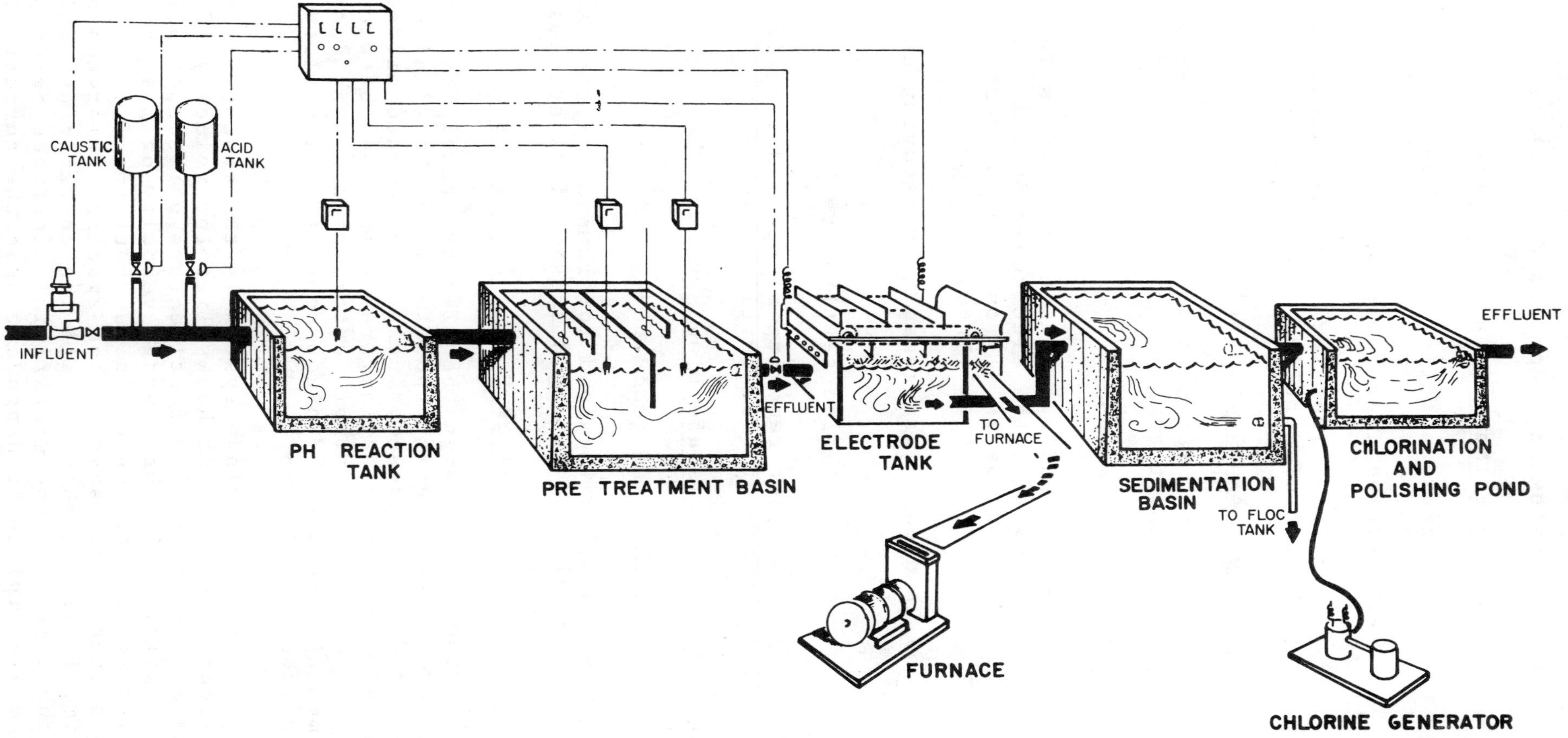

EFFLUENT

CHLORINATION AND POLISHING POND

CHLORINE GENERATOR

SEDIMENTATION BASIN

TO FLOC TANK

ELECTRODE TANK

TO FURNACE

FURNACE

EFFLUENT

PRE TREATMENT BASIN

PH REACTION TANK

ACID TANK

CAUSTIC TANK

INFLUENT

FIG. 1 - Basic System

A Domestic Waste Water Treatment

A Pollution Engineering International Process

These component systems are comprised of the following parts:

Group A - Primary Treatment System

1) A primary settling tank.

2) Screening if required.

3) Pump and lines for removing sludge from
    bottom of settling tank.

Group B - Electroflotation System

1) Baffling for addition of floc in preliminary
    treatment tank.

2) Electrode tank.

   a) Depending upon the pollutants in the
       water, the electrodes in this tank
       may be of all one type or combination
       of several types.

3) A paddle arrangement for removing generated
    foam.

4) A primary floc settling tank.

5) Pumps for removing sludge and floc from
    bottom of floc settling tank.

Group C - Disinfecting System

1) A chlorine generator.

2) A polishing pond for removal of excess
    chlorine before disposing into streams.

System Design Principles

The basic design theories or principles that are advanced here were those incorporated into the final electrochemical system design. These principles concerned different electrodes, electrode combination, retention time between plates, plate spacing and removal of contaminates. Each design criteria was aimed at obtaining both good operational utility and acceptable economic efficiency.

A parallel plate system is the most efficient electrode arrangement. Our physics books list the velocity of ionic movement in an electrolyte in centimenters per second per unit of potential difference. The average velocity of most of those contaminants we are interested in is about one (1) inch per hour per volt per inch. This says if the spacing is large and the retention time is short, that not all the contaminants will be treated. Since the amount of the sacrificial electrode being pumped out into the electrolyte is dependent upon the current (Faraday's

Law) the best economics from a power cost is that the voltage be as low as possible. This voltage has to be at least equal to or greater than the decomposition voltages (sum of the half-cell voltage) of the sacrificial electrodes. If, for instance, iron electrode pairs are used, the decomposition voltage would be twice 0.771, or 1.54 volts. We want to be sure that the voltage at the plate face is 1.5, so to take care of losses up to the face, we design the system for 2.5 volts. Now with 1.5 volts at the plate face, the average treatment time for plates spaced 10 inches apart would be

$$\frac{(10\ in)}{(\quad)} \times \frac{(hr \times volt)}{(1\ in \times in)} \times \frac{(10\ in)}{(1.5\ V)}\ or\ 67\ hours.$$

This would be a little hard to tolerate. One inch spacing would be 40 minutes and 3/4" spacing would be 30 minutes.

Now that we know what the plate spacing and configuration is, information about the waste water contamination will allow the rest of the electrode design. What should the area of the plates be? The electromotive series indicates the combining activity of the sacrificial electrode with waste water contaminants. Stoichiometric calculations indicate the amount of the sacrificial electrode needed. For instance, if the electrodes are aluminum, we know that one part per million (ppm) is required to remove 3.5 ppm of phosphates and 1.35 ppm is required to remove one ppm of silicates, and so on. We generally need an extra 12 to 15 ppm to generate the floc to quickly precipitate these newly formed compounds. Thus, we have determined the amount of electrodes (ppm) needed to treat the water. Equation (1) will determine the size of the power pack needed since the total amperage has been determined.

$$Amp = 6.085 * Elect * Q/GEW * Eff \qquad (1)$$

Where: Elect = Sacrificial Electrode Needed in PPM
           Q = Flow Rate in Gallons per Minute
         GEW = Gram Equivalent Weight of Sacrificial Electrode
         Eff = System Efficiency

Knowledge of the electrolyte specific resistivity and Equation (2) will determine the plate area required.

$$A = 0.002735 * SR * d * I/E \qquad (2)$$

Where: A = Plate Area in Ft^2
        SR = Waste Water Specific Resistivity in Ohm-Cm
         d = Plate Spacing in Inches
         I = Total Current in Amps
         E = System Voltage

Example: Let's assume that 150 ppm will be needed to remove the colloidal and solute contaminants. We will use iron electrodes. The system efficiency is 82%. The flow rate is 50 gallons per minute and the specific resistivity is 500 ohm-cm. What is the amperage used and the plate area needed?

$$\text{Amp} = 6.085(150)(50)/28 \times .82 = 1988 \text{ amps}$$

$$\text{Plate Area} = .002735(500) \times .75 * 1988/2.5$$
$$= 816 \text{ Ft}^2$$

This means the 25 plates 4' * 8' will be needed.

Sewage Treatment

   Now that we have a concept of what the system looks
like, let us see what the system can actually do. The raw sewage
coming from the primary surge tank is laden with fine suspended
colloidal material. This imparts the BOD and COD and cannot be
removed by ordinary sedimentation means. In most cases, the
colloidal material is stable because it has negative electrical
charges surrounding each particle. By passing the water between
two sacrificial electrodes, the colloids' electrical equilibrium
is disturbed and now these colloidal particles will start to
collesce and form large lumps and chunks. At the same time
that this process is going on, fine bubbles of oxygen and hydrogen are being generated at the anode and cathode, respectively.
These fine bubbles are tending to escape and they are entrapped
in these colloidal particles and float them to the surface as a
foam. The more the colloidal particles, the denser the foam that
is formed. Another process is being activated while the removal
of the colloidal particles is being effected. The current that
is being passed between the sacrificial electrodes is pumping
electrode ions out through the electrolyte and these ions start
looking around for something with which to combine electrically
and chemically. If the water is laden with phosphates, silicates,
magnesium, and calcium--and sewage water generally is--the sacrificial electrode will combine chemically with these constituents forming insoluble precipitants and can be removed in a settling chamber. The amount of the sacrificial electrode that is
pumped into the water will determine how much of the contaminants
are removed. In most cases, if the sacrificial electrode is iron
or aluminum, there will be some hydroxide formed as a floc and
will be advantageous for removing turbidity and these newly
formed chemical compounds.

   Reaction with the sacrificial electrode ions with the
contaminants is selective to a certain extent. Those contaminants that appear lower on the electromotive series and furtherest
away from the sacrificial electrode will react first. If any of
the sacrificial electrode ions are left, then those contaminants
higher on the electromotive series will be removed, and so on,
until all of the sacrificial electrode ions have been consummed.
This selective reaction sequence is not entirely true. There are
millions of these ions that are being pumped out through the
water and they can come in contact with ions of contaminants that
are higher on the electromotive series and react before they have
an opportunity to contact some of the contaminants that are lower
on the electromotive series. Some of all the contaminants will
be removed. There will always be some electrode hydroxide (floc)
formed. The amount of sacrificial electrode pumped into the system will reflect how much of the contaminants have been removed.

Thus, the system will remove both organic contaminants and inorganic contaminants.

The contaminants that the basic system will not readily remove are the highly soluble salts. Those contaminants such as sodium chloride, ammonia, etc., are not easily removed. However, by proper selection of the correct electrodes, even these can be reduced while the flotation and precipitation reactions are occurring.

Let us look at some typical examples of the treatment of domestic sewage. At our Houston offices, we have several different sizes of equipment that we use to treat water under dynamic conditions to evaluate the need for a particular type of sacrificial electrode. We have units that will treat anywhere from four gallons per hour to as much as five gallons per _minute_. We also have a trailer-mounted rig which can treat from 25 to 50 gallons per minute of waste water. For demonstration purposes, it is relatively easy to borrow some raw sewage from the City of Houston, haul it to our plant in tank trucks, and treat the water. The water is exposed between the electrodes for about a thirty minute time period. We have found that this is extremely adequate and will remove all the BOD to less than five ppm; there will be no turbidity and the color will be less than five Jackson units. All the phosphates, silicates, and magnesium will be removed from these waters and a good part of the calcium ions will be removed. This water will be rendered very desirable for industrial reuse and, after disinfecting with chlorine, the water would be potable. Table No. 1 is an analysis of what this process can do for domestic sewage treatment.

<u>Table No. 1</u>

<u>Laboratory Test Results</u>

Samples Taken: Sewage Water from Sims Bayou Treatment Plant

	Before Treatment*	After Treatment*
COD	490	26
Total Solids	602	401
Suspended Solids	73	7
Settleable Solids	21	5
Total Hardness	127	11
Alkalinity	257	11
pH	6.88	7.02
IOD	0.98	<0.1
BOD	220	9
Coliform	318,000/ml	0
Phosphates	38	0

* - Values shown in parts per million except for pH and Coliform

Economics

Operating Cost

The cost to treat water in this system is directly proportional to the contaminants that must be removed. If a water had no inorganic contaminants in it, then the system would remove colloidal suspended particles by putting approximately 12 to 15 ppm of the sacrificial electrode into the water. However, most waters contain organic compounds as well and they will use up ions themselves in forming insoluble particles. So, the knowledge of the magnitude of the organic contaminants must be known as well as the inorganic contaminants, and since Mr. Farada told us how the sacrificial electrode could be used to put ions into the water, we can predict what the operating cost for a system will be. The total cost per thousand gallons of treated water is derived in Equation (3).

$$\text{Total Cost/Kgal} = \frac{ELT}{120} \left(\frac{4.79 * SR * CD * C_E}{GEW * Eff} + C_L \right) \quad (3)$$

Where: ELT = Electrode Consumption in PPM
       SR = Water Specific Resistance in Ohm-Cm
           (SR = 415,000/ppm * * 9735)
    PPM = Resistance of Water in PPM of NaCl
     CD = Current Density in Amps/in^2
   GEW = Gram Equivalent Weight of Electrode
     Eff = Electrode Consumption Efficiency
      C_E = Cost of Electricity in \$/Kw-Hr
      C_L = Cost of Electrode in \$/#
    PPM = Water TDS as NaCl

A close examination of Equation (3) reveals that the first part of the equation in the parenthesis is the cost of electricity and C_L in electrode cost. For the previous example the current density is 0.017 amps/in^2. The cost of power for 1¢/kw electricity is 2.22¢/kg. The cost of iron installed in the electrode box is 15¢/# and the operating cost would be 18.75¢/kg. The total operating cost would be 21¢/kg. From the above, it is apparent that the cost is fairly sensitive to the cost of the sacrificial electrode--more so than it is to the power cost. Whenever possible, we use iron as the sacrificial electrode.

The cost of using the same amount of chemical to do the same job will cost more than an electrochemical system. We also find out that electrically we do a better job for the reactions than can be done with chemicals. So, when comparing this system's costs with a secondary plant effluent cost + chemical-coagulant cost, the electrochemical treatment is more economical.

Capital Investment

The capital investment for electrochemical systems is comparable to the cost of secondary effluent systems. Figure No. 2 shows what a typical sewage treatment plant would look like.

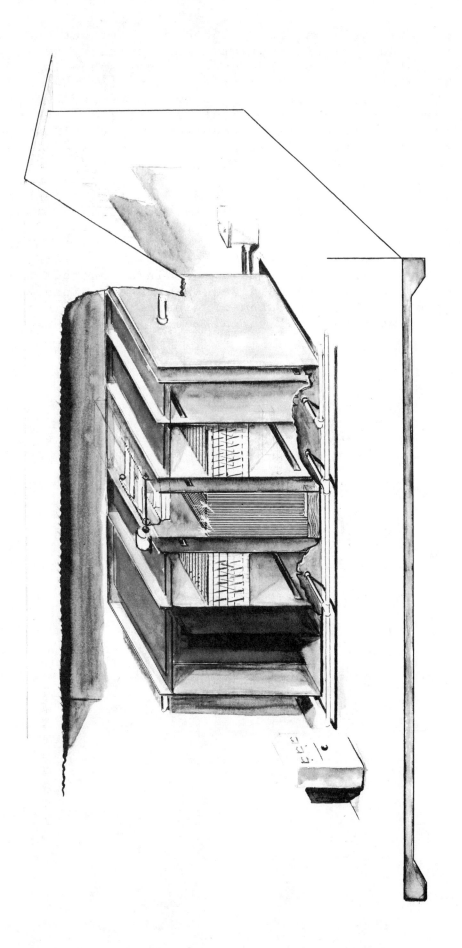

FIG. NO. 2
TYPICAL PLANT

The system cost will be equal to a secondary plant but much less than a secondary and a tertiary plant. But, tertiary quality water is being delivered from this system. Less space is required for this system since the retention time for the electrodes is generally thirty minutes or less, no extensive air blower system is needed, and in general, about 1/10 of the space needed for a secondary effluent system is required. With this system, you can get a tertiary treated water for the same capital investment that secondary plants are costing.

I believe Mr. Thomas Crapper's flush toilet is here to stay. Any other system would require the user to have his own independent disposal system. I believe the public is willing to pay the price to have someone else handle this particular problem. Our solution is to take this contaminanted water, concentrate the contaminants, and remove them in a controlled environment, where they will be oxidized and rendered innocuous and inert. The solution to the problem of generating pure water from sewage waters can be economically realized.

ULTRAFILTRATION AND MICROFILTRATION MEMBRANE PROCESSES FOR

TREATMENT AND RECLAMATION OF POND EFFLUENTS IN ISRAEL

G. SHELEF, R. MATZ AND M. SCHWARTZ
Technion - Israel Institute of Technology, Hydronautics-Israel Ltd. and
Hebrew University of Jerusalem.

Introduction

Most of the work done in the U.S. on the use of membrane systems in waste-water treatment and reclamation, has been concentrated on the application of reverse osmosis for the removal of salinity as well as other wastewater conta-minants - notably organic and nitrogenous compounds (1,2). Tertiary, secondary and primary effluents were used, in addition to raw sewage. Most of the membranes used were within the ionic size range i.e. between 0.5×10^{-3} to 8×10^{-2} microns.

Although these membranes proved to be excellent for the rejection of TDS and COD, there are still serious problems to be contended with such as reduced flux rates, flux decline, membrane durability, poor rejection of various nitrogenous compounds and high applied hydraulic pressure. The problems of brine disposal and overall costs still limit the feasibility of reverse osmosis systems on a large scale.

Cost analyses indicated that the earlier the stage of wastewater treatment the membrane processes are introduced into, the more economically feasible the system becomes (1). Thus, the current trend toward membrane treatment of primary effluent or even raw sewage can eliminate several expensive steps in the chain of processes such as the secondary and/or the tertiary steps. On the other hand, the earlier the step membrane processes are applied, the more difficult are the problems of membrane fouling.

The work described in this paper departs from the general trend of using the small pored reverse osmosis membranes for the removal of both salinity and other contaminants. It concentrates on using the ultrafiltration - micro-filtration membranes of pore size ranging between 2×10^{-3} to 10 microns, emphasizing the rejection of molecules of molecular weight over approximately 500 rather than the rejection of ionic salinity. Obviously this approach can be useful only under conditions in which reclamation and recycling of wastewater can be accomplished without the removal of the increased salinity in municipal wastewater effluents.

Figure 1 illustrates the size ranges of various membrane separations, indicating the extensive range of molecular sizes covered by ultrafiltration and microfiltration membranes.

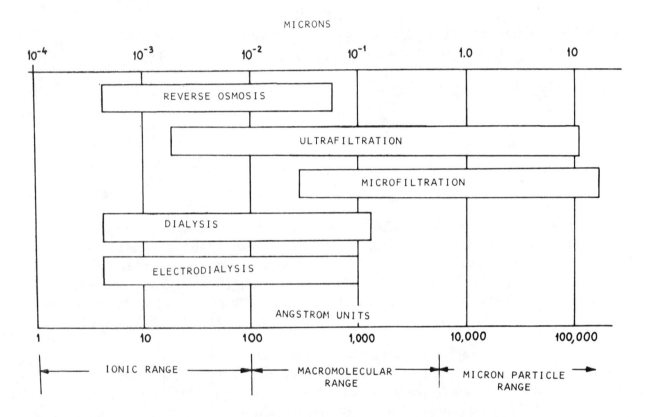

FIG. 1

Size Range of Various Membrane Separations.

It can be shown that precisely in a country like Israel, in which it is well accepted that reclaimed wastewater will play an important part in the National Water System in the coming decades, conditions in many instances do not require the removal of salinity from reclaimed and recycled wastewater. The first part of the paper, therefore, analyzes the exact conditions which allow wastewater reclamation and recycling without the necessity of desalination.

The second part of this paper describes experiments of ultrafiltration and microfiltration of oxidation ponds effluents. These effluents were used because of the following reasons: (a) oxidation ponds are still the most prevelent method of wastewater treatment in Israel; (b) the cost of treatment is minimal, thus allowing the introduction of membrane processes in an earlier stage; (c) it was assumed that the residual organic contaminants in such effluent would be of larger molecular size, particularly whenever ponds of increased detention periods are used.

In the experiments various membranes are used ranging from the large pore size microfiltration membranes to the smaller pore size ultrafiltration membranes. The cut-off point of the membrane with regard to the molecular size of the solute is, therefore, reduced in the course of the experiments. Only the experiments with the larger pore size membranes have been completed and are described in this paper, and although they are limited in practical value, they do show the direction of the current research.

Conditions Affecting The Need For Desalination of Wastewater

Wastewater reclamation is of first priority in Israel, due to the fact that Israel already utilizes over 90 percent of its natural water resources. The cost of new water is currently around 25 ¢/1000 gal at source and the cost is rapidly increasing. The coming 15 years will be marked by the incorporation of both reclaimed wastewater and desalinated seawater into the overall National Water System. This System presently incorporates most water resources in the country and provides maximum flexibility of supply, recharge and benefical uses as well as planned dilution of various waters according to their quality and salinity.

The role of reclaimed wastwater is visualized in the following three ways:

A. Segragation of reclaimed wastewater from the National System for use in agricultural irrigation.

B. Injection of reclaimed wastewater into the National Water System, 85% of which is used for irrigation while the remaining 15% is used for domestic and industrial use. The reclaimed wastewater, after dilution with other sources should meet all Drinking Water Standards.

C. Direct recycling of the reclaimed wastewater into the domestic water supply of the community.

The last category is the most problematic with respect to the need for removal of the increased salinity in the effluent. Such need, however, is affected by a few factors as follows: (a) the recycling ratio of wastewater effluent into the community water supply denoted by r; (b) the salinity of the incoming water supplied to the community before any mixing with recycled waters, C_f; (c) the increased salinity in the effluent in each passage through the system, C_i; (d) the tolerable salinity of the mixed water supply (which in Israel was set to around 1000 mg/l TDS).

The resulting effluent salinity Ce after n cycles with a recycling ratio r, is given by -

$$Ce(n) = Cf + Ci \ (1 + r + r^2 + r^3 \ldots \ldots + r^{n-1} + r^n) \ldots \ldots \ldots (1)$$

When n approaches infinity, i.e. when recycling is continuous, Equation 1 becomes an expanded form of a series which results in the following:

$$Ce = Cf + \frac{Ci}{1-r} \ldots \ldots \ldots \ldots \ldots \ldots \ldots \ldots (2)$$

The salinity of the mixed water supply Cm, can be calculated as follows:

$$Cm = Cf + \frac{rCi}{1-r} \ldots \ldots \ldots \ldots \ldots \ldots \ldots \ldots (3)$$

Figure 2 illustrates the calculated increase of salinity (expressed in mg/1 of TDS) under three sets of conditions of Cf and Ci, as it is affected by the recycling ratio, r. It can be shown that when the initial salinity of the incoming water is moderate and if the increased salinity is kept at a relatively low level, no desalination of the recycled wastewater will be needed, once the recycling ratio is kept below 0.6.

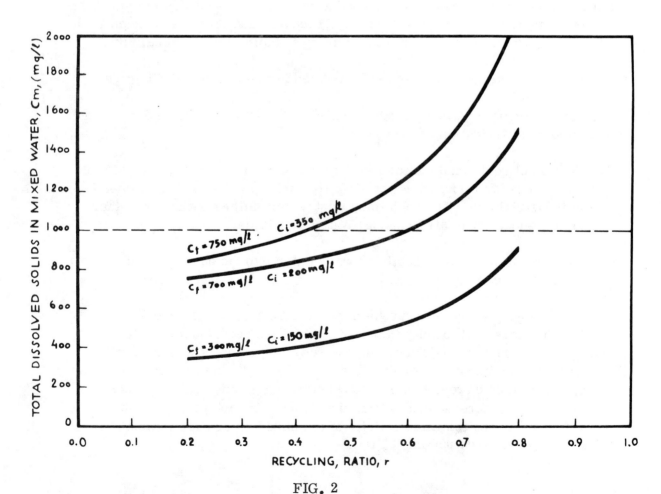

FIG. 2

Salinity Increase in Mixed Water Supply as a Function of Recycling Ratio of Various Waters (calculated).

Table 1 gives salinity values (expressed as TDS) in supplied water and in raw effluent of various communities within the Dan Region (Greater Tel-Aviv) in Israel.

This Region has adopted a major wastewater reclamation program and is currently using oxidation ponds as the treatment process.

TABLE 1

Salinity Increase of Wastewater Effluent in Various
Dan Region Communities (3)

	Salinity Concentration as TDS, mg/l			
	Water Supply	Wastewater	Increase	Recycled Supply (calculated at r=0.6)
	C_f	$C_f + C_i$	C_i	C_m
Holon	238	312	74	349
Bat-Yam	153	347	198	450
Tel-Aviv North	613	909	296	1057
Tel-Aviv South	647	893	246	1016

Both Holon and Bat-Yam are supplied with low salinity waters and the increased salinity of Bat-Yam is due partly to sea water intrusion into the city sewers and partly to industrial wastes. Initial salinity in Tel-Aviv is higher and the marked increase is due primarily to spent regenerant from water softners, which is the main cause of high salinity increase in most cases in Israel (4,5). Table 1 indicates that even with recycling ratio of 0.6, the values of C_f and C_i will require desalination of the effluent, once direct recycling of wastewater will be applied. It is assumed, however, that reduction of C_f can be accomplished by supplying water with lower TDS to communities in which direct recycling of wastwater will be practiced. This can be facilitated by using the inherent characteristics of flexibility and planned dilution of the National Water System, particularly when the System will incorporate desalinated seawater low in TDS, as one of its main new water sources, in addition to reclaimed wastewater. It is also assumed that reducing C_i can be accomplished by an extensive program aimed at diverting spent softeners regenerants away from municipal sewers or even by replacing ion exchange softners by central lime or lime-soda softening plants. This program should take place in communities in which direct recycling of wastewater is planned.

Reducing the recycling ratio, r, to levels below 0.6 is practical since only between 65 to 75 percent of the water supply flow reaches the sewage treatment plant. Similarly the community can use directly some of its effluent for irrigation of green belts, public parks and near-by farm land, thus keeping the recycling ratio at or below 0.6.

Methods

Pond System

Effluent from a 280 sq.m. experimental oxidation pond at the Environmental
Health Laboratories of the Hebrew University in Jerusalem served for the experiments. The pond is operated at high rate using the method of Accelerated
Photosynthetic System (APS) and treats between 30 to 50 cu.m/day of raw
sewage from a Jerusalem residential area. The pond is 45 cm deep, is fully
aerated at night-time while some agitation is provided during day-time, in
order to increase photosynthetic efficiency of the algal biomass. The detention
period at the course of the experiment was 3.5 days. Effluent was withdrawn
only at day-time and it contained a high concentration of unicellular algae, mostly
Chlorella, Scenedesmus and Euglena.

Membrane System

The experimental system is shown in Figure 3; it consists of two 90 cm long
12 mm (inside diameter) porous tubes, into which an ultrafiltration membrane had
been directly cast. The effective membrane area was 680 sq.cm. and about 40
liters of pond effluent were recirculated at a mean linear flow rate of approximately 40 cm/sec.

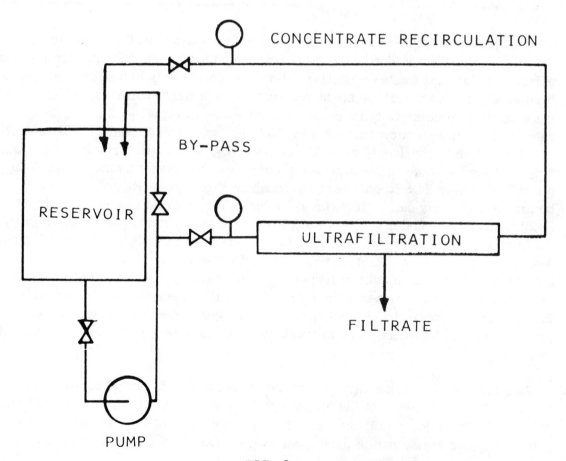

FIG. 3
Experimental Arrangement

Three ultrafiltration membranes produced by Hydronautics-Israel Ltd. were used and they are designated as CA-2010, CA-1310 and CTA-6. Only the experiments with the CTA-6 membrane has been completed and reported in this paper. The typical rejection data of the membranes is given in Table 2, while the cut-off points of the membranes is given in Figure 4.

TABLE 2

Properties of HIFIL Ultrafiltration Membranes *

Type of Membrane		CA-2010	CA-1310	CTA-6
Normalized Water Flux $\dfrac{ml}{min \cdot cm^2 \cdot atm.}$		0.002	0.10	0.20

	Solute	Mol. Wt.			
	Urea	60	< 30	0	0
	Sucrose	342	~ 50	0	0
	Bacitracin	1,400	~ 50	~ 30	0
	Dextran T10	10,000	> 90	~ 50	0
Rejection of Solutes	Lysosyme	14,400	> 90	~ 60	0
	Dextran T40	40,000	100	> 90	0
	Ovalbumin	45,000	100	> 85	10
	Bovine Serum Albumin	67,000	100	100	> 95
	Dextran T70	70,000	100	> 90	> 90
	Dextran T150	150,000	100	> 95	100
	ϒ-Globulin	160,000	100	100	100
	Dextran T250	250,000	100	100	100

* The sign ~ implies possible deviation of ± 10%
  Selectivities have been determined in 0.1% aqueous solutions and describe
  membrane characteristics after 30 min. of filtration.
  Excluding the CA-2010, which was tested at 50-100 psi, filtration through
  all other membranes were performed at 5-10 psi.
  The membranes from the UF series are autoclavable.

CTA-6 membrane is noted by the sharp cut-off point at molecular weight of about 70,000 and by increased flux per unit of pressure. It serves, therefore, for the rejection of large organic molecules, inorganic and organic colloids, bacteria and algae.

Results and Discussion

The concentrations of the various substances in Jerusalem raw sewage, in the APS pond effluent and in the CTA-6 membrane effluent (product) are given in Table 3. Effective fluxes during the experiments decreased during

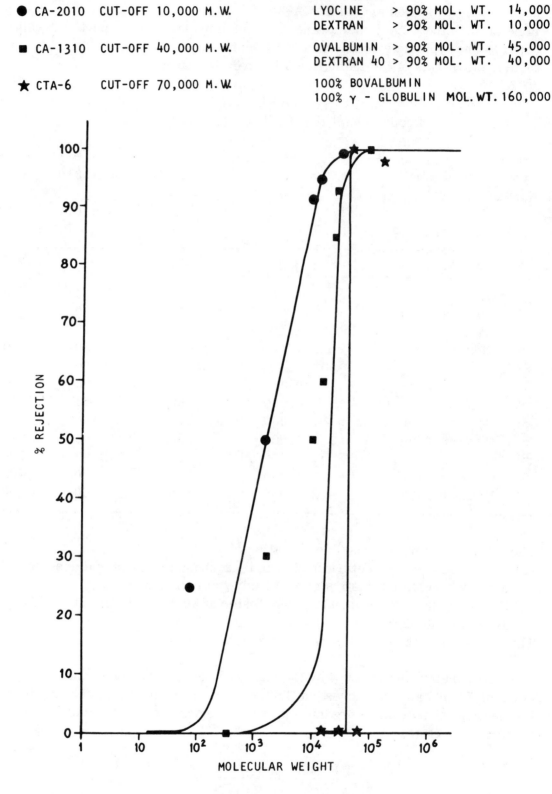

FIG. 4
Cut-Off Points of Ultrafiltration Membranes
Produced by Hydronautics-Israel Ltd.

the first hour from 19 gal/sq.ft/day to a steady value of about 10 gal/sq.ft./day at a constant operating pressure of approximately 30 psi. During the experiments, temperatures were kept constant at 30°C by using a cooling system to remove the heat generated by the recycling pump. Samples of filtrate were accumulated for 4 hours and subsequently analyzed.

TABLE 3

Effluent Composition and Rejection of Various
substances by CTA-6 Membrane Using Pond Effluent

Item	Raw Sewage (Jerusalem) mg/l	Pond Effluent mg/l	Membrane Effluent	
			mg/l	% Rejection
BOD (5 days) - Total	510	85	7.5	91.2
BOD (5 days) - Filtrate	-	9	7.5	16.7
COD - Total	1140	512	61.0	88.1
COD - Filtrate	-	133	61.0	54.0
Total Nitrogen	72	41	7.1	82.7
Total Nitrogen - Filtrate	-	11.6	7.1	38.8
NH_4^+ - Nitrogen	-	6.4	2.6	59.4
NO_3^- - Nitrogen	-	0	0.3	-
Suspended Solids	348	274	3.2	98.8
Total Dissolved Solids	1080	1038	812	21.8
Chlorides	320	312	298	4.5
Coliform Bacteria (per 100 ml)	3×10^8	1.6×10^6	30	99.99
Algae Count (per mcc)	-	9800	0	100.00

The raw sewage of Jerusalem is exceptionally strong due to the reduced water consumption caused by the shortage of water and the high cost of water in the City. Values of BOD, COD and suspended solids in the pond effluent remain relatively high due to the presence of algal biomass in the effluent.

After removing the algal biomass by glass filtration for example (hence filtrate), the effluent quality is excellent, primarily in regard to removal of BOD (over 95%) and nitrogenous compounds (84% removal).

Treatment of the pond effluent by the CTA-6 membrane produced effluent which is superior to the one produced after glass filtration under laboratory conditions. This is particularly marked by further rejection of COD (54%), total nitrogen (39%) and TDS (22%), as compared to the filtered pond effluent. This can be explained by the rejection of organic macromolecules, some containing nitrogen, which would otherwise pass through most filtration processes. It is also assumed that rejection of molecules, smaller than the nominal 70,000 m.w. for CTA-6 membrane, took place under continuous membrane use, due to accumulation of filter cake on the membrane surface. This might explain the substantial reduction of TDS (22%) by this relatively large-pored membrane.

As indicated above, the pond was operated at high rate of 3.5 days detention. It is assumed that in ordinary ponds, in which detention periods range from 15 to 80 days, the proportion of macromolecules will increase while the proportion of the more biodergradable small molecules will decrease. This assumption is supported by works regarding the accumulation of large molecular weight proteins, tannins and fluvic and humic acids in natural waters and in wastewaters which were exposed to a prolonged biodegradation (6, 7).

By using smaller pored ultrafiltration membranes, such as the CA-1310 and CA-2010, it is assumed that further rejection of COD, nitrogenous compounds and TDS can be accomplished.

It should be noted that the residual bacterial count of the membrane effluent was probably caused by contamination of the sample after the experiment. The residual suspended solids can be explained by some after-growth of bacteria in the sample.

Due to the high level of residual COD, the CTA-6 membrane produced effluent, which is not yet suitable for recycling into domestic water supplies, unless substantial dilution and/or further treatment will be applied. Furthermore, such effluent can be produced by the more conventional train of tertiary processes, which include ammonia stripping, multi-media filtration and activated carbon sorption. Even if the smaller-pored membranes will prove their expected rejection efficiency, a rigorous cost analysis should be made in order to determine the economical competetivness of ultrafiltration processes, as compared to other tertiary processes.

Probably the most encouraging results in applying the concept of larger size range membranes, with only partial rejection of electrolytes, is the recent development of the Dynamic Membranes. Electrolytes rejection of between 30% to 66% and COD rejections of up to 95% were accomplished at fluxes of 50 gpd/sq.ft. (8,9,10). As indicated above, even partial removal of salinity can be adequate for wastewater recycling in most cases, even at high recycling ratios.

Summary

(a) Recycling of reclaimed wastewaters into community water supplies can be accomplished,in many cases,without removal of electrolytes salinity. The initial salinity content, the increased effluent salinity and the recycling ratio are the major factors determining whether, and to what a degree,desalination will be required. These factors can be controlled by the use of a well planned water system, by appropriate regulation regarding handling of spent softeners regenerant wastes and by optimal beneficial use of wastewater effluents,other than by recycling.

(b) Membranes of the ultrafiltration - microfiltration size range can substantially upgrade the quality of pond effluent, in regard to removal of soluble COD, nitrogenous compounds and some dissolved solids. The experiments indicate that further increase in rejection values can be accomplished by using smaller size ultrafiltration membranes.

(c) A rigorous cost analysis is required in order to examine the economical feasibility of ultrafiltration membrane processes, in comparison to other tertiary treatment processes.

References

(1) Sachs, S.B. and E. Zisner "Reverse Osmosis for Wastewater Reclamation" Proceedings of the Beitan Aharon Symposium (Israel) on Brackish Water Desalination (1972).

(2) Middleton, F., Various reports on advanced wastewater treatment by reverse osmosis, Robert Taft Research Center, Cincinnati, Ohio (1965-1972).

(3) Folkman, Y., "Dan Region Survey" Tahal Co. (1971).

(4) Rebhun, M., Int. J. Air Wat. Poll. $\underline{9}$, 253 (1965).

(5) Feinmesser, A., Proc. of the 5th Int. Wat. Poll. Conf. Pergamon Press, p. I-33/1 (1971).

(6) Ghassemi, M. and R.F. Christman, Limn. Oceon, $\underline{13}$, 583 (1968).

(7) Rebhun, M., and J. Manka, Envir. Sci. Tech. 606 (1971).

(8) Savage, H.C., Bolton, N.E., and H.O. Phillips, Wat. Sewage Works, $\underline{116}$, 192 (1969).

(9) Sachs, S.B., Baldwin, W.H. and J.S. Johnson, Desalination, $\underline{6}$, 215 (1969).

(10) Johnson, J.S. et al, Quarterly Report for OSW, June 15-Sept. 15 (1970).

Applications of New Concepts of Physical-Chemical
Wastewater Treatment
Sept. 18-22, 1972

MEMBRANE EQUIPMENT SELECTION
FOR THE MUST HOSPITAL WATER RECYCLE SYSTEM

James A. Heist
AiResearch Manufacturing Company
of Arizona

In 1963, the U. S. Army Medical Research and Development
Command awarded AiResearch Manufacturing Company of Arizona a
contract to develop the MUST (mobile unit, self-contained, transportable) Hospital. As these hospitals were used in Viet Nam,
it became apparent that a system to insure an adequate potable
water supply was needed. Since there are potential areas of
operation where no ground water sources are available, the decision was made to develop a system to recycle the water used in
the hospital for potable use.

In 1967, a second contract was awarded to AiResearch to develop the Water and Waste Management System (WWMS) to provide the
potable water supply for the hospital. The requirements of the
system were laid out in the Qualitative Materiel Requirement
(QMR)[1]. The potable water product was specified to be of a purity
meeting standards established in Solog 125, "Minimum Potability
Standards for Field Water Supply," with a desired quality meeting U. S. Public Health standards. Based on the QMR specifications, a first generation system was developed and delivered to
the Army Test Center at Fort Lee, Virginia. This system has been
described in previous literature[2] and the results of testing of
the system by the Army's General Equipment Test Activity has

---

[1]Letter, CDCMR-O, HQ USACDC, 29 August 1969, subject: Revised
Department of the Army Approved Qualitative Materiel Requirement
(QMR) for Medical Unit, Self-Contained, Transportable (MUST).
Appendix II.

[2]A. Gouveia and K.A.H. Hooton, "Potable Water from Hospital
Wastes by Reverse Osmosis," Chemical Engineering Progress,
Symposium Series, Vol. 64, No. 90, Pages 280-284, 1968.

been published[3].

The first generation system consisted of two separate units which could be used separately or in series. The first section, the Water Treatment Unit (WTU), was designed to clarify the waste stream for further purification. It could also be used separately to clarify and chlorinate the waste for non-polluting discharge to the environment. The second unit in the system is the Water Purification Unit (WPU), which purifies the WTU effluent for distribution in the MUST hospital potable water system. The WPU is also capable of processing brackish ground waters without the support of the WTU.

The aforementioned GETA Test Report found certain deficiencies in the first generation WWMS. In 1971, two more contracts were awarded to AiResearch to develop a second generation WWMS. The first of these contracts was to evaluate the possibility of a membrane system for primary treatment of waste streams. The second contract was to investigate the state-of-the-art advances made in physical-chemical waste treatment since development of the first generation WWMS, and incorporate these advances in a second generation prototype unit.

Waste Characterization

The GETA test plan for evaluation of the first generation WWMS characterized each of the waste stream components generated in a MUST hospital. Seven streams are generated in various areas of the hospital with a composite of these wastes producing an eighth discernible waste stream:

    1)   Shower Waste
    2)   X-Ray Waste
    3)   Kitchen Waste
    4)   Laboratory Waste
    5)   Operating Room and CMS Waste
    6)   Composite Waste (combination of 1) thru 5))
    7)   Type I Laundry Waste
    8)   Type II Laundry Waste

The characteristics of these waste streams were determined by professional personnel thoroughly familiar with the activities that generate the waste in each area. Synthetic formulations for each of the waste streams were defined using only the items found in use in the MUST hospital. These synthetic wastes have been used in all testing during this program. The only waste stream not included in the recycle system is the human waste, which is incinerated. The composite waste stream is an average of the other hospital waste streams. It is approximately the influent to the recycle system. Table 1 shows the chemical

---

[3]Letter, USAEHA-ES, Office of the Commander, USAEHA, Sept. 1970-Feb. 1971, subject: Water Quality Engineering Special Study No. 99-003-71, Evaluation of the Water Processing Element, MUST.

TABLE 1
Composite Waste Analysis

Alkalinity	104.	Lead	.32
Surfactant	75.	Magnesium	15.
Arsenic	.001	Manganese	.03
Barium	.12	Phosphate	156.
Cadmium	.02	Potassium	34.6
Calcium	15.	Sodium	360.
Chromium	1.05	Sulfate	28.
COD	870.	Organic Carbon	192.
Copper	.04	Dissolved Solids	1100.
Greases & Oils	43.4	Suspended Solids	400.
Iron	.33		

analysis of the average composite waste stream. As can be seen from the analysis data, the composite hospital waste is roughly twice as concentrated as most municipal waste streams.

In a high recovery system such as the MUST water recycle system, the size and economics dictate that there be either a high recovery per module or a recirculation system. The recovery per module per pass in tubular modules is typically not high enough to allow single pass systems. In recirculation systems the waste applied to the membrane is concentrated to some extent, depending on the recovery of the system. In the development of reverse osmosis for primary treatment of hospital wastes, eight times concentrated wastes were generally used in evaluating the membrane process, representing the stream entering the first of a series of modules in a 90% recovery recirculation system.

Primary Treatment by Membrane Systems

In the development of the first generation Water and Waste Management System (WWMS), the philosophy was to provide a system which would treat the waste sufficiently to allow purification by reverse osmosis. Specifically, the idea was to remove all of the possible fouling constituents of the waste before applying it to the membrane. At the time this philosophy represented a fair summation of the state-of-the-art of membrane processing. Operations of the membrane unit process to minimize fouling was largely undeveloped. Similarly, washing techniques for removing foulants from a membrane were not well developed.

In the last five years, advances have been made allowing a minimal fouling during operations and cleaning of the membrane where fouling has occurred. However, essentially all of the advances to date merely eliminate the effects of a certain foulant or are to remove that specific substance from the membrane. Very little work has been done to determine the nature of fouling constituents or to ascertain the necessary operating methods for minimizing fouling. At this time, very little work is being done to use a membrane system directly for the primary treatment of waste waters.

This program has developed membrane processing to the point of presenting a viable alternate to other physical-chemical primary treatment systems for the rigorous demands of an Army field water recycle system. The findings of this study, pointed out in this paper, dictate many of the requirements of the membrane equipment used for this application. The findings and limitations of the study will, it is believed, yield some insight into the direction required for future study in the application of membrane processing to primary treatment of all types of waste waters.

Membrane processing was considered for use as the primary treatment method basically because of the consistent quality of its product and its basic stability to wide variations in the nature of the influent waste streams. A membrane system is inherently capable of producing a high quality product even during the dynamic changes in the waste stream to the processing unit. The worst thing that can happen is that the membrane becomes fouled. So, the task was easily defined--eliminate the fouling as much as possible, and develop ways of removing the foulants where fouling does occur.

Module and Membrane Selection

The only requirement of the primary treatment process is that it clarify the waste stream for purification in further processing. Thus, dissolved solid rejection is not a necessary characteristic of the membrane and high flux ultrafiltration membranes would be suitable if fouling could be controlled.

Tubular modules were selected for use in the primary treatment section. The desire for simplicity and a minimum number of pretreatment steps dictated that only a gross solids separation step precede the membrane process. This meant that the colloidal material and much of the fine suspended matter would be left in the feed to the modules. Hollow fiber and spiral wound units do not allow passage of much of this material. Further, tubular units are necessary to provide the storage volume in the flow path to hold the highly concentrated suspended material.

The tubular modules chosen for evaluation were all manufactured by Calgon-Havens. They were selected because they were representative of the state-of-the-art in both membrane capability and module configuration. Originally, two membranes were selected for evaluation in the tubular modules. A low pressure (300 psig) ultrafiltration membrane, the Calgon-Havens 215, was the first choice because of its relatively high flux. A high flux reverse osmosis membrane, the Calgon-Havens 410X-1, was included in the test apparatus to provide a point of comparison. Initial testing showed that the short term potential for development of a system was greater with the 410X-1 module because of the lesser fouling rate. At this point, the 215 module was replaced by a Calgon-Havens 620 module, feeling that the 620 offered a potential to be the membrane with the least tendency to foul.

Initial Flux Versus Fouling

The unfouled flux capability of the Calgon-Havens 215 module is about four times that of the 620 module and twice that of the

410X-1 modules. Initial tests were carried out only on the 215 and 410X-1 modules. Eight times concentrated (8X) composite hospital waste was used for this phase of the testing. The brine effluent flow rate was 2.0 gpm, and the pH was adjusted to 5.0 to 5.5.

Both the 215 and 410X-1 modules experienced severe fouling during this testing. While the 215 had a higher initial flux, after several hours its flux rate had fallen below that of the 410X-1. Further, its flux level remained below that of the 410X-1 until some measure was taken to clean the membranes of the fouling material. The same trend was observed when the 410X-1 and 620 modules were compared. Based on this evidence, major emphasis was shifted from the 215 ultrafilter to the 620 reverse osmosis membrane.

This change in emphasis should not be interpreted to show any indication of reverse osmosis having higher potential fluxes than ultrafiltration when used for primary treatment. Basically, the membranes have certain capabilities; specifically, their unfouled flux rate. The problem becomes one of minimizing fouling in order to take full advantage of the inherent differences in the membranes. The initial results indicated, however, that the fouling processes are relatively more severe in the looser membranes. Thus, the shift in emphasis from the 215 to the 620 was merely a response to the belief that the potential for the short, limited study we were to perform was greater with the tighter membrane.

This tendency of the higher flux membranes to foul more severely continued throughout the testing. Generally, for any given waste, the flux rate of the 410X-1 was more unstable than that of the 620. Changes in any condition during the testing which tended to hasten or emphasize the fouling process caused a greater change in the 410X-1 than in the 620.

There appear to be two periods in the fouling process. The first period begins with the initial application of waste to a clean membrane and lasts for a few hours afterward. The rate of fouling of the membrane is higher during this phase than in the second phase. The length of this period varies with the membrane and the waste, as does the rate of fouling (as measured by the decline in flux). The second phase lasts through the remainder of the test period, until the membrane is washed. This has been a period of up to sixty hours. If the flux is plotted against the time that the waste has been applied to the membrane, the slope of the curve is an indication of the degree of fouling of the membrane. In the second phase, the slope is less than in the first phase, indicating that the fouling process is more pronounced in the first phase. The point at which the transition takes place varies with the waste. Generally, the more severe fouling wastes result in the transition occurring during the first 1 to 3 hours. With less severe fouling wastes, this transition occurs somewhere in the first ten hours.

Flow Turbulence

Higher brine flow rates were tried in an attempt to lessen the

fouling treands on the membrane. The reasoning was that this
would increase turbulence, decreasing the rate of deposition of
the fouling material on the membrane. It also seemed likely that
the effects of the concentration polarization phenomenon would
be decreased, effectively reducing the solids concentration of
the waste that was applied to the membrane.

The higher brine flow rate had two effects on the flux curve.
These effects are shown in FIG. 1. The first portion (the period
with a high rate of fouling) was altered, the degree of alteration dependent on the membrane. The rate of fouling during this
period was lower at the higher flow rate, and the rate was more
influenced with the 620 membrane than with the 410X-1 membrane.
Also, the point at which the flux curve changed slopes occurred
at a greater flux rate. Once again, the second portion of the
flux curve had less slope than the first portion. Unlike the
first portion fo the flux curve, whose slope was clearly altered
at the higher brine flow rates, the slope of the second portion
of the curve does not seem to change noticeably at the higher
brine flows. It should be noted here that the consistency of
the slopes in the second portion of the flux curves were only
noticeably unaltered as long as the only alterations to operating
parameters was the higher brine flow rate. When other changes
were made in operating procedures to lessen the effect of fouling
the distinction between the two portions of the curve became less
pronounced.

Depending on the nature of the waste, a higher brine flow rate
may not always do an adequate job of enhancing the flux of a
membrane system. Turbulence does not prevent fouling. It does
slow the rate of the fouling process, thereby allowing a more
efficient use of the membrane between cleaning operations.
Where the solids loading is so high as to create an excessive
osmotic pressure, other techniques must be used to lower these
factors to a level effectively and efficiently handled by membrane processing. Where the significant fouling material in a
waste is ineffectively restrained from the membrane by the greater turbulence in the flow, other techniques must be developed
for keeping the material from coating the membrane.

Chemical Additives

In the early stages of adapting a membrane system to primary
treatment of hospital wastes, much of the work centered around
finding a chemical additive to alleviate any severe fouling.
The first generation MUST WWMS used a constant addition of sodium triphosphate (STP) in the membrane section to prevent fouling by any substance left in the waste stream at that point.
This made STP a natural starting point in looking for a dispers-
ant to use in primary treatment.

The effectiveness of STP in slowing the fouling rate is best
demonstrated in a waste stream of 8X composite hospital waste.
In this test, additions of 0, 25, 100, 250, and 500 ppm STP were
compared at conditions otherwise identical. The brine flow rate
was 3.0 gpm. The flux curves at the various addition levels of
STP (FIG. 2) can be compared to the flux curves for the 620

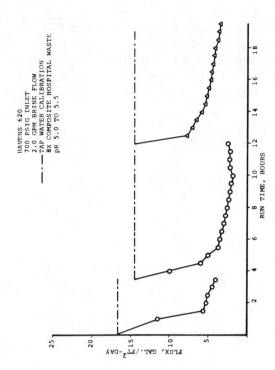

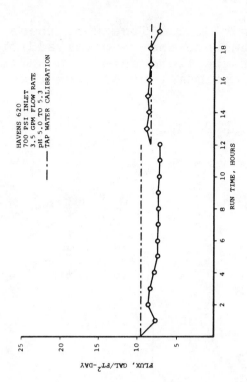

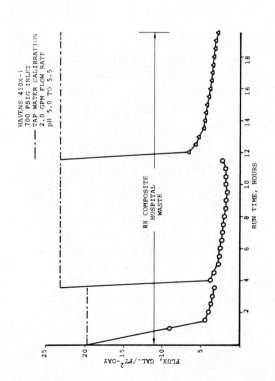

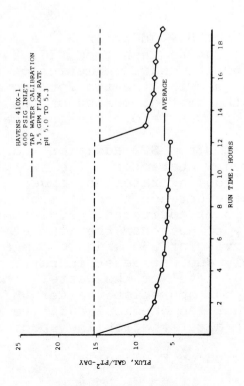

FIG. 1
Flux Curves

module in FIG. 1. Together, these curves show the effect of
flow turbulence and chemical addition on the fouling characteristics of the 620 module. An addition of 100 ppm seemed to provide the optimum flux characteristics for both the 620 and 410X-1.

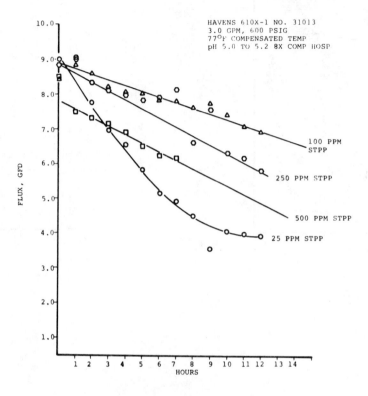

FIG. 2
Effect of Chemical
Addition on the Flux Curve

The 25 ppm STP addition had a definite effect on the flux
curve. As can be seen from FIG. 2, the 25 ppm addition level
has altered the characteristics of the flux curve somewhat. The
two phases showing the two fouling rates are still apparent.
However, the initial period of fouling does not seem to be as
severe as it was before the chemical addition.

At the 100 ppm STP addition and above, the flux curve assumes
a totally new character. The two phases of fouling with the
different fouling rates no longer exists. The new fouling rate
is somewhere between the rates observed in the first and second
phase at lower additions. Additions above the optimum level
serve only to decrease the flux at any given time, as indicated
by the curves for 250 and 500 ppm STP addition. This lower flux
is probably due, at least in part, to an increase in the effective solids level of the waste. The fact that excessive addition does not appreciably alter the slope of the flux curve indicates that the chemical additive above a certain level does
nothing to alter the fouling species in the waste.

Effects of pH on Fouling

The importance of pH on the fouling characteristics of a particular waste was first noted with the composite hospital waste stream. The natural pH of this stream is usally slightly basic, averaging around 7.5. Adjustment of the pH to 5.5 to 6.0 had been provided to minimize membrane hydrolysis. This strict regulation of pH may not be necessary, at least with some waste streams. Modules in operation for over six months at the City of San Diego Point Loma facility at a natural pH of 7.0 to 7.2 have shown no signs of hydrolysis. The 620 modules used in testing in the MUST development program have over 750 hours at the higher pH (6.8 to 7.2) and show no signs of rejection degradation.

The only waste in which detailed pH effects have been obtained is the composite hospital waste. In the range of 5.0 to 7.0, there is a noticeable improvement in permeation performance of the module with higher pH levels.

The mechanism which pH exhibits on membrane fouling has not been determined. The major constituents of fouling with composite hospital waste appears to be oils and greases (or similar organic materials). This makes it difficult to determine if the change in solubility of some dissolved material(s) is the cause of the difference in fouling rate. In a waste stream of the complexity of this composite waste, the effects of singular components are very difficult to define. The difference in fouling rates on the various membranes and at various pH levels indicates that the fouling phenomenon may be partially electrical in nature. If so, the existence and density of polar groups in the membrane and the degree of polarity of the fouling element may in part determine the intensity of the fouling process.

Membrane Washing

One of the major questions posed when it was determined to look at membrane processing for primary treatment was whether or not it was possible to effectively wash the membrane of foulants. Generally, other applications of membrane processing indicated that this was possible. Two wash solutions have been used predominately in testing to date and have proven generally effective in restoring the permeation capabilities of the membranes.

The two wash solutions used are (1) a solution of Biz and ammonium citrate, and (2) a sodium dithionite solution. The Biz-ammonium citrate seems to be particularly effective in removing organic foulants. The sodium dithionite wash is used to remove precipitated inorganic salts. The effectiveness of these washes on membranes fouled by the various MUST hospital waste streams varied quite drastically. However, neither wash is, by itself, always effective. Together they do seem to completely restore the flux capabilities of the membrane.

There were times with certain of the wastes when system depressurization provided at least some restoration of flux capabilities to the membrane. This simple method of defouling varied in effectiveness from nearly complete restoration with operating room wastes to very little effectiveness with x-ray wastes. This merely serves to point out the great variability

in the nature of the fouling mechanism. Some of the fouling materials seem to be held in position only by the flow of water to
and through the membrane. Other foulants seem to have an actual
attraction for the membrane and must be literally pried loose
from the membrane by either the cleaning agent or procedure.

Summary

Membrane processing has been shown to be technically capable
of treating MUST hospital waste waters in field operations. Flux
capabilities of the Calgon-Havens 41OX-1 and 620 modules, used
in primary treatment of the MUST hospital waste streams, are
sufficient to allow favorable comparison with other physical-
chemical primary treatment systems under consideration for use in
the Water Treatment System. Rejection capabilities of the 620
modules produce a product of suitable quality to meet U. S.
Public Health potable water standards with only a minimum of
polishing.

Three operating parameters were found to affect the rate of
fouling of the modules---flow velocities, chemical treatment, and
pH of the waste. Of these, the flow velocity and pH seem to have
the greatest effect. The characteristics of the individual membrane also affects the fouling rate. Finally, the characteristics of the individual wastes determine, in the end, the fouling
rates.

Fouling of the membrane is basically a phenomenon of the
waste. That is, it is the elements in the waste which cause the
fouling process and proper pretreatment of the waste can minimize the fouling. Membranes are limited to a few that presently
exist or are likely to be developed. The great diversity in
types of fouling and mechanisms by which fouling occurs precludes
the foreseeable development of a "foul-less" membrane. However,
the obvious importance of the membrane in fouling mechanisms can
be used to select the best membrane for any given application.
Before effective pretreatment of wastes can be performed to minimize fouling, the mechanisms of fouling must be defined.

The modules used in the primary treatment tests have been
operated for over 1000 hours. Permeation and rejection characteristics of the wastes indicate that there has been no hydrolysis to the membranes. As yet, there is no indication that the
life of a module is limited by hydrolysis. Further, it has been
shown in this study that the fouling inherent in this type of
operation is reversible. That is, the fouling material can easily be removed from the membrane by certain washing techniques.

The membrane processes incorporated in the MUST water recycle
system, when used with minimal polishing procedures, produces
a water meeting all Public Health Service standards. In a recycle system of this nature there is always concern for buildup
of toxilogical constituents. Before this system is used to provide potable water, there will be a thorough toxicity evaluation
program.

COMPARATIVE COST OF TERTIARY TREATMENT PROCESSES

Roy F. Weston and Robert F. Peoples
Roy F. Weston, Inc., West Chester, Pennsylvania

The title of this paper is "Comparative Cost of Tertiary Treatment Processes."
The title is ambiguous in that the phrase "tertiary treatment" frequently is
used interchangeably with "advanced wastewater treatment." Since the latter
term may identify a single or multi-stage system, it has much broader application and provides the area for comparison of "tertiary treatment" processes
among the presently available "advanced wastewater treatment" systems. In
this context, the paper will discuss the alternatives and present results of
several evaluations done for specific applications where an "advanced wastewater treatment" system was required.

"Advanced wastewater treatment" itself must be defined to assure a clear understanding of the topic under discussion. As used herein, the term is meant
to define a treatment system capable of providing effluent quality significantly better than conventional techniques can provide. That is, the system
will remove 95% or more of BOD_5 and suspended solids, will remove 85% or more
of phosphorus and total nitrogen, and will provide a disinfected effluent.

Table 1 shows a summary of treatment requirements and applicable treatment
schemes. It is readily apparent that "tertiary treatment" is but one of several alternatives depending on the required degree of treatment, the presence
of existing facilities, and anticipated future needs. This paper will compare
the costs of alternative solutions as they are directly related to needs.

TABLE 1

Treatment Requirements
vs.
Applicable Treatment Systems

Treatment Systems	Treatment Requirements				
	A	B	C	D	E
1. Conventional Systems (Pre-treatment, primary clarification, single-stage biological treatment)	X	X[1]			
2. Conventional System With Chemical Addition	X	X			
3. Tertiary System (Conventional System followed by <u>one or more</u> polishing steps other than biological)	X	X	X	X	
4. Multi-stage Biological System (Pre-treatment, primary clarification, followed by 2- or 3-stage biological system and polishing treatment)	X	X	X	X	X
5. Independent Physical and/or Chemical System (Pre-treatment followed by a combination of one or more of filtration, adsorption, chemically-assisted coagulation, and precipitation)	X	X	X		X

Treatment Requirements:

A - High-Level Secondary Treatment, 90-95% BOD, 90% SS Removal
B - 90% BOD Removal, 85% P Removal
C - 95+% BOD Removal, 95+% P Removal
D - 95+% UOD and P Removal (with N oxidation to NO_3)
E - 95+% UOD and P Removal, 85+% total N Removal

[1]Not always applicable.

High-Level Secondary Treatment

While conventional secondary treatment is not directly a topic for this discussion, it plays an important role since many "advanced" systems will be
built on this base. This is because regulatory agencies will probably upgrade
effluent quality in a typical systematic pattern -- first they will want improvement with respect to oxygen demand removal, then with respect to suspended solids removal, and lastly with respect to nutrient removal. Thus, a
treatment system designed for 85% BOD_5 and suspended solids removal may be
upgraded step by step to achieve "advanced wastewater treatment" results.

The costs of high-level secondary treatment would be essentially the same as
a conventional system, but with a higher degree of reliability in operation.
This treatment cost can be expected to vary from 40¢/1,000 gal. for a 1-mgd
plant to 10¢/1,000 gal. for a 100-mgd plant at 6% interest and 25-year
amortization (1).

High-Level BOD Removal, Moderate Phosphorus Removal

When moderate (85%) phosphorus removal is added to a 90% BOD removal requirement, all of the systems shown in Table 1 may still apply. However, the conventional Secondary Treatment System is applicable only in certain circumstances, and pilot work is required to determine if it will work in any given
instance. Several researchers (2,3) have demonstrated moderate removal of
phosphorus simply by an improved control of existing activated sludge systems.
The exact mechanisms for this phenomenon are not yet defined, and removal cannot be assured without pilot or full-scale testing. For this reason, back-up
chemical feed equipment is normally provided to assure phorphorus removal.
Because of biological phosphorus removal, the chemical requirements may be
only a fraction of the requirements for chemical treatment alone. The capital
cost is small ($\approx$ $10,000 per million gallons), and operating costs are only for
the chemicals actually needed to meet the effluent requirements (2-3¢/1,000
gal.) (1). The remaining systems listed in Table 1 (Items, 3, 4, and 5) can
provide a higher quality of treatment than needed in this case.

95+ Percent BOD and Phosphorus Removal

Once treatment requirements reach this level, conventional secondary processes
cannot assure reliable compliance. Therefore, the conventional secondary system must be supplemented with a polishing treatment system. This may be a
"tertiary system", a multi-stage biological system, or an independent
physical-chemical system. There are various combinations of unit processes in
each of these alternatives which may be utilized to meet specific needs.
Tertiary systems include chemical precipitation, filtration, and carbon adsorption, or a combination of these processes. Capital costs for each of
these can be expected to be in the range shown in Table 2. These systems provide relatively low-cost means of meeting regulatory requirements for high-
efficiency BOD and phosphorus removal when existing secondary facilities have
adequate capacity, but not adequate efficiency.

TABLE 2

Capital Cost of Tertiary Unit Processes[1]

	5-mgd	30-mgd
1. Chemical Precipitation and Clarification	$1.4 M	$3.5 M
2. Filtration	$0.25 M	$1.0 M
3. Adsorption	$1.5 M	$4.8 M

[1] ENR = 1680 (1972)

Other tertiary processes are retention ponding and lagoons. However, these have limited application when high-quality effluents are required.

Another method is spray irrigation or effluent spreading. However, this may require a considerable amount of land. Since land costs and land area requirements are highly variable, costs cannot be generalized. These methods provide an excellent choice where no adequate receiving stream is available and soil conditions permit their use. However, the effluent quality needed to satisfy regulatory agency requirements for irrigation or effluent spreading may make advanced wastewater treatment necessary. In such cases, these techniques would be considered ultimate disposal methods rather than tertiary treatment.

95+ Percent UOD and Phosphorus Removal

The basic difference between this and the previous case is the stipulation of UOD (ultimate oxygen demand) as contrasted to BOD. UOD is defined as carbonaceous BOD plus oxidation of ammonia and organic nitrogen (TKN) to nitrate. Only two basic types of systems are applicable to this requirement -- Tertiary systems and Multi-Stage Biological systems. Tertiary systems would be applicable only if nitrification were achieved reliably in the secondary treatment step. Costs for this type of treatment would be about 50% higher than conventional secondary treatment to provide two-stage activated sludge with chemical addition for phosphorus removal and effluent filtration.

It should be noted that the systems mentioned above provide nitrogen oxidation but not nigroten removal. Nigroten removal is discussed below in a broader category because nitrogen removal without nitrification is possible.

95+ Percent UOD Plus 85 Percent Nitrogen Removal and 95+ Percent P Removal

These limits are thought of today as total treatment, or essentially the best that can be assured using today's effluent treatment technology. Actually it is better to think of this degree of treatment in terms of allowable effluent concentrations rather than percent removals. A tertiary system can meet these requirements, but it would be essentially a conventional system followed by the equivalent of an independent physical-chemical system. Therefore, the real comparison should be limited to: a multi-stage biological system (MSB) with effluent polishing, and an independent physical-chemical system (IPC).

Evaluations of these alternative concepts are discussed in the following sections of this paper.

Maximum Removal of BOD, Nitrogen, and Phosphorus

Future treatment requirements for a large municipal authority include maximum removals of organics (BOD), suspended solids, nitrogen, and phosphorus. This particular situation is complicated by a rather commom problem -- namely, that a well-functioning conventional secondary treatment system having the full required capacity already exists. In attempting to determine the most economical means of accomplishing the objective in such a situation, the cost of existing facilities is a significant factor.

Preliminary evaluation reduced the alternatives to: 1) a multi-stage biological system with effluent polishing; and 2) an independent physical-chemical system. Schematic drawings for these two systems are shown in Figures 1 and 2. The estimated capital costs for these complete systems are summarized in Tables 3 and 4.

TABLE 3

Summary of Capital Costs for
Three-Stage Biological Treatment[1]

Pre-Treatment	$ 1,500,000
First Stage	4,500,000
Second Stage	4,600,000
Third Stage	3,700,000
Filtration	1,400,000
Chlorination	200,000
Sludge Thickening and Supernatant Treatment	500,000
Sludge Conditioning and Dewatering	700,000
Sub-Total	$17,100,000
Piping, Electrical, Instrumentation, Site Work	6,900,000
Sub-Total	$24,000,000
Construction Contingency	3,600,000
TOTAL CAPITAL COST	$27,600,000

[1]ENR = 1680 (1972)

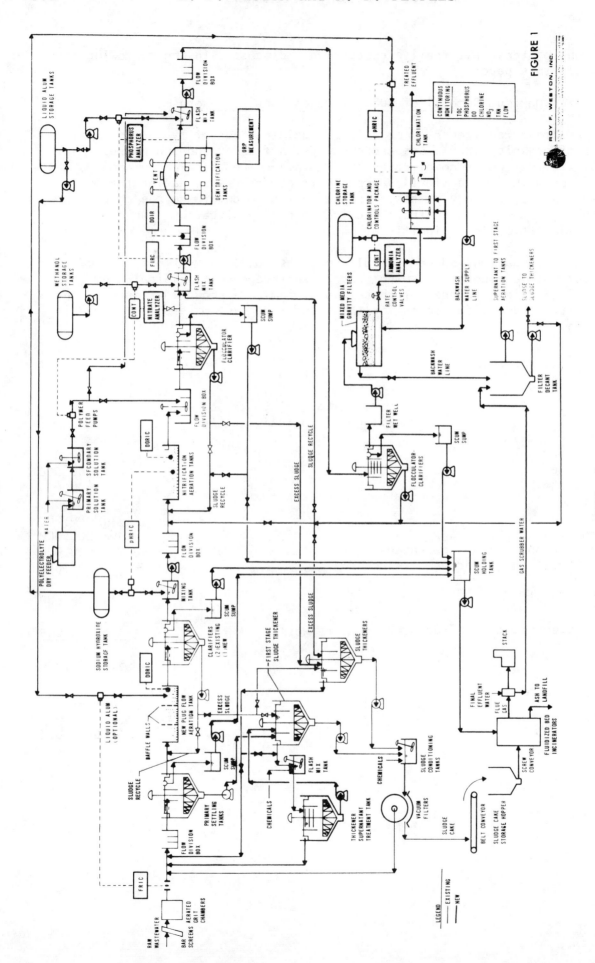

FIG. 1

Process Flow Diagram

Three-Stage Biological-Chemical Treatment

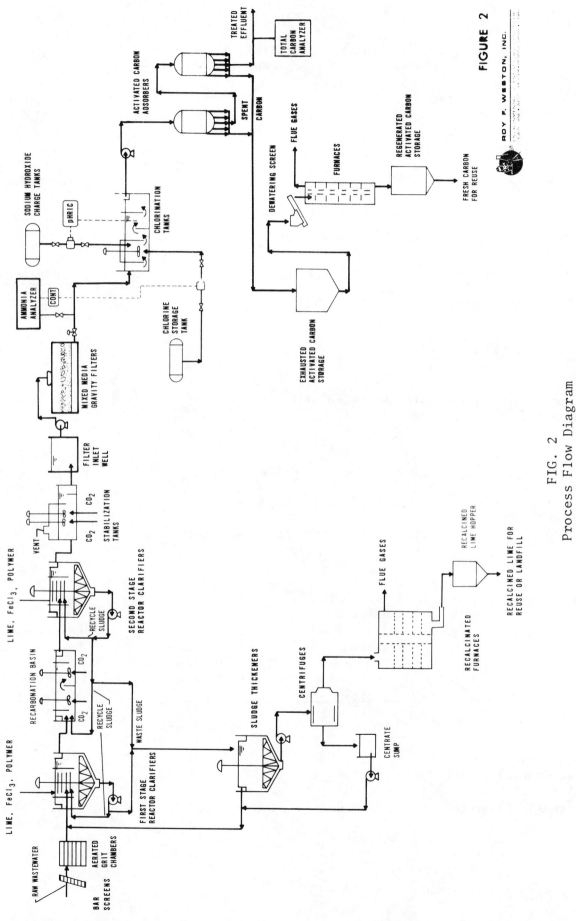

FIG. 2
Process Flow Diagram
Independent Physical-Chemical Treatment

TABLE 4

Summary of Capital Costs for
Independent Physical-Chemical Treatment[1]

2-Stage Lime Treatment	$ 5,400,000
Activated Carbon	2,600,000
Chlorination	300,000
Sludge Handling	5,700,000
Sub-Total	$14,000,000
Piping, Electrical, Instrumentation, Site Work	6,000,000
Sub-Total	$20,000,000
Construction Contingency	3,000,000
TOTAL CAPITAL COSTS	$23,000,000

[1]ENR = 1680 (1972)

The comparison of the alternative solutions presented in Table 5 illustrates
several noteworthy points.

TABLE 5

Cost Comparison
for
Total Treatment of Municipal Wastewater[1]
(30 mgd)

	Complete Multi-Stage Biological System with Effluent Polishing (MSB)	Independent Physical-Chemical System (IPC)	Conversion of Existing Adequate Secondary System to MSB System
	Millions of Dollars	Millions of Dollars	Millions of Dollars
Capital Cost	$27.6	$23.0	$18.0
Annual Operating Cost	2.9	4.3	2.9
Total Annual Cost 30 years, 6 percent	4.8	6.0	4.2
	45 cents/ 1,000 gallons	56 cents/ 1,000 gallons	39 cents/ 1,000 gallons

[1]ENR = 1680 (1972)

The first concerns capital costs. If no significant facilities exist at the outset of the project, the capital cost of the IPC system is about 10% less than that of the MSB system. If adequate secondary treatment facilities exist, the MSB has the lower cost. Clearly, a critical evaluation of the suitability of existing facilities for incorporation into an advanced waste-water treatment facility has a significant impact on capital cost.

However, the overriding factor in this evaluation can be seen to be annual operating costs.

As shown in Figure 2, the nitrogen removal in the IPC System is by breakpoint chlorination. The cost of breakpoint chlorination is approximately 7¢/1,000 gallons which nearly equals the total chemicals costs for the MSB system. It should also be noted that the cost of breakpoint chlorination is almost entirely an operating expense since capital facilities are nominal. Thus, when the cost of carbon and other chemicals are added, the total chemical costs account for the considerable difference in annual operating costs. For this situation, the annual costs obviously favor the MSB System. Even in the case of a "new facility", the capital savings of the IPC System cannot over-come its high chemical costs. A similar study for a 309-mgd plant serving Washington, D.C. indicated costs of 27.6-29.6¢/1,000 gal. for MSB and 33.8¢/1,000 gal. for IPC, but noted that pre-treatment and primary sedimentation were not included in the comparison since they were common to both alter-natives. Capital costs again were within 10%, with the MSB having the higher capital costs (4).

The preceding comments indicate that operating costs for the IPC system are too high if nitrogen removal is required. However, if nitrification or total nitrogen removal is required only seasonally, the annual costs would be much closer and could shift enough to favor IPC. For example, if nitrogen removal is practiced only four months of the year, the savings at 7¢/1,000 gal. would be about one-half million dollars annually.

Ammonia removal by ion exchange may some day be less costly than breakpoint chlorination, but present estimates are about 10¢/1,000 gal. Furthermore, the process results in a brine or waste chemical (ammonium sulfate) disposal problem.

Organic Removal from Industrial Wastewater

The application of advanced wastewater treatment techniques to an oil refinery wastewater disposal problem may be of interest.

This oil refinery was required to provide stringent organic removal measured in terms of first-stage ultimate oxygen demand, but with no nutrient removal requirements. This situation demanded efficient oil removal plus a relatively high degree of removal of residual BOD_5.

Preliminary evaluation reduced the alternatives to an activated sludge system and an activated carbon system. Pilot plant work demonstrated that either could do the job. Also, pilot tests and preliminary cost estimates showed that deep-bed sand filtration was clearly preferable to the more conventional API Separator for oil removal. Thus, it was included as pretreatment in both cases. This use of filtration as the first step of a treatment process chain serves as another example of the broad application of the "advanced wastewater treatment" terminology, as opposed to a "tertiary treatment" concept.

The plant has an average design capacity of 2.9 mgd. Table 6 shows a comparison of the costs for the alternative systems. If is readily apparent that the activated carbon system which might be thought of as an expensive and "advanced" type alternative to the more conventional activated sludge treatment has the economic advantage for both capital and operating costs. Of course, it also provides an effluent quality which far exceeds present-day requirements.

TABLE 6

Comparison of Costs
for
Treatment of Refinery Wastewaters[1]

	Activated Sludge System	Activated Carbon Treatment
Capital Cost	$2,500,000	$2,000,000
Annual Operating Costs	$220,000	$180,000
Cash Flow-Annualized Costs[2]	45¢/1,000 gallons	36¢/1,000 gallons

[1] ENR = 1680 (1972)
[2] 20 years at 8%

It is not possible to generalize as to the relative economics of industrial wastewater treatment because the activated sludge process was the more economical for another oil refinery that had to meet the same effluent requirements. The cost differences were due to the availability of needed land areas.

Tertiary Treatment of Municipal Wastewater

In this case, tertiary treatment consisted of removal of residual BOD, suspended solids, and phosphorus after treatment in a conventional activated sludge system. The design effluent quality was 5 mg/L BOD_5, 0.5 mg/L SS, and 1.0 mg/L P. Since the primary purpose of the project was to demonstrate the feasibility of a specific technical approach and to develop the real costs for operation of this approach as a Model Plant sponsored by EPA, there was no comparison of alternatives. However, the cost data are significant in that they place the tertiary treatment concept in perspective.

Table 7 summarizes actual capital costs and estimated operating costs for the basic 5-mgd capacity project, and also shows the estimated cost for providing nitrogen removal with facility. Again, nitrogen removal adds significantly to the operating costs of such a system. The addition of breakpoint chlorination would increase the operating costs nearly 20%. Also, it must be noted that this system would remove ammonia nitrogen, but would not remove nitrate nitrogen. The costs are considerable in any case when it is noted that conventional secondary treatment costs of approximately 15¢/1,000 gal. must be added to those shown in Table 7.

TABLE 7

Cost of "Tertiary Treatment"
of
Municipal Wastewater[1]

	Capital Cost	Annual Operating Costs
BOD, Suspended Solids and Phosphorus Removal	$3,600,000	$600,000 (38¢/1,000 gallons)
Nitrogen Removal by Breakpoint Chlorination	$100,000	$110,000 (7¢/1,000 gallons)

[1]ENR = 1680 (1972)

Summary

This discussion has attempted to show a few direct cost comparisons of alternative means of providing a high level of treatment. A summary of costs for the various treatment systems is presented in Table 8. The emphasis has been to highlight the need to be extremely clear on:

1. The required degree of treatment;

2. Utilization of existing facilities; and

3. Consideration of all alternatives.

The case studies used here show that the optimum result will vary with the need and with the capabilities of the existing treatment facilities. If all these factors are given proper weight, the most technically feasible and least-cost solution can be determined.

TABLE 8

Costs of Advanced Wastewater
Treatment Systems[1]

Treatment Systems	Capacity	Cost in cents/1,000 gallons Treatment Requirement				
	mgd	A	B	C	D	E
1. Conventional Systems						
(Pre-treatment, primary clarification, single-stage biological-treatment)	100 1	10 40				
2. Conventional System						
With Chemical addition	100 1		13 45			
3. Tertiary System						
(Conventional System followed by one or more polishing steps other than biological)	100 1			30 110		
4. Multi-stage Biological System						
(Pre-treatment, primary clarification, followed by 2- or 3-stage biological system and polishing treatment)	100 1				15 60	20 80
5. Independent Physical and/or Chemical System	100 1			43 93		50 100

Treatment Requirements:

A - High Level Secondary Treatment, 90-95% BOD, 90% SS Removal
B - 90% BOD Removal, 85% P Removal
C - 95+% BOD Removal, 95+% P Removal
D - 95+% UOD and P Removal (with N Oxidation to NO_3)
E - 95+% UOD and P Removal, 85+% Total N Removal

[1] ENR = 1680 (1972)

References

1. Swanson, C.L., "New Wastewater Treatment and Technology Is Available Now", Division of Facilities Construction and Operation, Office of Water Programs, EPA, Wash., D.C., June 1971.

2. Milbury, W.F., McCauley, D., and Hawthorne, C.H., "Operation of Conventional Activated Sludge for Maximum Phosphorus Removal", Journal WPCF, Washington, D.C., Vol. 43, No. 9, Pgs. 1890-1901, Sept. 1971.

3. Bargman, R.D, Betz, J.M., and Garber, W.F.. "Nitrogen-Phosphate Relationships and Removals Obtained by Treatment Processes at the Hyperion Treatment Plant," Presented at the 5th International Water Pollution Research Conference, July-August 1970.

4. Samworth, R.B., Ph.D., Bethel, J.S. "The Use of Pilot Plant Studies in the Design of a Major wastewater Treatment Plant." Presented at the 44th WPCF Conference, San Francisco, Calif., Oct. 1971.

Applications of New Concepts of Physical-Chemical
Wastewater Treatment
Sept.18-22, 1972

PROCESS SELECTION AND COST OF ADVANCED WASTEWATER TREATMENT
IN RELATION TO THE QUALITY OF SECONDARY EFFLUENTS
AND QUALITY REQUIREMENTS FOR VARIOUS USES

L.R.J. van Vuuren and M.R. Henzen
National Institute for Water Research of the South
African Council for Scientific and Industrial Research

Introduction

During the past decade, the importance of pollution control and water reuse has been emphasized during several conferences and workshop sessions. New terminology has been formulated to define or describe new approaches to the complex field of wastewater treatment. Water reclamation, tertiary or advanced treatment, renovation, reuse, independent physical-chemical treatment, mineral addition, denitrification and numerous sophisticated technological processes have been described by a multitude of authors from various parts of the globe. Conclusions drawn have confirmed that conventional sewage treatment processes have their limitations, that pollution abatement should be tackled at source and that conventional water treatment systems are no longer a safe practice to cope with the deteriorating quality of surface waters attendant to industrial activity. However, it is also evident that available technology can be harnessed successfully to solve these problems at a cost which has to be assessed for each particular situation.

In South Africa, the pressure for more water rather than the pollution problem has instigated intensive research into water reclamation. The first plant for the reclamation of sewage for potable reuse on a permanent basis was put on stream in Windhoek, South West Africa, during the late sixties. This research also resulted in the construction of a full-scale plant (27 Ml/d) at Springs, South Africa, where renovated sewage is being used for the production of high quality pulp and paper since July, 1970. A third plant (4,5 Ml/d) was commissioned at Pretoria in 1971 to serve as a demonstration/research facility. The main objectives of the latter plant are to evaluate cost and quality and to formulate design criteria for future full-scale implementation of advanced treatment technology for water reuse and/or pollution abatement in South Africa.

This paper deals briefly with the process selection and economics of the above water reclamation schemes in relation to the quality parameters involved for domestic and industrial reuse.

371

The Windhoek Water Reclamation Scheme

Operational results and performance

The operational results, cost and performance data of the pilot and full-scale plants at Windhoek, were reported upon in earlier publications (1,2,3). Subsequent research at Windhoek was directed towards a continuous monitoring programme, particularly with respect to pathogenic and biochemical quality as well as epidemiological surveys in relation to the reclaimed and other water supply sources in the area.

The nominal capacity of the works (Fig. 1) is 4,5 Ml/d. During the first two years of operation, the reclamation plant has contributed an average of 13,4 per cent of the total water consumption (4). The average inflow of raw sewage during this period was 6,2 Ml/d which is equivalent to a capacity utilization factor on the nominal design output of a little over 50 per cent. The low utilization performance is attributed largely to interruptions in the operations caused by mechanical failures, process losses, recharging of carbon filters and cessation of operations during the winter months due to high ammonia concentration levels where existing chlorination facilities were inadequate to ensure breakpoint treatment.

The Windhoek water reclamation plant operated since October 1968 until the end of 1970. Towards the beginning of 1971, the loading on the conventional sewage treatment works had increased to such an extent that the quality of the maturation pond effluent did not comply with the water quality specifications of the reclamation plant. Reclamation of purified sewage effluent was therefore stopped temporarily pending completion of expansions to the conventional sewage treatment facilities. In addition the incidence of plentiful rain in this area, eliminated the urgent need for water reclamation. The plant will be commissioned again upon completion of the conventional treatment works.

Costs

During the entire period of operation the quality of the product water complied with WHO Standards. Hence the unit costs of reclaimed water were higher than would be the case for water of lower quality.

The costs obtained for the first two years of operation are given in Table 1.

TABLE 1

Costs for Reclaiming Potable Water from Maturation Pond Effluent at Windhoek

Cost Item	South African cents/kl	U.S. cents/1000 U.S. gallons
Capital costs	3,88	19,40
Labour	0,89	4,45
Chemicals	2,87	14,35
Activated carbon	2,40	12,00
Specialized supervision	1,50	7,50
Total	11,54	57,70

The above costs do not include that of conventional sewage treatment and are based on the actual utilization factor of 50 per cent. For an 80 per cent utilization factor, the total unit cost would drop to 9,19 c*/kl (45,95 U.S. c/ 1000 U.S. gallons). The latter figure compares favourably with the unit cost of 10,7 c/kl for conventional water treatment of supply sources originating from natural run-off in this area. It should be taken into account that for geographic reasons unit costs for chemicals and power (2,5 c/kW.h) are relatively high in the Windhoek area.

The combination of the individual processes of: algae flotation using aluminium sulphate as flocculant, foam fractionation, breakpoint chlorination and granular activated carbon adsorption followed by post-chlorination proved to be an economic system for the treatment of this type of effluent, provided the ammonia concentrations in the influent are maintained at levels below 7 mg/l. The low utilization factor referred to above as well as the inability of the conventional sewage treatment facilities and attendant maturation ponds to provide an effluent fit for reclamation particularly from a chlorination point of view, impose a constraint on the cost evaluation.

Refinement of Purified Sewage for use in the Manufacture of Fully Bleached Pulp

Background information

The South African Pulp and Paper Industries (SAPPI) is situated approximately 100 kilometres from the nearest pumping station of the Rand Water Board water supply authority. Water usage is made up of 16 Ml/d from this source and a further 27 Ml/d of purified sewage from the City of Springs.

During previous years the purified sewage (conventional treatment) received only limited tertiary treatment by sand filtration and low level chlorination. The demand for a higher brightness paper required further refinement of the available purified sewage. A quality survey confirmed the presence of heavy metals, particularly iron, manganese and copper and organics which are known to affect paper brightness. Research was then conducted using various adsorbents, oxidants and flocculants such as lime and aluminium sulphate to improve the water quality. These studies culminated in the design and construction of a full-scale plant which was commissioned in July, 1970 (5).

The full-scale plant consists essentially of the following units (Fig. 2): Flotation tank of 750 kl capacity; feeder line (0,6 m) with booster pump; aeration vessel with high speed disperser and air compressor operating at 10 psig; storage tanks and dosing equipment for aluminium sulphate, sodium hydroxide and chlorine; and auxiliary equipment such as pH and flow recorder controllers.

Aluminium sulphate is dosed at 75 mg/l into the feedline at a point succeeding the aeration stage. Approximately 1,4 mg/l of polyelectrolyte is added for improved flocculation and flotation and sodium hydroxide (10 mg/l) is dosed in the effluent launder to adjust pH value. After addition of 3 mg/l of chlorine as a measure against algal growth, the effluent is passed through gravity sand filters. Typical operational results are shown in Table 2.

---

* All costs are expressed in S.A. cents except where stated otherwise.

TABLE 2

Quality of Purified and Reclaimed Sewage
Effluent from Full-scale Plant at SAPPI
(mg/l where applicable)

	Springs Purified Sewage	Reclaimed water
pH	7,2	6,7
Conductivity	1100	1200
Colour (Hazen)	40	10
Chemical Oxygen Demand	75	40
Total Phosphate (as P)	2,6 - 6,7	0,6 - 1,2
Methylene Blue Active Substances	1,0 - 1,5	0,7 - 0,9
Iron (as Fe)	0,26	0,06
Manganese (as Mn)	0,55	0,50
Copper (as Cu)	0,45	0,02
Relative paper brightness in Elrepho Units	76,9	84,4
Distilled water in Elrepho Units	85,6	
Rand Water Board Water in Elrepho Units	82,4	

The reclaimed water was of such a quality that it could be used in all
sections of the mill without deleterious effect on the quality of the product
paper.   Very low turbidity (0 - 1 mg/l) was carried over from the flotation
unit which has a hydraulic retention of less than 30 minutes.

Costs

Average costs over almost 2 years of continuous operation are recorded in
Table 3 below in comparison with costs for water as supplied by the Rand Water
Board.

TABLE 3

Treatment Costs at SAPPI's Water Reclamation Plant

	South African cents/kl	U.S. cents/1000 U.S. gallons
Rand Water Board water	3,52	17,60
Springs Purified Sewage	0,22	1,10
Operating Costs (chemicals, maintenance and supervision)	1,21	6,01
Capital Expenditure		
R159 000 at R14 000 per annum, 25 Ml/d for 325 days per annum	0,17	0,85
Total cost reclaimed water	1,60	8,00

From these figures, the annual savings by using reclaimed water instead of Rand Water Board water amounts to R156 000 which is of the same order as the initial capital expenditure.

The excellent performance record of this plant, the relatively low capital expenditure and operating costs and the production of a quality water suitable for pulp and paper production have confirmed the suitability of the selected processes to meet the target requirements.   A cost saving of more than 50 per cent in relation to conventional supply sources is achieved.

Stander Water Reclamation Plant

Description of Plant

The design of this plant was based on extensive pilot-scale studies during the period 1966 to 1969.   These were reported on in earlier publications (6,7).

The full-scale plant was completed towards the end of 1970.   During the commissioning stages, several problems were experienced which required major modifications and adjustments to some of the process units.   The subsequent operational programme was directed towards optimization studies during which period further refinements had to be made.

The flow diagram illustrating the major process units (Fig. 3) is briefly described below, and details on the capacity of the units are given in Table 4.

TABLE   4

Details of Retention and Loading Rates of the Various Units as Shown   in Fig. 3

Process Unit	Retention time in min	Loading in kl/min
Foam fractionator	9,9	–
Chemical clarification tank	76	0,0367
Ammonia stripper	1,35	0,065
Mixing tank No. 1	17,2	–
Mixing tank No. 2	5,7	–
Mixing tank No. 3	5,7	–
Stabilisation tank	176	0,0182
Sand filter (2 in parallel)	8,5	0,119
Breakpoint chlorination tank	40	–
Granular carbon filter (2 in parallel)	12,9	0,231

Foam fractionation

During the pilot-scale investigations, the need for foam fractionation was considered an important pretreatment stage as a means to utilize powdered activated carbon more fully for the removal of organic carbon in a subsequent stage. By the time the full-scale plant was commissioned, it was evident that detergent concentrations in the secondary effluent was of such a low level as a result of the increased use of 'soft' detergents that foam fractionation was hardly warranted.   This process unit was subsequently by-passed.

Excess lime treatment

The flotation unit, the design of which was based on the pilot plant, gave disappointing results as only about 60 per cent of the solids could be separated by dispersed air flotation.   The cause of this failure was inter alia ascribed to the drop in detergent concentrations.   It then became necessary to convert the flotation unit to a solids contacting clarifier.   With a solids recirculation rate of 10 per cent at 35 gm/l and controlling the sludge blanket at a low level, the converted unit performed satisfactorily.   It became evident that prolonged sludge age in the clarifier had an adverse effect on dissolved COD in the effluent from this unit, probably as a result of chemical hydrolysis.

Optimization studies were aimed at achieving low turbidities in the effluent after excess lime treatment in order to improve phosphorus removal. Supersaturated calcium carbonate removal in this unit was also considered to be of major importance as a safeguard against excessive scaling in the ammonia stripping tower.

Diurnal variations in raw water alkalinity required a fluctuating lime demand which could eventually be controlled quite effectively by automatic pH control.

Ammonia stripping tower

This unit is of a longitudinal rectangular shape, comprising a multiple modular replica of the pilot plant.   Four induced draught fans (15 hp) are situated in parallel across the tower, each serving one quarter of the tower capacity.   A manifold system for distribution of liquid is situated on top of the tower.   Ammonia removals of 50 and 80 per cent were recorded during winter and summer months respectively.   However, the relatively high incoming ammonia during the winter months had a very marked influence on breakpoint chlorine demand.

The scaling problem appears not to be of grave concern after one and a half years of intermittent operation.

Recarbonation

Two submerged burners (one standby) using kerosene as fuel, are being used for recarbonation purposes.   The gases are dispersed mechanically in a counter-flow system.   As the design of these absorption units was based on rather limited criteria, the absorption efficiencies are only about 50 per cent. Recarbonation costs are, however, still of a low order (Table 5).

Carbon contacting and stabilisation

Activated carbon contacting and contact stabilisation are integrated using a three-stage mixing system followed by a solids contacting clarifier (Fig. 4). The three mixing stages comprise carbon contacting, sludge recirculation and chemical mixing using chlorinated ferrous sulphate (20 mg/l) and polyelectrolyte (0,5 gm/l).   The sludge drawn from the stabilisation tank is recirculated at a rate of 10 per cent containing 25 mg/l of solids in suspension.   The bed volume is kept low by back-mixing of the excess $CaCO_3$/carbon sludge to the excess lime clarifier.   Turbidities of less than 0,5 JTU's are obtained in the stabiliser unit continuously.

Pressure filtration

Three pressurised sand filters are operated in parallel, one as a standby.

Breakpoint chlorination

Breakpoint chlorination is maintained in tank with $1\frac{1}{2}$ hrs detention. Control of chlorine dosage would be improved by on-site monitoring of the ammonia concentration prior to chlorination.

Activated carbon

The chlorination tank is succeeded by two granular carbon columns in parallel. These serve essentially as dechlorination units to enable better control of free residual chlorination as a final treatment step before distribution.

Quality improvements

The quality of the conventional biofilter effluent is subject to extreme diurnal and seasonal variations of effluent in terms of alkalinity, ammonia and COD in particular (see Fig. 5(a) and 5(b)). These constituents have a direct bearing on chemical treatment costs in terms of lime, chlorine and activated carbon dosages respectively. The average quality improvement through the various process stages is given in Fig. 6.

Treatment costs

Cost estimates based on the intermittent operation of the Stander plant have shown that chemical treatment and supervision are the major cost items for this type and size of plant. A prototype research facility of only 4,5 Ml/d would obviously require unrealistic supervision and staff costs relative to a full-scale plant of larger capacity. Supervision and labour costs are, therefore, projected for larger capacity plants. For scaling-up purposes, it was assumed that the six-tenths power factor (8) is valid for both staff and capital expenditure whereas chemical treatment costs are maintained constant, using a weighted average for seasonal chemical requirements.

Chemical treatment costs

Chemicals are received in bulk and costs include delivery to site using current prices. A break-down of the chemical costs is shown in Table 5. It is evident that carbon treatment constitutes approximately 50 per cent of the total chemical treatment costs. This cost factor is based on the relatively high price of 39 U.S. c/lb for powdered carbon in South Africa as compared to some 10 c/lb in the United States.

Power consumption

The site available for erection of the plant did not allow for gravity flows so that the effluent is pumped between stages. Projections for cost savings in terms of pumping costs have been adjusted for practical situations where gravity feed is possible. For scaling-up purposes, this figure was maintained constant at 0,406 c/kl using a tariff of 0,800 c/kW.h.

TABLE 5

Chemical Treatment Costs in the Stander
Water Reclamation Plant

Chemicals used	Unit cost S.A. c/kg	Average Dosage mg/l	S.A. c/kl	U.S. c/1000 U.S. gallons
Hydrated lime	2,08	330	0,69	3,45
Chlorine	16,40	50	0,82	4,10
Powdered carbon	63,70	30	1,94	9,70
Ferrous sulphate	2,40	40	0,09	0,45
Carbon dioxide	1,87	75	0,14	0,70
Polyelectrolyte	129,0	1,2	0,15	0,75
Total			3,83	19,15

Capital expenditure

Capital depreciation is calculated over 20 years write-off and $8\frac{1}{2}$ per cent interest. Instrumentation costs have been separated from the capital and is maintained constant for large-scale projection.

Total costs

The overall costs for the 4,5 Ml/d research plant are given in Table 6.

TABLE 6

Total Costs for Reclaiming Secondary Effluent in
Stander Water Reclamation Plant (4,5 Ml/d)

Cost item	Cents per kl	U.S. c/1000 U.S. gallons
Fixed costs		
Chemicals	3,830	19,150
Power	0,406	2,030
Instrumentation	0,156	0,780
Total fixed costs	4,392	21,960
Capital ($8\frac{1}{2}\%$, 20 years)	1,253	6,265
Supervision and labour	2,500	12,500
Total costs	8,145	40,725

Cost projections for large-scale reclamation plants

Calculations based on the above estimate are illustrated for plants ranging from 3,8 to 380 kl/d (1 to 100 million U.S. gallons per day) (see

Fig. 7).   It is evident that the total costs approach the fixed costs asymptotically.   The relatively high costs for chemicals which have been assumed constant is still subject to further reductions pending current research.   In this connection the need for balancing diurnal variations, the use of secondary effluent from an activated sludge plant, and the greater demand for chemicals such as activated carbon would certainly have a favourable influence on chemical treatment costs.

Sludge disposal of the combined sludges in the Stander research plant is handled via drying beds and dumping on site.   The reuse potential of this high calcium carbonate sludge is another avenue which could be explored to effect a saving in chemical treatment costs.

Discussion and Conclusions

The abovementioned cost estimates for the three different reclamation plants currently in use in South Africa clearly illustrate how costs are dependent on locality, process selection, size of plant and quality parameters involved.   For industrial reuse such as for pulp and paper production, costs can be reduced by employing a one-stage aluminium sulphate flotation treatment process for removal of organic colloïds, phosphorus and heavy metals from a secondary effluent.   Treatment costs of 8,00 U.S. c/1000 U.S. gallons indicate a 50 per cent savings in cost relative to conventional water supply.   The favourable costs are primarily due to the simplicity of the process which does not require ammonia removal or dissolved organics adsorption with activated carbon.

The Windhoek plant has demonstrated that a high quality potable water can be produced in conjunction with other supply sources using maturation pond effluent as intake source.   The ponds were, however, found to have a limited capacity for ammonia removal and the success of the scheme is thus dependent on efficient functioning of the conventional sewage purification plant, particularly with respect to reductions in ammonia during the winter months. Treatment costs of 57,70 U.S. c/1000 U.S. gallons is of the same order as that for conventional water supply in this area for which the cost structure is relatively high.

The Stander reclamation plant, having the same nominal capacity as the Windhoek plant (4,5 Ml/d) produces potable water from secondary effluent at a more favourable overall cost (40,72 U.S. c/1000 U.S. gallons).   The major costs are in terms of chemicals and supervision, the latter cost showing a marked reduction when projected for plants having a capacity of 10×10^6 U.S. gpd (37,9 Ml) and higher.   The major problems encountered in this research facility are due to diurnal and seasonal quality variations of influent, especially with respect to ammonia and dissolved organic carbon.   In the absence of maturation ponds, such as are being used at Windhoek, the need for a balancing system appears to be an important consideration.   High level chlorine demand for breakpoint chlorination constitutes a major cost item under the prevailing conditions.   The need for breakpoint chlorination has been deemed essential for the treatment of this type of water, although current research has already indicated that it may not be essential under specific circumstances.   Disinfection in the excess lime treatment stage is already highly efficient, and the presence of high concentrations of chloramines below breakpoint may provide satisfactory sterilisation for both virus and bacteria.

The occurrence of algal blooms and hyacinths in polluted streams and impoundment reservoirs have given rise to grave concern from an aesthetic point of view.   The stage has been reached where due consideration should be given to a more strict control of pollution at source.   In terms of costs, the environmental aspect may be of greater significance than is generally realised today.

References

1.   L.R.J. van Vuuren and F.A. van Duuren, J. Wat. Pollut. Control Fed. 37, p. 1256 (1965).

2.   G.G. Cillie, L.R.J. van Vuuren, G.J. Stander and F.F. Kolbe, 3rd International Conference on Water Pollution Research, Munich, 5 - 9 September, 1966 (Section II, Paper 1).

3.   L.R.J. van Vuuren, M.R. Henzen, G.J. Stander and A.J. Clayton, 5th International Water Pollution Research Conference, San Francisco, 2th July - 5th August 1970, pp. 1 - 32/1 - 1 - 32/9.

4.   A.J. Clayton and P.T. Pybus, The reclamation of potable water from sewage effluents in the City of Windhoek, S.W.A. (In press).

5.   L.R.J. van Vuuren, J.W. Funke and L. Smith, 6th International Conference on Water Pollution Research, Jerusalem, 18 - 24 June, 1972.

6.   L.R.J. van Vuuren, G.J. Stander and G.L. Dalton, Wat. Pollut. Control, 70, p. 213 (1971).

7.   G.J. Stander and L.R.J. van Vuuren, Wat. Pollut. Control, 68, p. 513 (1969).

8.   J.H. Perry, Chemical Engineers Handbook. Fourth Edition, McGraw-Hill New York (1963).

Acknowledgements

The authors wish to acknowledge the assistance of the research team of the National Institute for Water Research who were actively engaged in the operation and evaluation of the Stander Water Reclamation Plant.

Thanks are also due to personnel of the South African Pulp and Paper Industries, the Windhoek City Council and the various consultants who played an active role in the planning and design of these plants.

The permission of the Director of the National Institute for Water Research to read this paper is acknowledged.

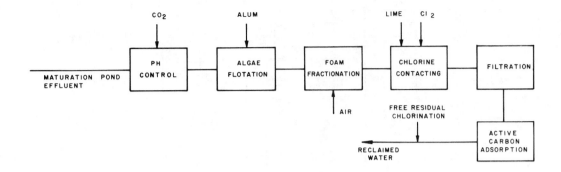

FIG. I   *Schematic Flow Diagram of Windhoek Water Reclamation Plant*

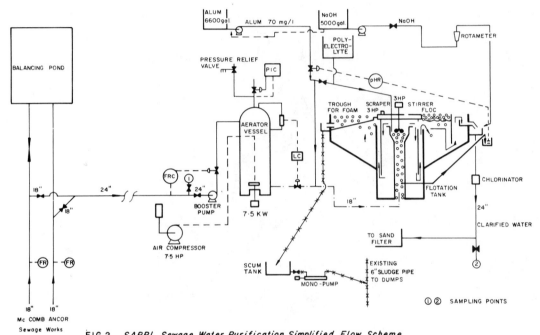

FIG. 2   *SAPPI Sewage Water Purification Simplified Flow Scheme.*

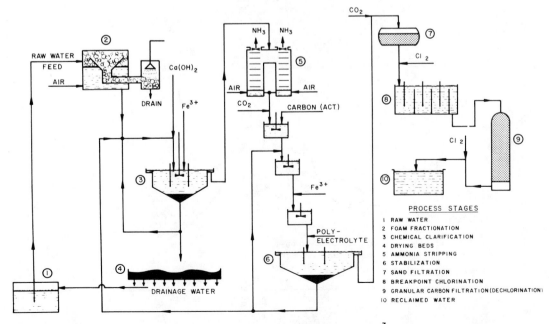

PROCESS STAGES

1 RAW WATER
2 FOAM FRACTIONATION
3 CHEMICAL CLARIFICATION
4 DRYING BEDS
5 AMMONIA STRIPPING
6 STABILIZATION
7 SAND FILTRATION
8 BREAKPOINT CHLORINATION
9 GRANULAR CARBON FILTRATION (DECHLORINATION)
10 RECLAIMED WATER

FIG. 3   *Flow Diagram of the Stander Water Reclamation Plant (4500m³/ Day)*

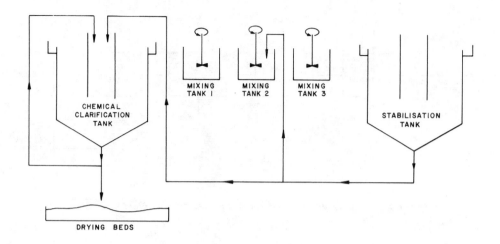

FIG. 4   *Integrated Carbon Contacting Stabilisation System.*

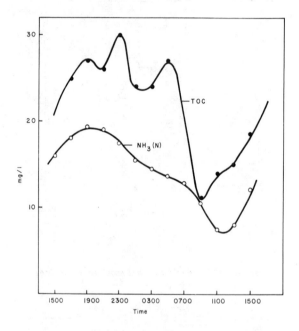

FIG. 5 (a)   *Diurnal variations in quality*

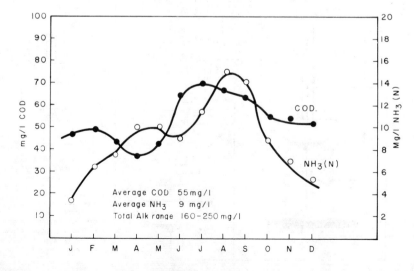

FIG. 5 (b)   *Seasonal variations in quality.*

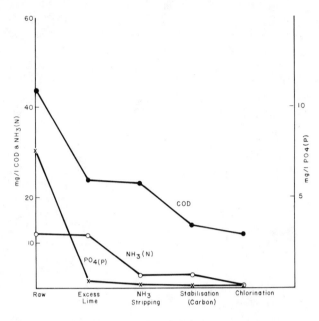

FIG. 6   *Quality Improvement through various process stages*
(Stander plant)

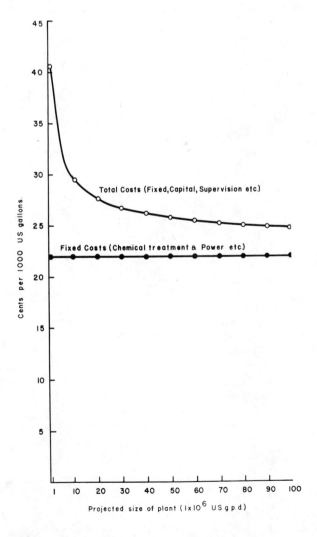

FIG. 7   *Reclaimed water costs based on Stander Plant.*

Applications of New Concepts of Physical-Chemical
Wastewater Treatment
Sept.18-22, 1972

INTERRELATIONSHIP BETWEEN FRESH WATER
AND WASTEWATER PLANTS

by
Lawrence K. Cecil, P.E.
Consulting Chemical Engineer
Tucson, Arizona

We can no longer think of fresh water and wastewater systems
as belonging in two different categories. Both involve water.
Both fresh water and wastewater are polluted. The pollutional
load difference is quantitative, not qualitative. Ordinarily
both belong to the same entity, usually a municipality or a district covering a metropolitan area.

Increasing demands for water caused by metropolitan population growth and individual affluence are creating supply problems
for many cities. Our affluence enables us to purchase more products, all of which require some water in manufacturing. Also we
are not as careful in the amount of water we use as we were in
the old hard times. As demand increases we are looking for additional sources of supply. If the nearby sources cannot be expanded, we must go farther afield, often at great expense. Our
wastewater is a source of fresh water for downstream cities.
There is some purification in the transport system, both by
natural biological processes and dilution with water from other
sources. As we look for new sources of fresh water, consider the
nearest, our treated wastewater. Can we treat our wastewater to
compensate for the purification in the river?

The American Water Works Association is beginning to do
exactly that. In recent years its annual meeting programs have
had an increasing number of papers dealing with sewage reuse.
This has culminated in the AWWA policy statement "On the Use of
Wastewater as a Public Water-Supply Source" (1). This starts,
"AWWA believes that the direct reuse of wastewater must be actively investigated with a view to realizing its full potential,"
and ends with, "The Association believes that the use of reclaimed wastewaters for public water supply purposes should be
deferred until research and development demonstrates that such
use will not be detrimental to the health of the public and will
not affect adversely the wholesomeness and potability of water

supplied for domestic use."

Denver, Colorado (2) is making an elaborate fifteen-year study evaluating successive water use as a potential means of supplemental available water supplies. The study covers the legal, aesthetic, economic, and technical practicality of treatment methods for reuse. There is the hope that by 1985 it will be possible to put 200 mgd treated wastewater into the mains.

Consider the plaintive remark made to the author a couple of years ago by a City Commissioner in a small Eastern Illinois city, "By the time we have our sewage treatment system upgraded to meet the State requirements for 1975, it will be of higher quality than the water in the fresh water lake draining an agricultural area which we use for our city supply." It would be an ideal situation if we could combine our waste and fresh water into one treating plant to supply all of the municipalities water needs.

It is an economic axiom in the water field that water should not be purified to a higher quality than needed for its use. In addition to supplying water for domestic use, industry and recreation in the form of parks and golf courses take a lot of water. If the cost of transportation from the treatment plant is not too high, these needs can be supplied at a lower treatment cost than the residual wastewater that must be upgraded to water main quality.

Tucson, Arizona, where plants grow all year, uses 3 mgd for irrigation at Randolph Park located in the center of the city. The city is currently installing an "in transit" sewage treatment system that will take sewage from a nearby main and upgrade it to park irrigation quality. This takes a 3 mgd load off both the sewage treatment plant, and the domestic water supply. This same type of operation can be applied to a water using industry located inside a city.

Not every city can be so fortunate as Odessa, Texas (3) where a petrochemical manufacturing complex takes the entire sewage plant effluent for industrial use. Dallas, Texas (4) is trying and is operating an elaborate pilot plant to establish effluent quality while it negotiates with industry for reuse.

Fresh water plants have been designed to produce a clear, palatable, and bacteriologically safe water supply for domestic use. The manufacture and widespread use of non-biologically degradable organic and metallo-organic compounds that may have carcinogenic qualities and perhaps genetic damage effects, and trace quantities of toxic metals bring into question the adequacy of present design.

Wastewater treatment plants have been designed to reduce suspended solids to the point sludge banks will not form in the receiving water, the BOD will be reduced to the point the dissolved oxygen in the receiving water will sustain fish life, and pathogenic bacteria are destroyed. Recently eutrophication problems have in some cases required phosphate removal, and the

realization that both ammonia and chloramines are toxic to fish
will require nitrogen removal. These upgrading requirements can
be handled by adding suitable unit processes to existing design,
although sometimes at considerable cost. The new effluent
quality criteria proposed by the Environmental Protection Agency
cannot be met by the conventionally designed wastewater treatment
plant. A new, adequate design must be developed.

   With new designs needed for fresh and wastewater treatment
plants, consideration should be given to a single design that
will meet both fresh and wastewater treatment criteria. If these
qualities can be incorporated into one design, there are definite
economic advantages. Fresh water treatment plants are usually
built at the edge of the city closest to the water source. Waste
water collection systems have gravity flow. The treatment plant
location is at the low point, and this is often on the other side
of the city from the fresh water plant. Very little water is
actually consumed in a municipality, except for irrigation and
special industries. A study at San Antonio, Texas (5) disclosed
that about 35% of the water main input did not reach the sewage
plant during the lawn irrigation season, but that there was
little loss during the rest of the year.

   If our wastewater becomes our principal source of water for
our domestic system, our single treatment plant can be located
where the sewage treatment ordinarily would be located. There
will be only one plant, and one set of operators. The small
amount of fresh water make-up can be piped through or around the
city to this one plant if the fresh water source is on the opposite side of the city.

   There are several routes we can go for our plant design.
Stander (6) reports on several years of good operation using
sequential conventional treatment methods. Heist (7) reports
a study demonstrating that the newer membrane processes can remove suspended matter, and most of the other contaminants in a
single operation. Such a system, if supplemented by massive
chlorination and subsequent dechlorination in activated carbon
columns can be the basis for design of a single treatment plant
that will use municipal wastewater as just another supply source.

   We must proceed with great caution in designing our single
plant. Each process in the sequence must be evaluated for its
effect on each of the pollutants, its economics, and the effect
on downstream processes if it fails temporarily. Our system
must contain some redundancy if we are to have the confidence of
our users that they can be assured they will always get the
quality of treated water they need.

References

(1) AWWA Policy Statement. On the Use of Wastewaters as a
     Public Water-Supply Source. Jour. AWWA, 63:609 (October
     1971).

(2) Linstedt, K.D., K. J. Miller, and E.R. Bennett. Metro-
     politan Successive Use of Available Water. Jour. AWWA,
     63:610 (October 1971).

(3) Smythe, Frank. Multiple Water Reuse. Jour. AWWA 63:623
     (October 1971).

(4) Graeser, Henry J., Dallas' Wastewater-Reclamation Studies.
     Jour. AWWA 63:634 (October 1971).

(5) Alamo Area Council of Governments. Basin Management for
     Water Reuse. EPA Water Pollution Control Research Series
     16110EAX02/72.

(6) Stander, G.F., Tertiary Treatment-The Cornerstone of Water
     Quality Protection and Water Resources Optimization. Proc.
     Application of New Concepts of Physical-Chemical Wastewater
     Treatment. Vanderbilt Univ., Nashville, Tenn. Sept. 1972.

(7) Heist, James. Membrane Processes Applied to Hospital
     Wastewaters. Proc. Application of New Concepts of Physical-
     Chemical Wastewater Treatment. Vanderbilt Univ., Nashville,
     Tenn. Sept. 1972.

Applications of New Concepts of Physical-Chemical
Wastewater Treatment
Sept.18-22, 1972

HOW USABLE IS PRESENT TECHNOLOGY FOR
REMOVING NUTRIENTS FROM WASTEWATER

by
Lawrence K. Cecil, P.E.
Consulting Chemical Engineer
Tucson, Arizona

Nutrient requirements for various forms of living organisms
vary. Our concern here is for plant growth, particularly algae.
Accelerated eutrophication of surface waters, especially lakes,
is one of the prime concerns of environmentalists. All lakes are
subject to eutrophication, but the enormous speed up caused by an
excess of nutrients from man's activities is something we should
stop. We are dealing with a complex relationship. Complete absence of nutrients for plant growth means no fish food, and so no
fish. This saddens the fisherman segment of the environmentalists. Complete removal of all nutrients by present biological or
physical treating methods is not possible with present technology.
We can treat to reduce the concentration of nutrients to a level
that makes improbable excessive plant growth.

The best application of present technology in slowing eutrophication is complete reuse of wastewater. This technology is
adequately developed, so that it is a management problem. The
next best system is the so-called "ring" technic, which carries
the wastewater around the lake to discharge into a flowing stream
where eutrophication is not a problem. If we must discharge our
wastewater into the lake we depend on treating processes. It is
not necessary, although it would be desirable, to remove all nutrients. Each is essential to growth. If we can reduce one to
a low enough concentration the food chain is broken, and growth
is controlled.

The relative quantitative requirements for growth of algae
are, on a descending scale, carbon dioxide, nitrogen, phosphorus,
potassium, magnesium, calcium, sulfur, iron, and a number of
trace materials including growth hormones, vitamins, amino acids,
boron, manganese, zinc, copper, molybdenum, cobalt, and probably
others.

Carbon dioxide is by far the principal nutrient. Biologi-

389

cally degradable carbon compounds will be oxidized by bacteria in the lake to furnish additional carbon dioxide. Our waste water should have low BOD. If the receiving water is low and our wastewater is high in bicarbonates, these should be reduced by treatment with lime, free mineral acids, or mineral acid salts.

Nitrogen is next in the algal nutrient scale. Nitrogen removal methods are adequate, although expensive and clumsy. Municipal wastewaters contain an average of 20 mg/l ammonia nitrogen. Some industrial wastewaters may be very high in nitrogen. Petroleum refineries, where ammonia is used for corrosion control, may contain 1000 mg/l ammonia nitrogen. Other industrial wastes may be very high in nitrate nitrogen.

Nitrification with activated sludge and anaerobic denitrification are the preferred methods for municipal wastes. Oxidation of ammonia to nitrogen with chlorine under pH control is effective. If the 8-1 chlorine to nitrogen ratio, with attendant increase of chloride salts is not objectionable, it is worth considering. It is an effective polishing treatment after the biological nitrification-denitrification process, which does not dependably approach zero nitrogen in the effluent.

Air stripping after the ammonia has been converted to the free form by raising the wastewater pH to 11.0 or higher, is both cheap and effective. Air pollution control authorities may object, but the ratio of 350-500 cubic feet of air to each gallon of wastewater gives very high dilution. This process is temperature dependent. As the temperature of either air or water decreases so does the ammonia removal efficiency. At very low air temperatures the desorption system freezes.

Ion exchange with clinoptilolite, a siliceous cation exchange material selective for ammonia, is effective on relatively low ammonia concentrations as in municipal wastewater, and is just beginning to be used. It is probable that ongoing studies of the mechanism of the ammonia selectivity of clinoptilolite will permit the synthesis of high capacity resinous cation exchangers that will overcome the disadvantage of the low capacity of the clinoptilolite.

High nitrate industrial wastes are effectively handled by biological denitrification if supplied with organic carbon compounds needed for cell synthesis.

Breaking the food chain by nitrogen reduction is not particularly favored because of the ability of some algae, especially the very objectionable blue-greens, to fix atmospheric nitrogen. Although phosphorus is not high in the algal nutrient chain, the carbon-nitrogen-phosphorus requirement being something on the order of 100-20-1, its removal is the preferred method of breaking the food chain.

Well developed treatment methods can reduce phosphorus consistently to very low concentrations. The cost is reasonable. These methods also reduce the suspended solid and biologically

oxidizable content of the wastewater. If lime is used to precipitate the orthophosphate to calcium hydroxyapatite, the resultant high pH leaves the wastewater ready for ammonia stripping. The low cost acid salts of aluminum and iron (both ferrous
and ferric) are capable of reducing the phosphorus to acceptably
low concentrations. The U.S. Environmental Protection Agency
has published a number of reports on both R&D studies and full-
scale plant operation of phosphate removal by all of these chemicals.

Magnesium, which is the metal component of chlorophyll, is
effectively reduced by lime treatment. The probability of there
being enough magnesium in the receiving water to satisfy chlorophyll demands is so great that magnesium removal from the waste
water is not a reasonable method of breaking the food chain.

The trace, or micronutrients are needed in such small concentrations that ordinary chemical and biological treatment
technics are not depended upon to reduce any of them to such a
low concentration algal growth can be controlled. Probably some
of them are partially reduced by existing treatments, but no
real study of this type has been reported.

The Organization for Economic Cooperation and Development,
Paris, France has recognized eutrophication as an international
problem. It has established a Steering Group on Eutrophication
Control in its Water Management Sector Group. One of the sub-
groups is Experts of Working Group on Treatment Processes, which
has completed its report "a cost/effectiveness evaluation of the
best existing processes for the treatment of wastewaters for
nutrients (phosphorus and nitrogen) removal. It also gives consideration to ring canalization for the protection of water
bodies and to the treatment of tributaries." The Eutrophication
Control Report of the Water Management Sector Group, now in preparation, will consider all phases of the eutrophication problem.
It should be a landmark in the program to reduce man's contribution to eutrophication of lakes.

There is a real revolution in treating technics taking place
right now. We have been accustomed to the sequential process
system, in which when another compound is to be removed we add
on another unit operation. The Office of Saline Water of the
U.S. Department of the Interior has given massive support to
development of distillation, membrane separations, and freezing
as desalting technics for the production of fresh water. All of
these systems are directly applicable to wastewater treatment.
The costs are comparable to our usual sequential treating systems. All three can remove all of the contaminants in wastewater,
in our case we call some of them nutrients, to very low values.
If you have a nutrient removal problem, make a thorough investigation of these OSW technics.

The answer to the question, "How usable is present technology for removing nutrients from wastewater?" is: "Yes. It
is not only usable but acceptable!"

RECOMMENDED READING

Algal Culture, by John S. Burley. Carnegie Institute of
     Washington, D.C.

Eutrophication: Causes, Consequences, Correctives. National
     Academy of Sciences, 2101 Constitution Avenue, Washington,
     D.C. 20418.

Fresh Water Algae of the United States, Gilbert M. Smith.
     McGraw-Hill Book Company, New York, N.Y.

Physiology and Biochemistry of Algae, Ralph A. Lewin.
     Academic Press, New York, N.Y.

Algae, Man, and the Environment, Daniel F. Jackson, Editor.
     Syracuse University Press, Syracuse, New York.